Freeze Drying of Food Products

Freeze Drying of Food Products

Fundamentals, Processes and Applications

Edited by

Roji Balaji Waghmare
Department of Food Engineering and Technology, Institute of Chemical Technology, Mumbai, India

Manoj Kumar
Chemical and Biochemical Processing Division, ICAR - Central Institute for Research on Cotton Technology, Mumbai, India

Parmjit Singh Panesar
Department of Food Engineering and Technology, Sant Longowal Institute of Engineering and Technology, Longowal, Punjab, India

WILEY

This edition first published 2024

Registered Offices
John Wiley & Sons, Inc., 111 River Street, Hoboken, NJ 07030, USA
John Wiley & Sons Ltd, The Atrium, Southern Gate, Chichester, West Sussex, PO19 8SQ, UK

For details of our global editorial offices, customer services, and more information about Wiley products visit us at www.wiley.com.

Wiley also publishes its books in a variety of electronic formats and by print-on-demand. Somecontent that appears in standard print versions of this book may not be available in other formats.

Library of Congress Cataloging-in-Publication Data Applied for
[ISBN: 978111998206; ePDF: 9781119982074; ePub: 9781119982081; oBook: 9781119982098]

Cover images: © Penpak Ngamsathain/Getty Images,Liudmila Chernetska/Getty Images, Eugenia_Kuz/Getty Images, SpeedPhoto/Getty Images
Cover design: Wiley

Set in 9.5/12.5pt STIXTwoText by Straive, Pondicherry, India

Contents

List of Contributors

Rubén Agregán
Centro Tecnológico de la Carne de Galicia
San Cibrao das Viñas, Ourense
Spain

Shalini S. Arya
Department of Food Engineering and Technology
Institute of Chemical Technology
Mumbai
India

Kalyan Barman
Department of Horticulture
Institute of Agricultural Sciences
Banaras Hindu University
Varanasi, Uttar Pradesh
India

Surya N. Chaurasia
Division of Vegetable Production
ICAR-Indian Institute of Vegetable Research
Varanasi, Uttar Pradesh
India

Tanuva Das
Department of Food Engineering and Technology
Institute of Chemical Technology
Mumbai
India

Ruben Dominguez
Centro Tecnológico de la Carne de Galicia
San Cibrao das Viñas, Ourense
Spain

Noemí Echegaray
Centro Tecnológico de la Carne de Galicia
San Cibrao das Viñas, Ourense
Spain

Ertan Ermiş
Department of Food Engineering Faculty of Engineering and Natural Sciences
Istanbul Sabahattin Zaim University
Istanbul
Turkey

Yogesh Gat
Department of Food Engineering and Technology
Institute of Chemical Technology
Mumbai, Maharashtra
India

Papiha Gawande
Department of Food Engineering and Technology
Institute of Chemical Technology
Mumbai, Maharashtra
India

Mouandhelmamou Hassani
Amity Institute of Food Technology
Amity University Rajasthan
Jaipur
India

Hatice E. Kırtıl
Department of Food Engineering Faculty of Engineering and Natural Sciences
Istanbul Sabahattin Zaim University
Istanbul
Turkey

Hare Krishna
Division of Vegetable Production
ICAR-Indian Institute of Vegetable Research
Varanasi, Uttar Pradesh
India

Manoj Kumar
Chemical and Biochemical Processing Division
ICAR-Central Institute for Research on Cotton Technology
Mumbai
India

Shiv Kumar
MMICT&BM (HM)
Maharishi Markandeshwar (Deemed to be a University), Mullana
Ambala, Haryana
India

Yogesh Kumar
Department of Agricultural and Food Sciences
University of Bologna
Cesena, FC
Italy

Yasemin Ş. Küçükata
Department of Food Engineering Faculty of Engineering and Natural Sciences
Istanbul Sabahattin Zaim University
Istanbul
Turkey

Jose M. Lorenzo
Centro Tecnológico de la Carne de Galicia
San Cibrao das Viñas, Ourense
Spain

Área de Tecnología de los Alimentos
Facultad de Ciencias de Ourense
Universidad de Vigo
Ourense
Spain

Rahul Mehra
MMICT&BM (HM)
Maharishi Markandeshwar (Deemed to be a University), Mullana
Ambala, Haryana
India

Anuradha Mishra
MMICT&BM (HM)
Maharishi Markandeshwar (Deemed to be a University), Mullana
Ambala, Haryana
India

Arun S. Mujumdar
Department of Bioresource Engineering, Macdonald College
McGill University
Montreal, Quebec
Canada

Paulo E.S. Munekata
Centro Tecnológico de la Carne de Galicia
San Cibrao das Viñas, Ourense
Spain

Kyuya Nakagawa
Department of Chemical Engineering
Faculty of Engineering
Kyoto University
Kyoto
Japan

Donald Lyngdoh Nonglait
Department of Food Engineering and Technology
Institute of Chemical Technology
Mumbai
India

Arun Kumar Pandey
Amity Institute of Food Technology
Amity University Rajasthan
Jaipur
India

Parmjit Singh Panesar
Department of Food Engineering & Technology
S.L. Institute of Engineering & Technology
Longowal, Punjab
India

Mirian Pateiro
Centro Tecnológico de la Carne de Galicia
San Cibrao das Viñas, Ourense
Spain

Research and Development Department
Idaho Milk Products
Jerome, ID
USA

Devraj Rajput
Department of Food Engineering and Technology
Institute of Chemical Technology
Mumbai
India

Gunvantsinh Rathod
Food Science Institute
Kansas State University
Manhattan, KS
USA

Research and Development Department
Idaho Milk Products
Jerome, ID
USA

Suchismita Roy
Food Science Institute
Kansas State University
Manhattan, KS
USA

Swati Sharma
Division of Vegetable Production
ICAR-Indian Institute of Vegetable Research
Varanasi, Uttar Pradesh
India

Rajat Suhag
Faculty of Agricultural, Environmental and Food Sciences
Free University of Bolzano
Piazza Università
Bolzano
Italy

Roji Balaji Waghmare
Department of Food Engineering and Technology
Institute of Chemical Technology
Mumbai
India

Preface

Freeze-drying, also known as lyophilization, has been a significant breakthrough for the food industry, particularly for the dehydration of sensitive and high-value foods. This technique gained prominence after World War II as a method of preserving and storing food products without the need for refrigeration. Compared to conventional drying methods, freeze-drying offers several advantages, including the preservation of morphological, biochemical, and immunological properties, structure, surface area, and stoichiometric proportions. This book aims to serve as a reference for anyone interested in using freeze-drying to extend the shelf life of food, whether for academic or practical applications.

The book describes the fundamentals and practices used in freeze-drying areas with respect to different food products. It also discusses the applications of freeze-drying in the preservation of different food products such as fresh produce, meat and poultry, dairy products, functional foods besides the effect on nutrients and their stability. The summary also includes advances in the conventional freeze-drying process through the application of more advanced freeze-drying setups and other techniques that aim to reduce drying time and increase drying rate.

The book comprises 10 chapters, which are contributed by experts in their area of research. Chapter 1 deals with a fundamental understanding of freeze-drying, its principles, and the process involved. The mathematical modeling of freeze-drying, presenting a mathematical approach to the freezing and drying of food products, has been discussed in Chapter 2. Chapter 3 has been focused on the application of freeze-drying for preserving fruits and vegetables, highlighting its effectiveness in retaining the quality characteristics of fresh produce.

Chapter 4 discusses the use of freeze-drying to increase the shelf life of meat and seafood products, while Chapter 5 focuses on the freeze-drying of dairy products. Chapter 6 deals with the freeze-drying of probiotics for incorporation into functional foods, with a particular focus on the stability and viability of probiotics during and after the freeze-drying process.

Chapter 7 describes the effects of freeze-drying on vitamins, minerals, and bioactive compounds, while Chapter 8 discusses the criteria required for packaging materials for freeze-dried foods and the various types of commercially available packaging for such products. Chapter 9 sheds light on advances in freeze-drying technology to reduce energy consumption and improve energy efficiency, while the final chapter has been focused on the commercial applications of freeze-drying in the food industry.

In summary, this book provides a comprehensive overview of freeze-drying as a drying technique for food products, offering insights into the fundamentals, principles, and process involved, as well as highlighting its applications and advancements. The book would be highly useful for the faculty, research students, and industry personnel involved in food processing, dairy technology, food microbiology, food biotechnology, and allied fields.

Roji Balaji Waghmare, Manoj Kumar, Parmjit Singh Panesar
October 2023

Acknowledgments

We are indeed grateful to the contributors to all the chapters for their input and timely revision of the manuscript. We also acknowledge the various sources of illustrations used in the book that have been given in the form of figures, tables, and references.

We are especially obliged to Rebecca Ralf, Commissioning Editor, Kelly Labrum, and Ashik Melvin, Managing Editor, John Wiley & Sons Limited, UK, for all their support in bringing out this volume. We also extend our thanks to the production team at John Wiley & Sons Limited, UK, for their efforts in the production of this volume.

Last but not least, we acknowledge the support of our family members while preparing this volume.

Editors
Roji Balaji Waghmare
Manoj Kumar
Parmjit Singh Panesar

1

Freeze-Drying: Basic Principles and Processes

Roji Balaji Waghmare[1], Manoj Kumar[2], and Parmjit Singh Panesar[3]

[1] *Department of Food Engineering and Technology, Institute of Chemical Technology, Mumbai, India*
[2] *Chemical and Biochemical Processing Division, ICAR-Central Institute for Research on Cotton Technology, Mumbai, India*
[3] *Department of Food Engineering & Technology, Sant Longowal Institute of Engineering & Technology, Longowal, Punjab, India*

1.1 Introduction

It is possible that drying food is the earliest means of preserving it for an extended period of time, making it a crucial unit operation. Drying is most commonly used to turn a wet or liquid product into a powder, flake, or solid form without altering the product's physicochemical qualities (Maisnam et al. 2015; Assegehegn et al. 2019). Foods that have been dried have a longer shelf life because the drying process stops enzyme and microbial activity (Prosapio and Norton 2018). Traditional drying, on the other hand, degrades the physical and chemical properties of food, which can make it unpalatable to customers. As a result, enhanced drying methods are crucial for food and agro-products to ensure the production of high-quality goods with reduced drying times, increased capacities, enhanced process controls, reduced operational costs, and enhanced safety. Conventional drying processes can degrade nutrients, decreasing the overall quality of dried foods. In light of this, developing and researching innovative drying tools and methods are crucial.

> "Altmann used freeze-drying as early as 1890, his technique went unnoticed for over 40 years. ...

Freeze Drying of Food Products: Fundamentals, Processes and Applications, First Edition.
Edited by Roji Balaji Waghmare, Manoj Kumar, and Parmjit Singh Panesar.

> ... the optimism may have somewhat dimmed, the promise remains, and economic changes in the future may well stimulate another surge in development."
>
> *Meryman (1976)*

The above words are extracted from the abstract of the article "Historical recollections of freeze-drying," where author was optimistic about the future of freeze-drying (FD) or lyophilization of food.

The age-old Chinese and Peruvian Incas had some knowledge of FD. These people used to leave meat outdoor in the cold season to prolong storage and improve flavor. Prior to drying, the meat was first frozen. On the mountain tops above Machu Picchu, the ancient Incas kept potatoes and other harvests. The meal in the high altitudes' low pressure slowly sublimated the water inside due to the cold mountain temperatures that caused it to freeze. Food that has been freeze-dried is lighter and stays longer than other dried samples (Hua et al. 2010).

In the nineteenth century, FD was widely used and employed in research and technology. Nevertheless, the uses of first industrial FD were not seen until after the Second World War. Initially, the medical business was the only one to use the FD procedure to produce medicines, cells, and other related items. The food industry did not begin to use FD until the middle of the twentieth century (Garcia-Amezquita et al. 2015). This technique has quickly risen to prominence as a crucial method for the long-term storage of food items, even among the many other types of drying methods. FD is frequently used in the food business since it results in superior quality dried food. FD relies on the fact that a product's solvent will eventually evaporate. In this process, the solvent – which can be water or an organic solvent – crystallizes at low temperatures before rapidly changing phase from solid to gas. Because FD is carried out at lower temperatures, the qualitative attributes of the food are preserved, and the damage to thermolabile food components is limited (Martínez-Navarrete et al. 2019). Because of this, FD serves primarily to ship a product with a longer shelf life, and the quality of food is maintained after being reconstituted with water.

FD offers numerous advantages compared to conventional drying technology. The main advantages of FD are maintenance of morphological, biochemical, and immunological characteristics; high recovery of volatiles; maintenance of structure, surface area, and stoichiometric proportions; long time frame of realistic usability; and reduced weight for storage, shipping, and handling (Ciurzyńska and Lenart 2011; Isleroglu et al. 2018). Since FD is performed at low temperature, they present a lower risk for products labile to heat degradation. Therefore, FD can be applicable for valuable materials that are heat sensitive or samples sensitive to heat that cannot be treated using other processes involving high temperatures (Morais et al. 2016).

1.2 Principle of FD Process

To remove the solvent from a liquid formulation is the fundamental idea of FD, which is a method of preservation. As can be seen in Figure 1.1, there are typically three stages to the FD process: freezing, primary drying, and secondary drying. However, there are five crucial processes that must occur in order to fully comprehend the mechanics at play here: freezing, sublimation, desorption, vacuum pumping, and vapor condensation (Liu et al. 2008). At first, the liquid formulations are cooled to a low temperature, which freezes all of the water existing in the substance. In the sublimation drying process, the frozen solvent is first heated to a point where it leaves its solid state and enters the vapor phase (primary drying). After the initial freezing process, the unfrozen liquid is removed via desorption (secondary drying). Consequently, FD involves two equally important processes: freezing and drying (Tang and Pikal 2004).

The time and energy needed at each stage is highly dependent on the nature of the product, the dryer's design and configuration, and other process variables. In addition, moisture content standards of final product vary by type. While the necessity of proper storage of freeze-dried products may not immediately jump out, it has a profound impact on the durability and safety of the dried goods. Most FD is intended for long-term storage, which means that it must be packaged and stored in such a way as to prevent the deteriorating biochemical and

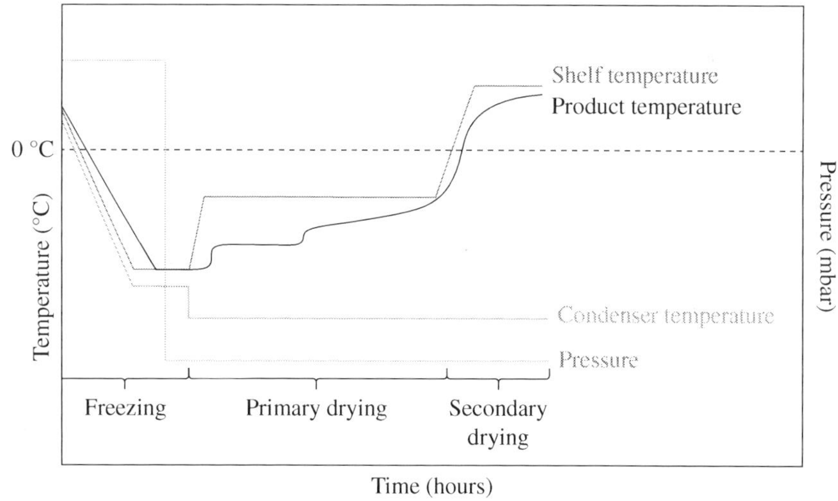

Figure 1.1 Freeze-drying process, presenting freezing, primary drying, and secondary drying stages. *Source:* Morais et al. (2016)/Reproduced with permission from Elsevier.

microbiological reactions that would otherwise occur. Storage stability has been estimated using appropriate shelf-life tests, which have included approaches incorporating accelerated shelf-life testing.

1.2.1 Freezing of Raw Material

Freezing is the first phase and one of the essential steps during FD. All the food products such as fresh produce, beverages, and coffee should be subjected to freeze before primary drying (Nowak and Jakubczyk 2020). During the preliminary phase of freezing, the sample is reduced to a temperature below its eutectic point or glass transition temperature. It is the lowest point of the combination of composition and temperature at which the sample freezes. In this step, the primary focus is on freezing any liquid water that may be present in the sample. In this step, it is advised to cool the sample to a temperature lower than the glass transition temperature for each humidity concentration. Over this point, sample is in unstable rubbery form, whereas, under this point, sample changes into glassy or amorphous phase (Garcia-Amezquita et al. 2015). This step transforms 70–90% of free liquid water present in food sample to solid frozen state. However, the remaining water is present in bound state (Bhushani and Anandharamakrishnan 2017). During FD, the suitable freezing rate relies on the type of food sample which could be solution, suspension, or biomaterials. Freezing rate depends on the temperature of the sample, temperature of the freezing medium, and also on the heat transfer resistance (Nowak 2017). To achieve the less primary drying time, it is necessary to maintain the big ice crystal size which results in less resistance to mass transfer in the final dried sample. However, development of several small ice crystals results in high resistance (Assegehegn et al. 2019). The temperature reducing the rate of cooling and freezing also has significant impact. Faster freezing results into the creation of limited glassy phase and inhibits extreme dehydration of food sample during freezing, whereas extremely fast freezing leads to the destructive changes, for example rupturing the cell wall and its other cellular components (Hua et al. 2010).

1.2.2 Primary Drying – Sublimation

Following the freezing stage and during primary drying, ice is removed by sublimation. Sublimation begins at the ice surface and proceeds across the frozen–dry matter interface, mainly dependent on the temperature–pressure combinations maintained. The drying chamber's pressure and the drying process's heat intensity are crucial factors in the FD process (Rambhatla et al. 2006). If the pressure in the compartment or the partial pressure of vapor in the compartment is lower than the vapor pressure of ice, then the frozen solvent can be evacuated from the solid state and entered straight into the vapor phase. This has been suggested that the compartment pressure should be maintained in between one fourth to one half of vapor pressure of ice to achieve controlled sublimation process (Bhushani and Anandharamakrishnan 2017). For contact heating, maintaining the suitable temperature of shelf is essential. However, during radiation heating, the space between the radiation source and sample and the intensity of radiation are important factors to be maintained (Nowak and Lewicki 2005). The primary drying begins from the top layer of the material and it progresses to the bottom layer. Thus, the vapor is evacuated by diffusion or convection through the void area (Ingvarsson et al. 2011). It is an essential step which has a significant influence on the mass transfer rate. As the sublimation step progresses, the water from the sample is decreased and therefore the glass transition temperature also differed (Fellows 2009). Freeze-dried products show low shrinkage and highly porous structures as the ice crystals in the samples get sublimated and the space which originally contained the ice is maintained (Jiang et al. 2017).

1.2.3 Second Drying – Desorption

During secondary drying, leftover moisture (in the unfrozen form) is removed by desorption, justifying higher time requirements. Desorption is performed by rising the temperature and decreasing the vapor pressure in the drying compartment of freeze-dryer. The secondary drying temperature should be maintained in a proper range to prevent denaturation of the material (Cheng et al. 2017; Liu et al. 2022). When compared to primary drying, the time required for desorption is only about 30–50%. Lower drying rates are achieved in secondary drying by reducing the amount of water present in the material, increasing barrier to heat and mass transfer, and binding water particles to dry particles (Millman et al. 1985; Tang and Pikal 2004). When the majority of the water has been eliminated throughout the FD process, leaving a structurally stable dried food material with ideally less than 0.5–3% leftover water, the process is said to have finished (Bhushani and Anandharamakrishnan 2017; Liu et al. 2022).

1.3 Basic Components of a Typical Freeze-Dryer System

A FD system is mainly composed of a drying chamber, a vacuum pump, a heat source, and a condenser. In order to achieve the maximum benefits of the FD system, these components must be properly selected and operated, and this depends on the specifications of every product material (Garcia-Amezquita et al. 2015). The basic components of a typical freeze-dryer system are shown in Figure 1.2. The food material must be placed in a drying chamber that can

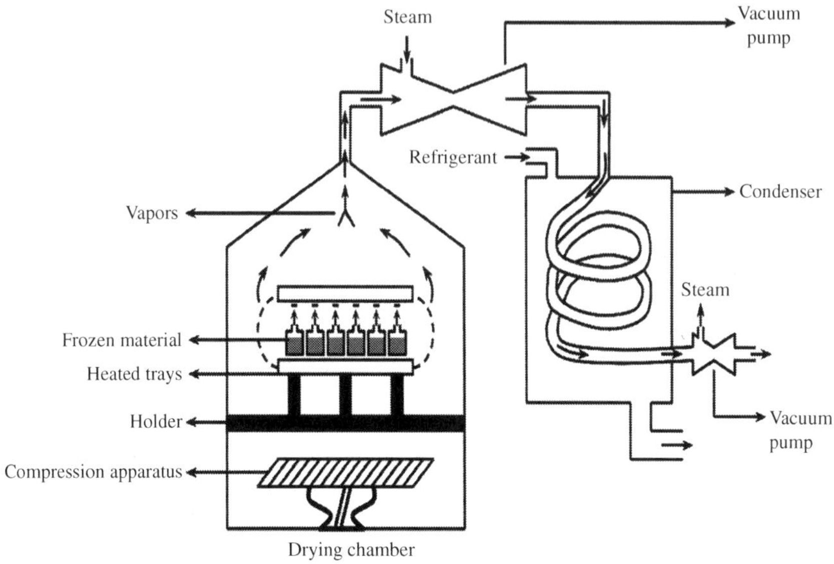

Figure 1.2 Basic components of a typical freeze-dryer system. (Freeze-dryers are equipped with a vacuum chamber that may chill and heat the containers and their contents, allowing for the drying of food products while they are stored on shelves. The vacuum chamber is linked to a vacuum pump, a refrigeration unit, and accompanying controls.)

maintain full vacuum and has shelves that are temperature controlled. It is required that the drying chamber to be completely sealed. The latent heat of sublimation is supplied by the heat supply for the shelves during the first stage of drying. Depending on the model of freeze-dryer, a number of different tray styles and heating options are available. Ribbed trays, for instance, are used in contact and conduction freeze-dryers and sit on heater plates. Due to the uneven distribution of heat, drying time increases. An accelerated freeze-dryer has superior heat transmission since the food is heated via expanding metal mesh, as opposed to being directly contacted by the dryer's heating element, as in a contact dryer (Rolfgaard 1987; Bhushani and Anandharamakrishnan 2017; Hua et al. 2010). The plates are often used to heat the materials inside the drying chamber. Conduction and radiation are the two methods used to heat food sample. Conduction is used to heat sample by the lower plate, whereas radiation is used to heat sample upper plate. It is possible to add an electric heater or heat exchanger inside the plates. A temperature control system can regulate the plates' temperature (Hua et al. 2010). For the best results from FD, it is important to control the drying stage's temperature so that it is not too high and denatures the product. The drying can be accelerated by raising the heating temperature, however, excessive heat will lead to denaturation of the food sample (Liu et al. 2022).

There can be two possible sources of the noncondensable gases in the FD unit. Air can enter a vacuumed unit in two ways: first, it can leak in from external environment; second, gases can escape from food sample as they dry (Hua et al. 2010). Through the use of the vacuum pump, the drying chamber is emptied of any gases that cannot be condensed. This is essential in order to reach the vacuum level of less than 0.61 kPa. The dryer system's total volume can inform the design of the vacuum system, which in turn will inform the size, kind, and number of vacuum pumps utilized (Mellor 1978).

1.4 Types of FD Processes

The high price tag of implementing FD means that it cannot be used for a wide variety of everyday food items. Market success has been achieved with this technology for biological formulations and high-value foods (Morais et al. 2016). However, with the development of cutting edge hybrid drying methods, FD is finding more and more uses in the food business. The following sections provide more detail on how FD is being utilized in tandem with a variety of other techniques, as well as some novel approaches that enhance FD procedures, to fulfill individual product needs. Common methods of FD food items include (1) vacuum FD, (2) atmospheric FD, and (3) spray FD.

1.4.1 Vacuum FD

Since the 1960s, the food industry has relied heavily on the vacuum FD technology to mass produce dried food items. This method has seen widespread use in recent years for drying a wide variety of high-value foods, including fruits, vegetables, coffee, milk, meat, and more. It is the gold standard for making high-quality dehydrated food items with minimum changes in texture, flavor, rehydration ability, nutritional value, and chemical makeup, and it is used all over the world (Ma et al. 2018). The process of vacuum FD combines the advantages of both vacuum drying and FD because it employs both of these methods simultaneously. This increases the rate at which water sublimates and helps to preserve the internal structure of particles (Shi et al. 2012).

To extract the water or solvent from a suspension using vacuum FD, the mixture is first frozen at a low temperature. Heat and mass are transferred concurrently in the vacuum FD process. Increases in heat and mass transfer rates during drying can be achieved through a number of different techniques (Patil et al. 2010). During the FD process, ice crystal size, drying rate, and heat and mass transfer rate can all be improved by annealing (Shi et al. 2012). For a set amount of time, a frozen liquid is kept above its final freezing temperature in the annealing process (Lim et al. 2018; Patil et al. 2010). Ice crystals form within and outside of the starch nanoparticles, thanks to the annealing process and the various cryoprotectants (Shi et al. 2012). Starch nanoparticles' glass transition temperature, particle size, and residual moisture content were all affected. To see how microbial transglutaminase (MTGase) might change the characteristics of gels made from vacuum freeze-dried silver carp surimi (Guo et al. 2019), proteins can be dried in a vacuum without losing their gelling properties during the process (Danial et al. 2016). On the other hand, Guo et al. (2019) found that the gel texture of the surimi was diminished due to the vacuum FD process denaturing the proteins to some degree. The gel texture of the dried surimi is enhanced by heating it to 40 °C and the addition of MTGase. Pretreatment followed by vacuum FD results in surimi powder with a better gel texture. Grass carp (*Ctenopharyngodon idella*) fillets are dehydrated using vacuum FD to create high-quality dried products with favorable nutritional quality, texture, flavor, color, little shrinkage, and high porosity. A partial least squares regression (PLSR) model was created for grass carp fillets to predict the percentage of mass loss during dehydration and the percentage of mass gain during rehydration (Ma et al. 2017).

1.4.2 Atmospheric FD

The process of atmospheric FD combines the benefits of both FD and convective drying. To facilitate the sublimation of the frozen solvent, a high vacuum must be applied to the frozen samples, which increases the energy requirements of the process and hence the cost. The diagram of atmospheric spray FD is shown in Figure 1.3. There is no need for a strong suction while drying frozen samples. Because of this, the original idea behind air FD was to reduce energy usage compared to vacuum FD while keeping the same high product quality. This can be done if the water vapor pressure in the drying chamber is less than that of the frozen sample. This state is attained by exposing a frozen sample to a stream of cold, dry gas (Ishwarya et al. 2015). Both atmospheric FD and vacuum freeze-dried products share similar rehydration kinetics and hygroscopic properties. Antioxidant levels were reduced by 15% when air-dried versus vacuum-dried (Rahman and Mujumdar 2012; Taylor et al. 2010). The antioxidant capabilities of eggplant (*Solanum melongena* L.) samples were examined by Colucci et al. (2018), who looked at how various drying process variables affected the eggplant. Antioxidant capacity, total phenolic content, and ascorbic acid in atmospheric FD eggplant samples were observed to decrease with increasing drying temperature and air velocity. On the other hand, the antioxidant content of the eggplant sample is unaffected by the ultrasonic-assisted atmospheric FD process.

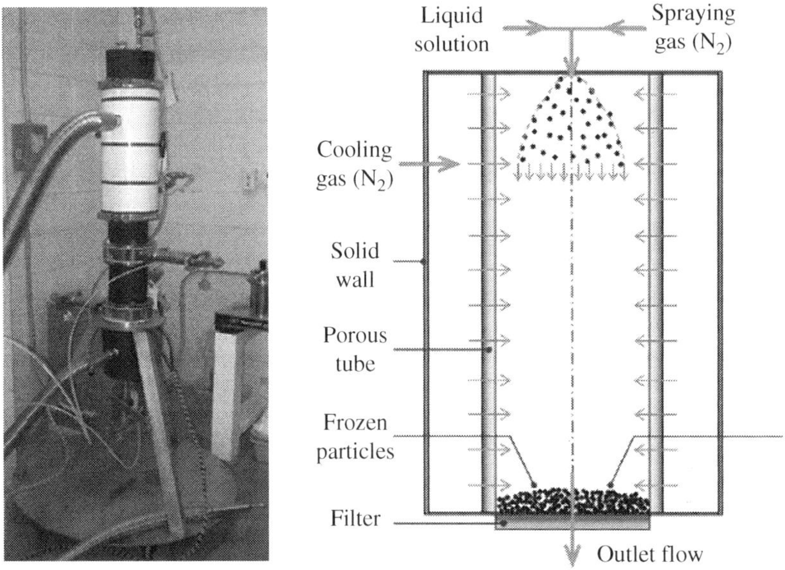

Figure 1.3 Atmospheric spray freeze-drying apparatus. *Source:* Israel Borges Sebastião et al. (2016)/Reproduced with permission from Elsevier.

1.4.3 Spray FD

With the advantages of both spray drying and FD, spray FD is a potential drying method that produces powdered materials (Anandharamakrishnan 2019). The spray FD methodology can be broken down into the following steps: (1) atomization, wherein the solution or suspension is broken up into minute droplets, resulting in a high surface-to-mass ratio; (2) solidification of the droplets (freezing), when the atomized liquid metals solidify upon coming into contact with a cryogenic medium (usually liquid nitrogen), and (3) sublimation of frozen particles at low temperature and pressure (Parthasarathi and Anandharamakrishnan 2016; Isleroglu et al. 2018). The flow diagram of spray FD is shown in Figure 1.4.

Due to the lower temperatures employed during the spray FD process, thermally sensitive materials can be processed with success. Microencapsulation of heat-sensitive probiotic microbes and chemicals including vitamin E, bovine serum albumin, docosahexaenoic acid, and fish oil has been a popular use for spray FD in recent years (Parthasarathi and Anandharamakrishnan 2016; Isleroglu and Turker 2019). Spray FD was reported to be an efficient way of encapsulating *Lactobacillus paracasei* utilizing a maltodextrin–trehalose mixture as a matrix, with final viabilities more than 60% (Semyonov et al. 2010). Scientists noted that the atomizer is crucial to the production of high-quality goods because it controls

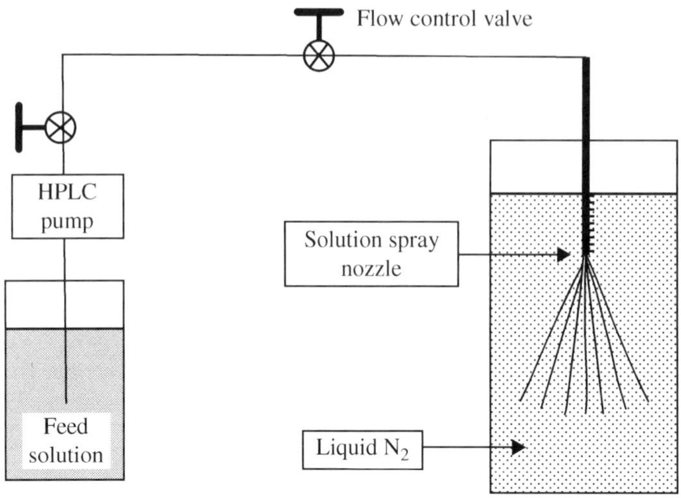

Figure 1.4 Flow illustration of spray freeze-drying. *Source:* Ishwarya et al. (2015)/ Reproduced with permission from Elsevier.

the spray's droplet size distribution. Ultrasonic-aided spray FD (USFD) is the most recent development in the industry, and it has been used to achieve controlled particle size and shorter drying times (Ishwarya et al. 2015; Isleroglu and Turker 2019). Ultrasonic nozzles were found to be responsible for regulating the particle size and aerodynamic properties of spray FD particles when using USFD to create large, porous particles of mannitol, lysozyme, and BSA protein (D'Addio et al. 2012).

Atomization can be accomplished with a variety of nozzle configurations, including those that use two fluids, three fluids, four fluids, and even ultrasonic waves. It has been reported that the ultrasonic nozzle provides very precise regulation of particle size, and that four fluid nozzles are well suited for medications with low aqueous solubility. Figure 1.5 depicts the recent use of four-fluid nozzles for atomization in spray FD (Anandharamakrishnan 2019). There are two compressed air feeds, one for each fluid supply, that atomize the fluids independently. A thin film of liquid is formed over a fast-moving gas when the streams collide. Around the epicenter of the impact, the shock wave disperses small droplets into a mist. Together with liquid nitrogen, these tiny droplets create a more uniform temperature field for heat transfer, which in turn encourages homogenous nucleation and the formation of fine ice crystals.

The comparative research on spray drying, FD, and spray–FD highlighted the significance of spray FD (Ishwarya and Anandharamakrishnan 2014). Authors compared the SFD-grown coffee's physical attributes and aroma quality to those grown using the FD and spray drying procedures. Spray FD aids in preserving coffee's signature low-boiling aromatic components, which are otherwise lost during the first stages of the FD and spray drying processes.

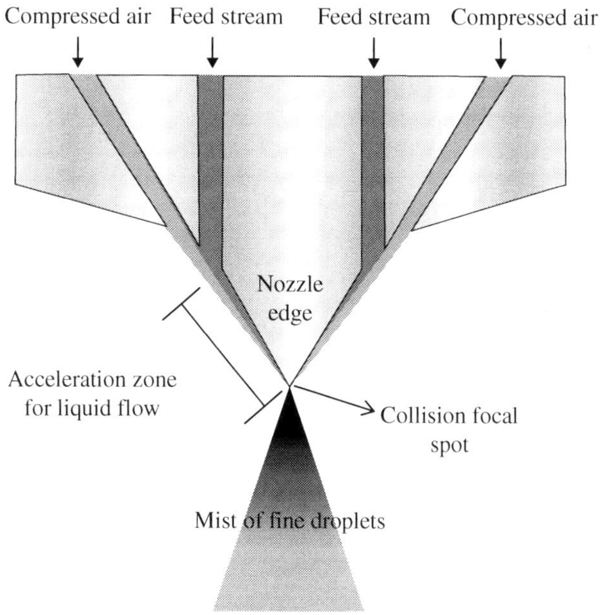

Figure 1.5 Diagram of four-fluid nozzle. *Source:* Anandharamakrishnan 2019/Reproduced with permission from Elsevier.

1.5 Characterization of FD Samples

To determine whether the process was effective and resulted in an appropriate freeze-dried matrix, it is crucial to characterize the systems after they have been freeze-dried. Additionally, it is important to determine if the process has not altered the sample's qualities (Morais et al. 2016). Freeze-dried products are often associated with reduced shrinkage, greater porosity, and the retention of the original cellular structure. The FD method also does not significantly alter the structure of the sample (Ciurzyńska et al. 2022).

1.5.1 Color

Consumers' perceptions of a product's quality are greatly influenced by its visual qualities, making color one of the most crucial indicators of food quality (Waghmare 2021). The quality attributes of FD powder and spray-dried bayberry powder can be determined using color evaluation (Cheng et al. 2017). The powder's browning index significantly enhanced during the period of the 50 days of storage at temperatures of 25 °C. ΔE for FD product was observed to be greater than spray-dried product. In this study, it was found that the FD is suitable and effective technique to retain the superior quality of dried bayberry fruits. Carvalho et al. (2017) investigated the effects of FD and air-drying methods on the color characteristics of whipped yogurt foam. Redness (a^*) was the only color characteristic that did not indicate a significant variation between FD and AD foams. When compared to AD foams, which displayed an average L^* of 79.7, FD foams displayed the lightest color, with L^* value of 93.3. FD aids in preserving whipped yogurt foam's original color and flavor, whereas air-drying promotes Maillard reactions, resulting in a browning exterior and toasted flavor.

1.5.2 Moisture

FD can decrease moisture level up to 2% (Deshwal et al. 2020). The effect of hot air-drying and FD on kimchi powder preparation has been documented (Park et al. 2016). The amount of moisture of kimchi dried powder varied from 10.82% to 7.86%. The amount of moisture present in market product is lower than 10%. Hence, hot air-drying and FD produces dried Kimchi powder as compared to market sample in moisture content. Drying methods for chives were investigated which included hot air-drying (HAD), FD, catalytic infrared drying in combination with hot air-drying (CIRD-HAD), and catalytic infrared drying in combination with FD (CIRD-FD) (Gu et al. 2022). The influence of catalytic infrared drying (CIRD) alone was also studied. The application of CIRD in addition to FD produces greater quality final dried product in less time and releases higher water from sample. It is difficult to preserve dried amorphous products in powdered form due to their ability for instinctive aggregation and caking, which lowers their quality. The damping of the particle surface results in plasticization and disintegration which leads to derivative variations in sample. Stępień et al. (2020) added greater molecular weight biopolymers such as maltodextrin and inulin to decrease hygroscopicity of freeze-dried avocado powders but it was observed that addition of these biopolymers is not required to enhance its stability.

1.5.3 Scanning Electron Microscopy

The ability of a freeze-dried sample to retain and preserve biologics, such as proteins, cells, and other fragile structures, depends on the stability of the sample's structure (Palmkron et al. 2023). The steps of FD are dehydration, freezing, and secondary drying. Powder's microporous surface is a result of the sample's water sublimating directly from ice to water vapor in the vacuum (Chen et al. 2023). Morphological variations were observed for FD and freeze-dried powder with cryoprotectants for *Lactobacillus rhamnosus* (Chen et al. 2023). In FD sample, the cells are extremely damaged, whereas FD powder with cryoprotectants showed smooth surface and entirely coated by the cryoprotectants. Zhou et al. (2021) observed shrinkage dissimilarity among vacuum FD, and hot air-dried garlic slices have presented in SEM images in Figure 1.6. In comparison to the vacuum

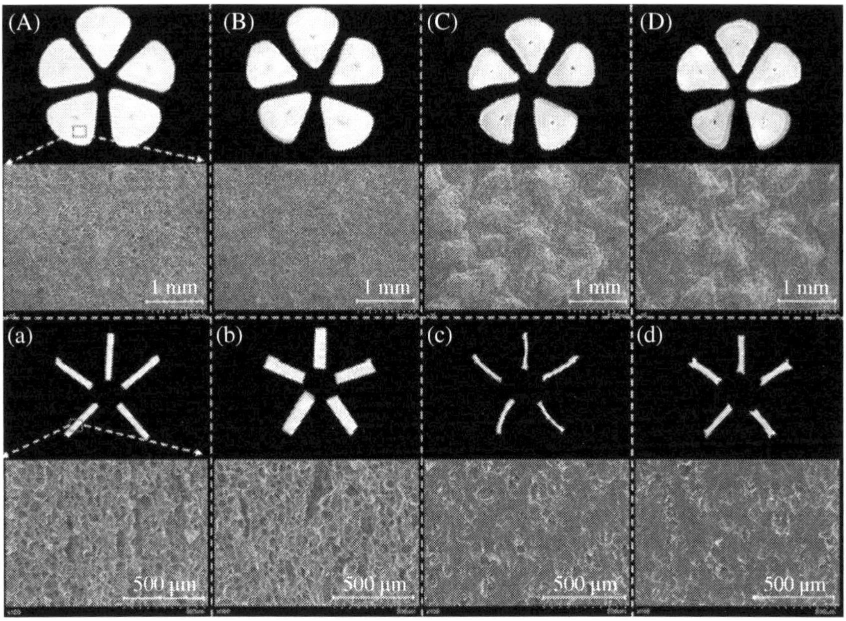

Figure 1.6 SEM pictures of the surface (A–D) and interior (a–d) of freeze-drying garlic with various slice thicknesses treated by vacuum freeze-drying and hot air-dried. (A) Vacuum freeze-drying – 3 mm sample. (B) Vacuum freeze-drying – 6 mm sample. (C) Hot air-dried – 3 mm sample. (D) Hot air-dried – 6 mm sample. *Source:* Cunshan Zhou et al. (2021)/Reproduced with permission from Elsevier.

FD slices, the surface of hot air-dried slices was irregular and significantly distorted. Vacuum FD developed extremely wide pores because of the development of ice crystals in freezing step. While the quick evaporation of water caused the garlic structure to extreme shrinkage and deform during hot air-drying, which resulted in the creation of a dense structure. In another research, the damaged surface layer with the invariable dispersion of minerals on the surface was visible with SEM images of FD bovine colostrum whey powder (Mehra et al. 2022).

1.5.4 Thermal Analysis

In some FD processes, freezing is followed by a further annealing phase. The material temperature is raised during the annealing process above the glass transition temperature (T_g'), but lower the melting point of the frozen state. Improvement in ice crystal area and improved homogeneity inside and between vials in a FD are the two main aims of annealing (Palmkron et al. 2023). Glass transition temperatures (T_g) indicate the conversion in amorphous form of material from glassy form to rubbery forms. Greater T_g value is related to the greater stable product (Waghmare 2021). Examining the behavior of materials as they undergo thermal decomposition is vital to ensure safety, and this assessment can be achieved through a range of tests such as differential scanning calorimetry (DSC), thermogravimetry (TGA), microcalorimetry (C80), and adiabatic calorimetry (ARC) (Ding et al. 2022). Calorimetry examines thermal characteristics of material by observing its particular physical characteristics in relation to temperature. Differential scanning calorimetry evaluates the amount of heat absorbed in response to variation in temperature. This method is frequently employed to monitor variations in phase transformation, including freezing, melting, glass transition, boiling, crystallization, and deterioration (Morais et al. 2016). In addition, this method is crucial for evaluation the recrystallization that can develop with few cryoprotectants, which may cause the process to become unstable (Zhang et al. 2008).

1.5.5 Rehydration Characteristics

Due to the ability to signify the physical and chemical modifications in samples during drying, the rehydration property of the dried sample was used as a quality index (Yali and Dongguang 2018). The quality of the dry sample improves with increasing rehydration. Freeze-dried pineapple, mango, guava, acerola, and papaya were analyzed by Marques et al. (2009) for their rehydration properties. These particular items were rehydrated in distilled water at a temperature of 25 °C over a time frame of six hours. In this study, it was observed that the rehydration of the FD fruits included an initial high water uptake rate, a subsequent low rate, and eventually a condition that was near to saturation. In contrast to acerola and guava, which had increased rehydration ratios at saturation, mango, papaya, and pineapple showed increased rehydration rates. The ability to absorb water was influenced by damages during moisture elimination and by structural disintegration brought about by the rehydration process. Aksoy et al. (2019) conducted research on dried minced beef and found that the rehydration ratio of FD samples was higher than that achieved by ultrasound-assisted vacuum drying (USV) and vacuum drying (VD). The rehydration ratio was shown to be greater for FD and CIRD-FD than for hot air-drying and CIRD-HAD by Gu et al. (2022).

1.6 Applications of FD in Food Processing

Freeze-dried foods are extensively used in various field activities for example aeronautics, armed forces, climbing, and research. These dried materials can be preserved at ambient temperature for significant time period after packaging (Liu et al. 2022). Major applications of FD in food products are presented in Figure 1.7.

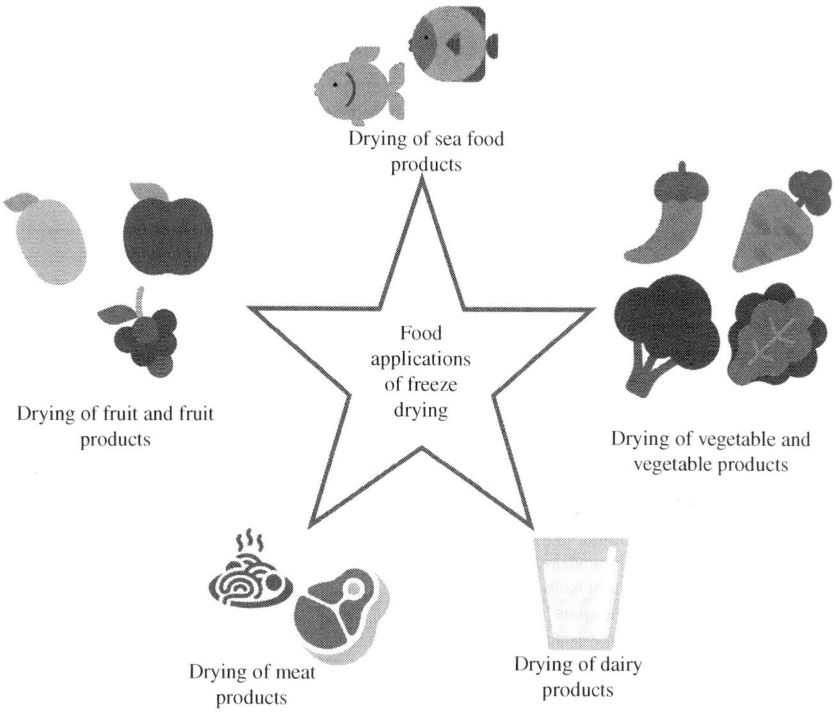

Figure 1.7 Major food applications of freeze-drying.

1.6.1 Fresh Produce

Fresh produce might not always be readily accessible for consumer and increased shelf life of fresh produce may be difficult due to presence of high water, lack of cold-storage devices (especially in developing and poor nations), and potential for quality degradation. As a result, drying of these fresh produce makes them easier to handle, transport, and increase shelf life. Fresh fruits have high water content and are crucial to dry using traditional drying methods because it causes considerable damages to the product (Bhatta et al. 2020). Feng et al. (2022) observed that the taro powder which primarily contains polysaccharides and apple powder which primarily contains mono-/disaccharides prevent structural damages in selected 12 freeze-dried fresh produce. It was also observed that the mono- and disaccharides, particularly fructose, enhances hygroscopicity, while polysaccharides decreased it in final FD samples. Wang et al. (2022) studied energy-conservation technique of vacuum FD kiwi fruit by providing changes in primary drying step. The method in which material temperature is greater than eutectic point temperature reduces energy use by 35% when compared to alternative procedures. Schössler et al. (2012) also reported decrease in period of drying needed to achieve final moisture of 10–11.5% (on dry matter basis) using ultrasonic treatment to FD for red bell pepper cubes. Ultrasonic treatment reduces drying periods and hence has the ability to decrease production cost of final dried powder.

1.6.2 Animal Products

Due to their high levels of important nutrients like protein, vitamins, minerals, and vital fatty acids, foods derived from animals are widely regarded as healthy and of high quality. Nevertheless, due to their large water activity and nutrient content, animal foods are prone to microbial spoilage. Therefore, animal products are acknowledged as foods that spoil rapidly and have a short storage period (Zouaghi and Cantalejo 2016). A series of extraction procedures were used to create a land snail meat substance with higher protein content by eliminating mucus, ash, and lipids (Pissia et al. 2022). A freeze-dried snail meat concentration was created from the remaining leftover part from the last extraction. However, the prepared snail meat powder needs further research to use it as a functional ingredient in food products. In another study performed by Cantalejo et al. (2016), four months storage life was achieved for FD raw chicken breast meat. The application of ozone on FD product presented a storage period of up to 8 months at $21 \pm 1\,°C$. Similarly, condition of modified atmosphere packaging along with ozonated treatment retains the textural and sensorial characteristics of FD chicken meat samples.

1.6.3 Dairy Products

Since spray drying can rapidly remove moisture from atomized milk droplets using hot air, it can be used with a wide variety of foods and is also a cost-effective method, it is mostly used to manufacture powders of dairy products. Dairy products receives several benefits from FD, since it inhibits protein denaturation and the Maillard reaction, increases the preservation of thermally sensitive vitamins, and produces stiff and porous products that are shelf stable (Deshwal et al. 2020). Over the past several years, the interest for dairy products that are good for human health has increased due to increase in consumer demand (Mehra et al. 2022). As a result of its high microbial load and low coagulation temperature, bovine colostrum spoils quickly. Bovine colostrum has a short shelf life because of the high number of bacteria in it and the low temperature at which it coagulates. FD is preferred drying method for converting bovine colostrum whey into bovine colostrum whey powder. It helps to reduce degradation of heat-sensitive compounds, mainly IgG, and increases storage period without changing its qualities. FD bovine colostrum whey powder presented a significant quantity of various nutrients such as IgG, protein, and total amino acids. Carvalho et al. (2017) investigated the impact of air-drying and FD on the characteristics of yogurt foam. Water activity was higher in air-dried samples than in freeze-dried samples. In addition, freeze-dried samples were much softer and less prone to cracking than their air-dried counterparts.

1.6.4 Miscellaneous Applications

Because of its popularity, coffee is one of the world's most traded agricultural products. Ishwarya and Anandharamakrishnan (2015) examined the efficiency of spray drying, FD, and spray FD method for producing soluble coffee. Spray FD enhances instant solubility and flow characteristics of coffee powder. The electronic nose study revealed that the aroma profiles of spray FD and FD coffee granules were similar. The researchers demonstrated that the spray FD method may be used to develop soluble coffee with enhanced characteristics. Park et al. (2016) compared the physicochemical characteristics of vacuum freeze-dried kimchi powder with those prepared by hot air-dried. It was found that the vacuum FD is the most effective for preparing superior quality kimchi powder. Furthermore, sensory score was higher for FD product than hot air-dried product. The water-related quality in freeze-dried multivegetable snacks was determined, specifically the effect of water activity on the structure and shrinkage of product (Ciurzyńska et al. 2022). Sodium alginate-infused FD snacks have a compressed structure and without having more prominent pore. The effect of FD on physicochemical properties of different food products is presented in Table 1.1.

1.7 Advantages of FD

FD offers several benefits over conventional drying techniques in the food area. Following are the few advantages:

- FD retains the original physical characteristics of food material as well as enzymatic and nonenzymatic reactions are also prevented.
- The amount of water in FD sample can reach up to 2–3%, which permit higher shelf life of food products.
- Significant benefits include high solubility and readily reconstituting dried product by rehydration, particularly for coffee or other instant foods. In this case, greater porosity and uniformly dispersed pores are also essential (Ciurzyńska et al. 2022).

Table 1.1 Effect of freeze-drying (FD) on physicochemical properties of different food products.

Food products	Sample	FD	FD model	Other drying methods	Key conclusion	References
Fresh produce	Garlic slices	Vacuum FD: prefrozen at −80 °C for 3 h; drying temperature − 18 °C; vacuum 0.518 mbar; cold-trap temperature of −85 °C	Epsilon 2-6D LSC plus, Martin Christ Ltd.	Hot air-drying: temperature 60 °C; air speed 2 m/s	Vacuum FD product presented less moisture than hot air-dried product. Vacuum freeze-dried product showed increased water absorption rate than hot air-dried product.	Zhou et al. (2021)
	Pineapple, mango, guava, acerola, and papaya	Pressure 13 kPa and temperature −30 °C	Model L4KR, Edwards, Sao Paulo, Brazil	—	Mango, papaya, and pineapple showed increased rehydration rates in comparison to acerola and guava.	Marques et al. (2009)
	Bayberry	Drying temperature −50 °C; drying time 48 h	FD-1C-50 (Boyikang, Beijing, China)	Spray drying: aspirator rate 100% (35 m^3/h), entry temperature 160 °C, exit temperature 90 °C	Higher amount of bioactive compounds were found in FD sample than SD. FD is an effective and appropriate technique to retain bioactive compounds.	Cheng et al. (2017)
	Chives	The prefreezing temperature: −20 °C; drying temperature: −80 °C	Epsilon 2-6D LS C plus, Martin Christ Ltd., Germany	Hot air-drying: temperature 70 ± 5 °C, catalytic infrared drying (CIRD): temperature 70 ± 5 °C	CIRD-FD most effectively retains quality characteristics. Highest retention of chlorophyll in CIRD-FD and freeze-dried samples.	Gu et al. (2022)

	Avocado	Temperature −20 °C; pressure 0.035 mbars for 48 h	Alpha 1-2LD plus Christ (United Kingdom)	—	Water activity of the FD powders including avocado, maltodextrin, and inulin were varying from 0.11 to 0.86.	Stępień et al. (2020)
	Red bell pepper	Temperature −30 °C	Edwards Modulyo II, Edwards High Vacuum Int., UK	—	Ultrasonic treatment to the sample enhances temperature which helps in decrease of energy consumption.	Schössler et al. (2012)
Animal products	Minced meat	Drying temperature: −55 °C and 1 hPa for 72 h	Martin Christ, Beta 1-8 LSC plus, Osterode am Harz, Germany	Ultrasonic vacuum drying: 25, 35, and 45 °C Vacuum drying: 25, 35, and 45 °C; pressure 60 mbar; pump speed 2 l/s	Drying period remarkably reduced with increment in drying temperature. FD method leads to increased rehydration ratios, lesser shrinkage, and color variation.	Aksoy et al. (2019)
	Dogfish skin gelatin gels	—	—	Spray drying	Maximum yield (8.67%) was found for oven-pretreated FD and least (3.06%) for oven-pretreated spray dried. Application as functional additive in various food production.	Salem et al. (2020)

(*Continued*)

Table 1.1 (Continued)

Food products	Sample	FD	FD model	Other drying methods	Key conclusion	References
	Snail meat	—	—	—	Snail meat concentrate powder is high in protein and less in lipids content. The production of snail meat concentrate influences the physicochemical characteristics of snail meat protein.	Pissia et al. (2022)
Dairy products	Bovine colostrum whey powder	Cut into 1 cm^3 sized cubes and dried with lyophilizer	Free zone, Labconco, Kansas City, MO, USA	—	Significant quantity of protein, amino acids, and trace minerals. A novel functional food or additive.	Mehra et al. (2022)
	Yogurt foam	Temperature: −80 °C; time 24 h Pressure: 0.1 mbar	Lyoquest, Telstar, Spain	Air-dryer: Temperature: 57 °C; time 24 h	FD retained foam structure however, complete damage was observed in air dried sample. Air-dried sample presented increased water activity freeze-dried samples.	Carvalho et al. (2017)

	Camel milk powder	Pressure: 1.60 mbar; Temperature: −55 to −60 °C	LABCONCO®, Model- 2.5 FreeZone, Kansas City MO, United States	Spray drying: speed of atomizer 22 000 rpm; inlet air temperature 170 ± 5 °C and exit air temperature 70 ± 5 °C	Maximum amount of calcium (15.33 g/kg) and iron (0.012 g/kg) in freeze-dried sample than spray dried. Higher water binding capacity in freeze-dried sample.	Deshwal et al. (2020)
	Raw donkey milk	Pressure: 0.3 mbar; time: 72 h; frozen temperature −80 °C	Zirbus Technology, Bad Grunt, Germany	—	UV-C radiation retains the protein's quality properties which is same as raw milk.	Tedeschi et al. (2023)
Microorganisms	*Lacticaseibacillus paracasei*	Pressure: 0.2 mbar; Temperature: −40 to −30 °C for 24 h	Lyophilization Systems, Inc., USA.	—	Probiotic FD method retained higher probiotic cell survivability (10^{11} CFU/g). Final dried sample presented low aw (~0.363), which proved satisfactory keeping standards.	Sornsenee et al. (2022)
	Lactobacillus rhamnosus	Temperature: −80 °C for 48 h	ALPHA 2–4 LD plus (Marin Christ, Osterode, Germany)	—	Freeze-dried sample can be stable at −20 °C for 60 d. Freeze-dried sample has smooth surface which is completely coated by cryoprotectants.	Chen et al. (2023)

(Continued)

Table 1.1 (Continued)

Food products	Sample	FD	FD model	Other drying methods	Key conclusion	References
Other food products	Multivegetable snack bars	Heating trays 30 °C; pressure 63 Pa; time: 24 h	Christ ALPHA 1–4 (Germany)	—	More porous structure of FD sample affects its water vapor absorption capacity.	Ciurzyńska et al. (2022)
	Kimchi powder	55 ± 2 °C for 12 h; 60 ± 2 °C for 10 h and 65 ± 2 °C for 8.5 h	Bondiro DC1316 Iishin Lab Co., LTD	Hot air-dryer: temperature 55 ± 2 °C, 60 ± 2 °C and 65 ± 2 °C	Higher water absorption index and water-soluble index of freeze-dried product in comparison to hot air-dried product. Highest retention of quality parameters in FD than hot air-drying.	Park et al. (2016)
	Coffee	Temperature ranges from −40 to 10 °C	Lyodryer-LT-5S; Lyophilization systems, Inc., USA	Spray FD: pressure 588.39 kPa; −25 to −10 °C under vacuum of 107 Pa Spray drying: inlet air temperature 150 ± 2 ° Coutlet air temperature 100 ± 2 °C; pressure 392.27 kPa	Spray FD leads to more free and tapped bulk densities followed by spray drying and FD. Spray FD preserved more aroma which were lost in freeze-dried and spray fried samples.	Ishwarya and Anandharamakrishnan (2015)

- For the encapsulation of food components, FD is least hazardous drying techniques, especially for bioactive substances.
- FD prohibits the loss of nutrients as well as it is highly appropriate drying technique for thermally sensitive food materials (Liu et al. 2022).
- Less residual humidity (10%) makes it simple to handle the FD material during transportation and storage (Cantalejo et al. 2016).

1.8 Limitations of FD

- FD is distinctive drying method used to create premium food materials. Notwithstanding, because of intensive energy needs for sublimation and vacuum creation, the drying technique is quite expensive as well as time-consuming (Schössler et al. 2012; Salem et al. 2020).
- Hygroscopicity and brittleness are the drawback of FD material which may result in restrictions during industrial operations such as choice of the proper packaging material and handling.
- Freeze-dried final products loses its physicochemical characteristics and also its appealing appearance when comes in contact with oxygen and light (Ciurzyńska et al. 2022).
- The intricate heat and mass transfer methods in FD have an effect on the final product. This facet should therefore be thoroughly investigated (Liu et al. 2022).

1.9 Commercial Status

1.9.1 Technological Applications in Food

The technique of FD is frequently used to remove moisture from various food products to improve their stability. Not all food samples are appropriate for FD, but some foods freeze-dry efficiently such as coffee, fruit juices, vegetables, herbs, meat, and eggs. It is required to cut big size food samples into smaller portions before being subjected to FD. Fruits are an excellent example of a food sample that freeze-dries effectively. The FD duration as well as the quality standards used to assess fruits that have been frozen dried are affected by the processing circumstances. Rehydration and texture are two important factors that are related to the final application of the freeze-dried product. These are now widely used as snacks and as components in other meals, like cereal bars and corn flakes. (Ciurzyńska and Lenart 2011; Freeze-Drying Fundamentals n.d.).

Furthermore, FD is generally employed for preservation of cultures and the formation of concentrated starter cultures. Severe climatic circumstances, like reduced temperature and reduced water activity, are experienced by the cells during FD, which reduces their viability (Palmfeldt and Hahn-Hägerdal 2000). In addition to this, therapeutic proteins are frequently enhanced in terms of shelf life by FD. Understanding the processes by which solutes preserves the protein from stresses brought on by FD as well as from injuries brought on by later storage is crucial if one is to produce a product with the best possible storage stability (Kreilgaard et al. 1998).

1.10 Conclusion and Future Perspectives

The process of FD consists of three distinct phases: freezing, primary drying, and secondary drying. It is the method of choice because it consistently yields higher quality products with longer storage lives than do other drying methods. For long-term storage and reconstitution, FD is an excellent method for preserving a wide range of heat-sensitive items. Furthermore, FD food product quality characteristics like rehydration and porosity are beneficial for producing different food products for example soup, quick beverages, and bakery products. In recent past, FD technical problems have been overwhelmed by adding pretreatments before FD or using FD with innovative techniques. Therefore, FD technology has immense market potential.

References

Aksoy, A., Karasu, S., Akcicek, A., and Kayacan, S. (2019). Effects of different drying methods on drying kinetics, microstructure, color, and the rehydration ratio of minced meat. *Foods* 8: 216. https://doi.org/10.3390/foods8060216.

Anandharamakrishnan, C. (2019). Spray-freeze drying of coffee. In: *Caffeinated and Cocoa Based Beverages* (ed. A.M. Grumezescu and A.M. Holban), 337–366. Elsevier https://doi.org/10.1016/B978-0-12-815864-7.00010-6.

Assegehegn, G., Brito-de la Fuente, E., Franco, J.M., and Gallegos, C. (2019). The importance of understanding the freezing step and its impact on freeze-drying process performance. *Journal of Pharmaceutical Sciences* 108: 1378–1395.

Bhatta, S., Stevanovic Janezic, T., and Ratti, C. (2020). Freeze-drying of plant-based foods. *Food* 9 (1): 87. https://doi.org/10.3390/foods9010087.

Bhushani, A, and C Anandharamakrishnan. 2017. Freeze Drying. In Handbook of Drying for Dairy Products Editors: C. Anandharamakrishnan https://doi.org/10.1002/9781118930526.ch6.

Cantalejo, M.J., Zouaghi, F., and Pérez-Arnedo, I. (2016). Combined effects of ozone and freeze-drying on the shelf-life of broiler chicken meat. *LWT - Food Science and Technology* 68: 400–407.

Carvalho, M.J., Perez-Palacios, T., and Ruiz-Carrascal, J. (2017). Physico-chemical and sensory characteristics of freeze-dried and air-dehydrated yogurt foam. *LWT - Food Science and Technology* 80: 328–334.

Chen, B., Wang, X., Li, P. et al. (2023). Exploring the protective effects of freeze-dried *Lactobacillus rhamnosus* under optimized cryoprotectants formulation. *LWT - Food Science and Technology* 173.

Cheng, A.W., Xie, H.-X., Qi, Y. et al. (2017). Effects of storage time and temperature on polyphenolic content and qualitative characteristics of freeze-dried and spray-dried bayberry powder. *LWT - Food Science and Technology* 78: 235–240.

Ciurzyńska, A., Galus, S., Karwacka, M., and Janowicz, M. (2022). The sorption properties, structure and shrinkage of freeze-dried multi-vegetable snack bars in the aspect of the environmental water activity. *LWT - Food Science and Technology* 171.

Claussen, I.C., Ustad, T.S., Str⊘mmen, I., and Walde, P.M. (2007). Atmospheric freeze drying—a review. *Drying Technology* 25 (6): 947–957. https://doi.org/10.1080/07373930701394845.

Colucci, D., Fissore, D., Rossello, C., and Carcel, J.A. (2018). On the effect of ultrasound-assisted atmospheric freeze-drying on the antioxidant properties of eggplant. *Food Research International* 106: 580–588.

D'Addio, S.M., Chan, J.G.Y., Kwok, P.C.L. et al. (2012). Constant size, variable density aerosol particles by ultrasonic spray freeze drying. *International Journal of Pharmaceutics* 427 (2): 185–191.

Danial, D., Seid, M.J., and Maryam, A. (2016). Influence of drying on functional properties of food biopolymers: from traditional to novel dehydration techniques. *Trends in Food Science & Technology* 57: 116–131.

Deshwal, G.K., Singh, A.K., Kumar, D., and Sharma, H. (2020). Effect of spray and freeze drying on Physico-chemical, functional, moisture sorption and morphological characteristics of camel Milk powder. *LWT - Food Science and Technology* 134: 110117.

Ding, X., Zhao, S., Ni, L., and Pan, Y. (2022). Thermal hazard and mechanism study of 5-(4-Pyridyl) tetrazolate (H4-PTZ). *Emergency Management Science and Technology*, 2(1), 1–8.

Fellows, P.J. (2009). Freeze drying and freeze concentration. In: *Food Processing Technology: Principles and Practice* (ed. P.J. Fellows), 441–451. Elsevier.

Feng, S., Bi, J., Yi, J. et al. (2022). Cell wall polysaccharides and mono–/disaccharides as chemical determinants for the texture and hygroscopicity of freeze-dried fruit and vegetable cubes. *Food Chemistry* 395: 133574. https://doi.org/10.1016/j.foodchem.2022.133574.

Freeze-Drying Fundamentals (2016). Available from: https://biotecharticles.com/Applications-Article/Fundamentals-and-Applications-of-Freeze-Drying-3557.html (accessed on 03 March 2023).

Garcia-Amezquita, L.E., Welti-Chanes, J., Vergara-Balderas, F.T., and Bermúdez-Aguirre, D. (2015). Freeze-drying: the basic process. In: *Encyclopedia of Food and Health* (ed. B. Caballero, P.M. Finglas, and F. Toldrá), 104–109. Elsevier Inc.

Gu, C., Ma, H., Tuly, J.A. et al. (2022). Effects of catalytic infrared drying in combination with hot air drying and freeze drying on the drying characteristics and product quality of chives. *LWT - Food Science and Technology* 161: 113363.

Guo, X., Shi, L., Xiong, S. et al. (2019). Gelling properties of vacuum-freeze dried surimi powder as influenced by heating method and microbial transglutaminase. *LWT- Food Science and Technology* 99: 105–111.

Hua, T., Liu, B., and Zhang, H. (2010). Introduction. In: *Woodhead Publishing Series in Food Science, Technology and Nutrition, Freeze-Drying of Pharmaceutical and Food Products* (ed. T.-C. Hua, B.-L. Liu, and H. Zhang), 1–17. Woodhead Publishing. ISBN: 9781845697464 https://doi.org/10.1533/9781845697471.1.

Ingvarsson, P.T., Yang, M., Nielsen, H.M. et al. (2011). Stabilization of liposomes during drying. *Expert Opinion on Drug Delivery* 8: 375–388.

Ishwarya, S.P. and Anandharamakrishnan, C. (2014). Spray-freeze-drying approach for soluble coffee processing and its effect on quality characteristics. *Journal of Food Engineering* 149: 171–180. https://doi.org/10.1016/j.jfoodeng.2014.10.011.

Ishwarya, S.P., Anandharamakrishnan, C., and Stapley, G.F. (2015). Spray-freeze-drying : a novel process for the drying of foods and bioproducts. *Trends in Food Science and Technology* 41 (2): 161–181. https://doi.org/10.1016/j.tifs.2014.10.008.

Ishwarya, S.P. and Anandharamakrishnan, C. (2015). Spray-freeze-drying approach for soluble coffee processing and its effect on quality characteristics. *Journal of Food Engineering* 149: 171–180. https://doi.org/10.1016/j.jfoodeng.2014.10.011.

Isleroglu, H. and Turker, I. (2019). Thermal inactivation kinetics of microencapsulated microbial transglutaminase by ultrasonic spray-freeze drying. *LWT - Food Science and Technology* 101: 653–662.

Jiang, N., Liu, C., Li, D. et al. (2017). Evaluation of freeze drying combined with microwave vacuum drying for functional okra snacks: antioxidant properties, sensory quality, and energy consumption. *LWT- Food Science and Technology* 82: 216–226.

Kreilgaard, L., Frokjaer, S., Flink, J.M. et al. (1998). Effects of additives on the stability of recombinant human factor XIII during freeze-drying and storage in the dried solids. *Archives of Biochemistry and Biophysics* 360 (1): 121–134.

Lim, J.Y., Lim, D.G., Kim, K.H. et al. (2018). Effects of annealing on the physical properties of therapeutic proteins during freeze drying process. *International Journal of Biological Macromolecules* 107 (PartA): 730–740.

Liu, Y., Zhao, Y., and Feng, X. (2008). Exergy analysis for a freeze-drying process. *Applied Thermal Engineering* 28: 675–690.

Liu, Y., Zhang, Z., and Hu, L. (2022). High efficient freeze-drying technology in food industry. *Critical Reviews in Food Science and Nutrition* 62 (12): 3370–3388. https://doi.org/10.1080/10408398.2020.1865261.

Ma, J., Jia Huan, Q., and Sun, D.W. (2017). Developing hyperspectral prediction model for investigating dehydrating and rehydrating mass changes of vacuum freeze dried grass carp fillets. *Food and Bioproducts Processing* 104: 66–76.

Ma, Y., Wu, X., Zhang, Q. et al. (2018). Key composition optimization of meat processed protein source by vacuum freeze drying technology. *Saudi Journal of Biological Sciences* 25 (4): 724–732.

Maisnam, D., Rasane, P., Dey, A. et al. (2015). Recent advances in conventional drying of foods. *The Journal of Food Technology and Preservation* 1 (1): 25–34.

Marques, L.G., Prado, M.M., and Freire, J.T. (2009). Rehydration characteristics of freeze-dried tropical fruits. *LWT - Food Science and Technology* 42 (7): 1232–1237.

Martínez-Navarrete, N., Salvador, A., Oliva, C., and Camacho, M.M. (2019). Influence of biopolymers and freeze-drying shelf temperature on the quality of a mandarin snack. *LWT - Food Science and Technology* 99: 57–61.

Mehra, R., Kumar, S., Singh, R. et al. (2022). Biochemical, dielectric and surface characteristics of freeze-dried bovine colostrum whey powder. *Food Chemistry* X 15.

Mellor, J.D. (1978). *Fundamentals of Freeze-Drying*. London: Academic Press.

Meryman, H.T. (1976). Historical recollections of freeze-drying. *Developments in Biological Standardization* 36: 29–32. PMID: 801137.

Millman, M.J., Liapis, A.I., and Marchello, J.M. (1985). Note on the economics of batch freeze dryers. *Journal of Food Technology* 20: 541–551.

Morais, A.R.D.V., do Nascimento Alencar, É., Júnior, F.H.X. et al. (2016). Freeze drying of emulsified systems: a review. *International Journal of Pharmaceutics* 503 (1–2): 102–114. http://dx.doi.org/10.1016/j.ijpharm.2016.02.047.

Nowak, D. (2017). *The Innovative Measurement System of the Kinetic of Freeze-Drying and Sorption Properties of Dried Products as a Tool for Controlling and Assessing the Course of Freeze-Drying*, 217. Warsaw, Poland: Warsaw University of Life Sciences Press.

Nowak, D. and Lewicki, P.P. (2005). Quality of infrared dried apple slices. *Drying Technology* 23: 831–846.

Nowak, D. and Jakubczyk, E. (2020). The freeze-drying of foods the characteristic of the process course and the effect of its parameters on the physical properties of food materials. *Food* 9 (10).

Palmfeldt, J. and Hahn-Hägerdal, B. (2000). Influence of culture pH on survival of *Lactobacillus reuteri* subjected to freeze-drying. *International Journal of Food Microbiology* 55: 235–238.

Palmkron, S.B., Bergenståhl, B., Håkansson, S. et al. (2023). Quantification of structures in freeze-dried materials using X-ray microtomography. *Colloids and Surfaces A: Physicochemical and Engineering Aspects* 658: 130726. https://doi.org/10.1016/j.colsurfa.2022.130726.

Park, H.J., Lee, Y., and Eun, J.B. (2016). Physicochemical characteristics of kimchi powder manufactured by hot air drying and freeze drying. *Biocatalysis and Agricultural Biotechnology* 5: 193–198.

Parthasarathi, S. and Anandharamakrishnan, C. (2016). Enhancement of oral bioavailability of vitamin E by spray-freeze drying of whey protein microcapsules. *Food and Bioproducts Processing* 100: 469–476.

Patil, V.V., Dandekar, P.P., Patravale, V.B., and Thorat, B.N. (2010). Freeze drying: potential for powdered nanoparticulate product. *Drying Technology* 28: 624–635.

Pissia, M.A., Matsakidou, A., Paraskevopoulou, A., and Kiosseoglou, V. (2022). Structural characteristics and physicochemical properties of freeze-dried snail meat. *LWT - Food Science and Technology* 155.

Prosapio, V. and Norton, I. (2018). Simultaneous application of ultrasounds and firming agents to improve the quality properties of osmotic + freeze-dried foods. *LWT - Food Science and Technology* 96: 402–410. https://doi.org/10.1016/j.lwt.2018.05.068.

Rambhatla, S., Tchessalov, S., and Pikal, M.J. (2006). Heat and mass transfer scale-up issues during freeze-drying, III: control and characterization of dryer differences via operational qualification tests. *AAPS PharmSciTech* 7 (2): E61–E70.

Rolfgaard, J. (1987). Freeze drying: processing, costs and applications. In: *Food Technology International Europe* (ed. A. Turner), 47–49. London: Sterling Publications International.

Salem, A., Fakhfakh, N., Jridi, M. et al. (2020). Microstructure and characteristic properties of dogfish skin gelatin gels prepared by freeze/spray-drying methods. *International Journal of Biological Macromolecules* 162: 1–10. https://doi.org/10.1016/j.ijbiomac.2020.06.033.

Schössler, K., Jäger, H., and Knorr, D. (2012). Novel contact ultrasound system for the accelerated freeze-drying of vegetables. *Innovative Food Science and Emerging Technologies* 16: 113–120.

Borges Sebastião, I., Robinson, T.D., and Alexeenko, A. (2016). Atmospheric spray freeze-drying: numerical modeling and comparison with experimental measurements. *Journal **of** Pharmaceutical Sciences* 106 (1): 183–192. https://doi.org/10.1016/j.xphs.2016.07.032. Epub 2016 Sep 28. PMID: 27692618.

Semyonov, D., Ramon, O., Kaplun, Z. et al. (2010). Microencapsulation of *Lactobacillus paracasei* by spray freeze drying. *Food Research International* 43 (1): 193–202.

Shi, A.M., Wang, L.J., Li, D., and Adhikari, B. (2012). The effect of annealing and cryoprotectants on the properties of vacuum-freeze dried starch nanoparticles. *Carbohydrate Polymers* 88 (4): 1334–1341.

Sornsenee, P., Chimplee, S., Saengsuwan, P., and Romyasamit, C. (2022). Characterization of probiotic properties and development of Banana powder enriched with freeze-dried *Lacticaseibacillus paracasei* probiotics. *Heliyon* 8 (10).

Stępień, A., Witczak, M., and Witczak, T. (2020). Moisture sorption characteristics of food powders containing freeze dried avocado, maltodextrin and inulin. *International Journal of Biological Macromolecules* 149: 256–261. https://doi.org/10.1016/j.ijbiomac.2020.01.154.

Tang, X. and Pikal, M.J. (2004). Design of freeze drying processes for pharmaceuticals: practical advice. *Pharmaceutical Research* 21 (2): 191–200.

Tedeschi, T., Aspri, M., Loffi, C. et al. (2023). Processing of raw donkey milk by pasteurisation and UV-C to produce freeze-dried milk powders: the effect on protein quality, digestibility and bioactive properties. *LWT - Food Science and Technology* 173: 114404. https://doi.org/10.1016/j.lwt.2022.114404.

Waghmare, R. (2021). Refractance window drying: a cohort review on quality characteristics. *Trends in Food Science & Technology* 110: 652–662. https://doi.org/10.1016/j.tifs.2021.02.030.

Wang, B.B., Sun, P.-Y., Chen, B.-L. et al. (2022). Influence of the process parameters on the energy conservation and quality of the kiwi fruit subjected to vacuum freeze-drying. *Case Studies in Thermal Engineering* 33.

Yali, W. and Dongguang, Z. (2018). Effect of pulsed electric field on collapse phenomenon of apple tissue in vacuum freeze drying. *IOP Conference Series: Materials Science and Engineering* 392: 052007.

Zhang, L., Liu, L., Qian, Y., and Chen, Y. (2008). The effects of cryoprotectants on the freezedrying of ibuprofen-loaded solid lipid microparticles (SLM). *European Journal of Pharmaceutics and Biopharmaceutics* 69: 750–759.

Zhou, C., Feng, Y., Zhang, L. et al. (2021). Rehydration characteristics of vacuum freeze- and hot air-dried garlic slices. *LWT - Food Science and Technology* 143.

Zouaghi, F. and Cantalejo, M.J. (2016). Study of modified atmosphere packaging on the quality of ozonated freeze-dried chicken meat. *Meat Science* 119: 123–131.

2

Mathematical Modeling of Freeze-Drying Process

Kyuya Nakagawa

Department of Chemical Engineering, Faculty of Engineering, Kyoto University, Kyoto, Japan

2.1 Introduction

The mathematical model of the freeze-drying process has great significance for the rational design of processes and operations. Because food drying often deals with products that are irregularly shaped and have a variety of physicochemical properties, a major challenge in modeling is how to include these complex properties into simplified formulas. Since the progress of drying is strongly dependent on the form and path of heat transfer to the product in the drying equipment, it is important to develop a model that is compatible with the mechanism of the drying equipment. Control of the freezing process, which affects the subsequent drying process, is also considered necessary to advance the process. This chapter first introduces the principles and characteristics of freeze-dryers, which are important to know for mathematical modeling. The formation of ice microstructures during freezing step that affects subsequent drying step is then described through a modeling approach of the freezing step. The derivation of mathematical equations and a simple simulation procedure are described for a conventional freeze-drying model applicable to a shelf-type freeze-dryer and a freeze-drying model proposed by the author applicable to a radiation-type freeze-dryer. Freeze-drying can enable the long-term storage of various emergency supplies, including food, medical resources, and other essential items. This can be crucial for disaster-prone areas where rapid response and accessibility to resources are critical for effective emergency management. Utilizing freeze-drying as a means to address emergency concerns and bolster preparedness efforts represents a potential stride toward mitigating the very challenges highlighted by Wang et al. (2021).

Freeze Drying of Food Products: Fundamentals, Processes and Applications, First Edition.
Edited by Roji Balaji Waghmare, Manoj Kumar, and Parmjit Singh Panesar.

2.2 Principle of Freeze-Dryer

2.2.1 Vacuum Freeze-Drying

Typical vacuum freeze-drying equipment consists of a drying chamber, condenser (cold trap), vacuum pump, and temperature control device. In shelf-type freeze-dryers, which are widely used in pharmaceutical processes, temperature-controllable shelves are installed inside to allow cooling, freezing, and heating of the product during freeze-drying (Figure 2.1). Drying involves the transfer of heat to the sublimation interface via conduction through a frozen layer. The shelf temperature is typically set between −30 and 0 °C to achieve an appropriate drying rate within which product collapse does not occur. A shelf-type freeze-dryer has the advantage of realizing precise product temperature control because the temperature of the product in contact with the shelf is controlled by conductive heat transfer.

The freeze-drying equipment that is widely used in food processing is equipped with a heater to heat the product during drying via radiation (Figure 2.2). As drying proceeds from the product surface, the dried product surface is heated by radiation heat transfer and heat is transferred to the drying interface via conduction heat

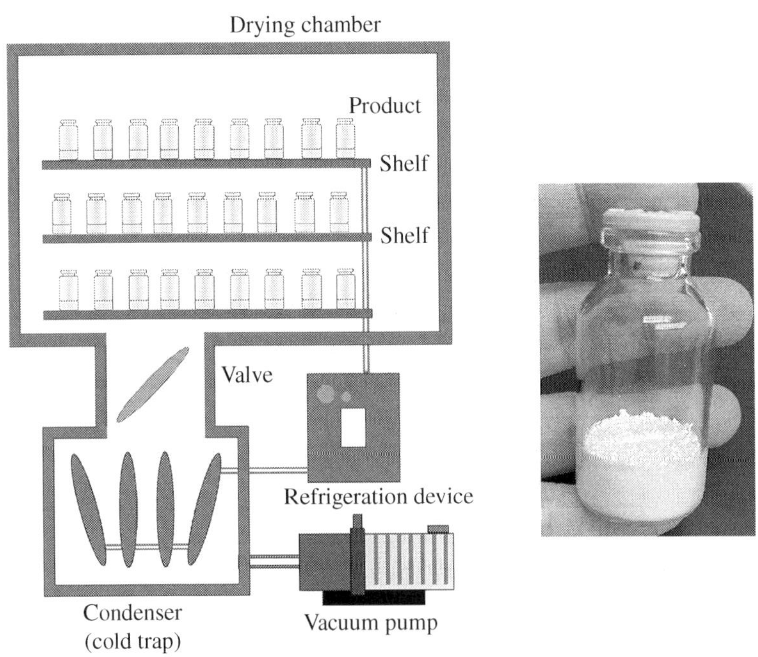

Figure 2.1 Shelf-type freeze-dryer equipped with temperature controllable shelf.

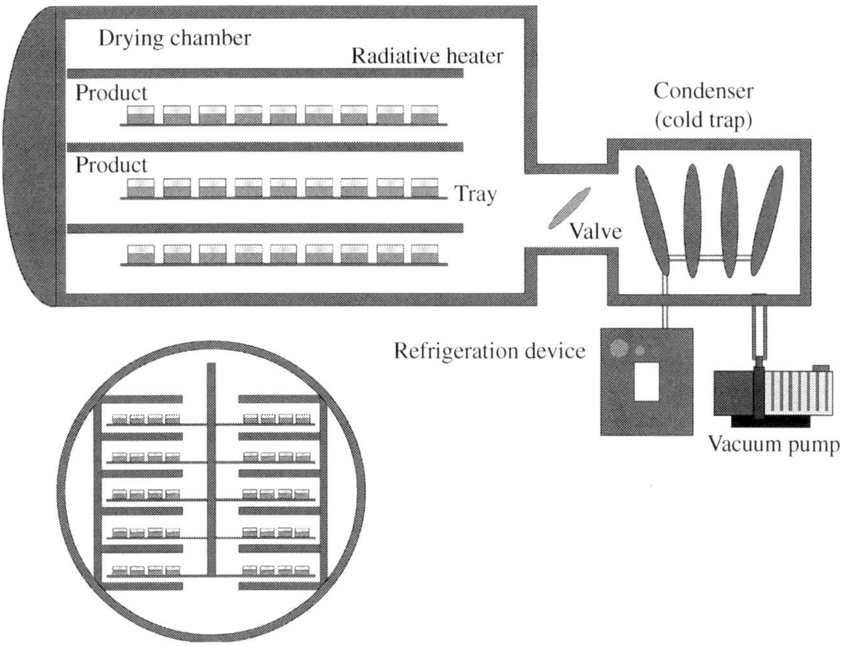

Figure 2.2 Radiation-type freeze-dryer.

transfer through the drying layer. Considering the low effective thermal conductivity of the drying layer, a large temperature gradient must be made inside the drying layer to deliver sufficient heat for drying. The heater temperature is typically set in the range of 50–100 °C.

The pressure in the freeze-dryer is reduced to a predetermined value by an exhaust pump in the initial stage of the drying process. Thereafter, as drying (sublimation) progresses, the water vapor discharged from the product is iced in a cold trap (condensing device/condenser). The cold trap is critical to the pressure in the equipment during the drying process. The ice condensed in the cold trap shows equilibrium water vapor pressure at the trap temperature; therefore, the pressure inside the equipment is ideally maintained equal to the saturated water vapor pressure of ice at the cooling plate temperature. In reality, the water vapor generated on the product surface moves through the equipment at a certain flow rate, which causes pressure distribution inside the equipment depending on the flow rate.

The driving force of the water vapor movement inside the product during the drying process is the pressure difference between the water vapor pressure at the sublimation surface and the pressure inside the drying chamber. The driving force of the water vapor released from the product to move inside the drying cabinet is the pressure difference between the water vapor pressure near the product and that in the cold trap. Therefore, the water vapor pressure at the sublimation surface of the product must be higher than that in the chamber, and the temperature in the cold trap must be maintained lower than the sublimation surface temperature. Thus, maintaining a large pressure difference is advantageous to improve the drying rate. The drying rate of the product is comprehensively determined by the dynamic balance between the mass transfer and/or fluid flow inside the product, the drying chamber, and the cold trap. Depending on the drying process, the rate-limiting step affecting the drying rate is the mass transfer (or fluid flow) inside the product or the heat transfer rate to the sublimation interface.

2.2.2 Atmospheric Freeze-Drying

Atmospheric freeze-drying (AFD) is an alternative freeze-drying technique that achieves ice sublimation at atmospheric pressure; herein, the water vapor partial pressure difference between the frozen zone and ambient gas is the driving force (Heldman and Hohner 1974; Meryman 1959; Wolff and Gibert 1988).

To perform drying at subzero temperatures, the air humidity must be maintained at an appropriately low level. A heat pump system is commonly used to generate low-humidity air (Perera and Rahman 1997). Additionally, the condenser temperature of the heat pump system must be set lower than the product temperature during drying, and the dehumidified air must be properly heated to realize an appropriate driving force for mass and heat transfer, as shown in Figure 2.3. Several studies on the AFD process have sought to optimize this heat pump drying system in terms of energy consumption and drying rate (Alves-Filho 2002; Claussen et al. 2007a,b; Wolff and Gibert 1990). The mass and heat transfer properties can be intensified by drying in a fluidized system and/or hybridizing drying components with external heating systems, such as microwave irradiation and ultrasound systems (Chou and Chua 2001; Claussen et al. 2007b; Colucci et al. 2017, 2018; Di Matteo et al. 2003; Garcia-Perez et al. 2012; Mumenthaler and Leuenberger 1991). The AFD process has easy operation owing to the use of simplified drying components. Additionally, the process above freezing temperatures, known as low-temperature drying, can be performed with the same drying device. Considering the long drying time required for AFD, both of these aforementioned conditions can be combined to reduce the drying time.

The dehydration mechanism of freeze-drying is primarily sublimation from the ice crystal phase (primary drying), which does not result in damage owing to the capillary force. Dehydration from the freeze-concentrated phase occurs by evaporation above the glass transition temperature of the maximally freeze-concentrated matter (T_g'). The T_g' of agricultural products is generally low; for example, that of strawberry ranges from −33 to −41 °C, and that of peaches, apples, persimmons,

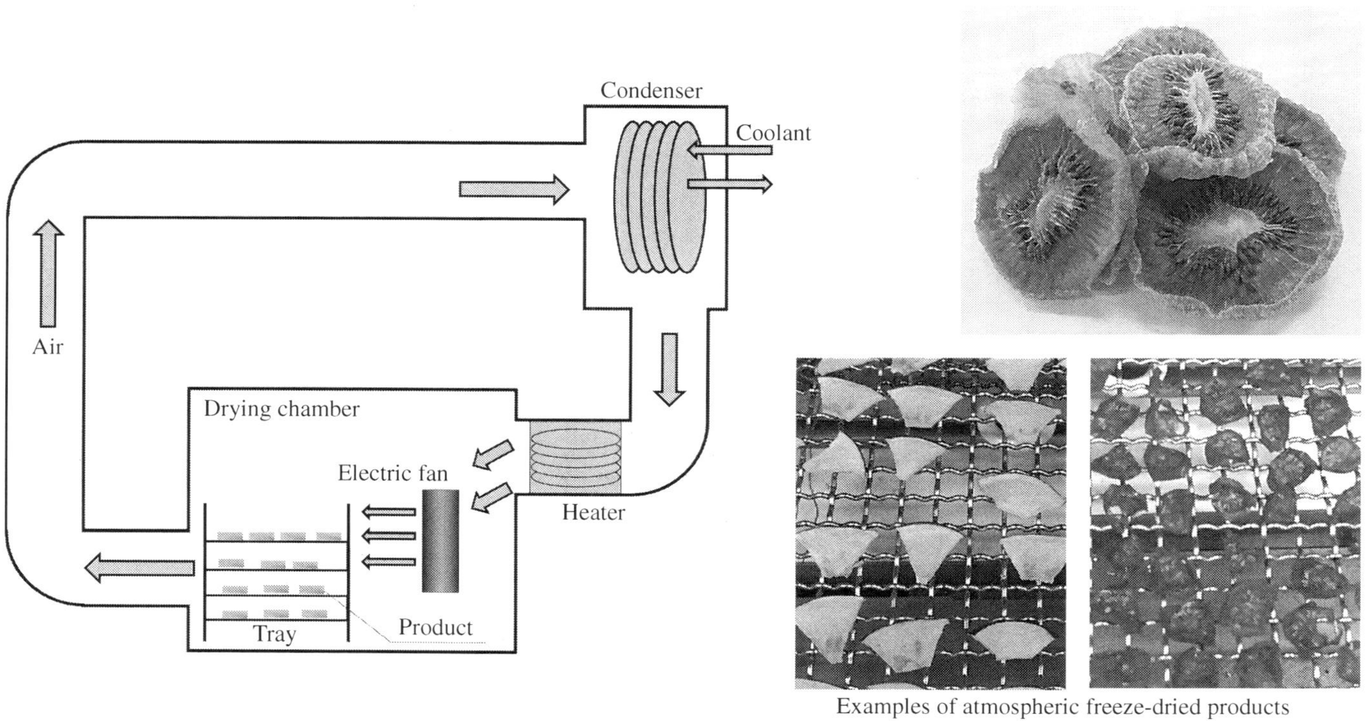

Figure 2.3 Atmospheric freeze-dryer (heat pump dryer).

pineapples, and tomatoes are approximately −36 °C (Slade et al. 1991), −58 °C (Bai et al. 2001), −55 °C (Sobral et al. 2001), −52 °C (Telis and Sobral 2001), and −59 °C (Telis and Sobral 2002) respectively. When a product is dried under typical AFD conditions, the product temperatures are approximately –20 °C or higher, depending on the setting air temperature. Therefore, the drying products are influenced by capillary forces caused by internal moisture because the product is maintained above T_g' for a long duration. Product deformations, such as shrinkage, melt-back, and/or structural collapse, occur as a consequence of glass–rubber–liquid transitions (Nakagawa et al. 2018a,b; Pikal and Shah 1990; Ratti 2001). Although the AFD process is considered to be advantageous for quality assurance owing to the use of low temperature and ice sublimation, the negative impact of quality loss due to structural collapse should be considered (Nakagawa et al. 2020a, 2021). Nonetheless, it exhibits significant potential as a drying method that creates new qualities derived from structural changes. In fact, dried fruits produced via the AFD process are very tasty, with a gummy texture and rich flavor.

2.2.3 Water Vapor Transfer in Freeze-Dryer

The water vapor generated by the product via sublimation is transferred inside the equipment and condenses as frost on the cold trap surface. In most freeze-drying cases, the vapor flow inside the equipment should be considered as a viscous or transitional flow. This can be determined from the Knudsen number (K_n) in the drying chamber, calculated as follows:

$$K_n = \lambda / D = \frac{3.11 \times 10^{-24} T}{d^2 p D}, \tag{2.1}$$

where λ is the mean free pass, D is the representative length (e.g. the diameter of the chamber duct), d is the molecular diameter, p is the pressure, and T is the temperature.

$$\begin{cases} K_n < 0.01 & : \text{Viscous flow} \\ 0.01 < K_n < 0.3 & : \text{Transitional flow.} \\ K_n > 0.3 & : \text{Molecular flow} \end{cases}$$

The velocity of the compressible fluid is determined by the pressure difference and conductance. Water vapor is trapped in the cold trap; however, the trapping performance and capacity gradually decreases as the amount of trapped ice increases. The movement of gas inside the dryer is illustrated in Figure 2.4. During the drying process, the rate of $\left(\frac{dm_s}{dt}\right)\left(\frac{RT_s}{M_W}\right)$ and that of frost formation on the cold trap $\left(\frac{dm_c}{dt}\right)\left(\frac{RT_c}{M_W}\right)$ are balanced, and the rate of gas leakage into the equipment

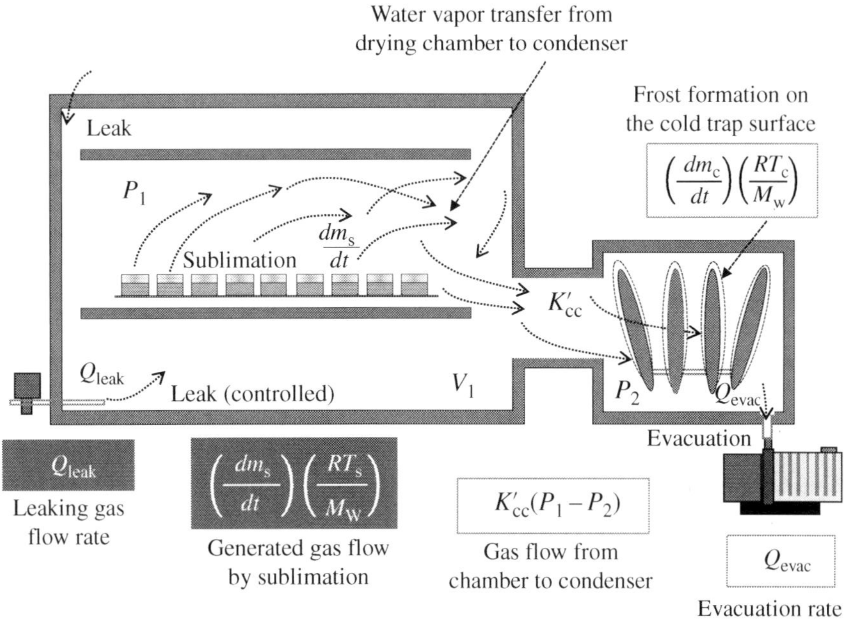

Figure 2.4 Mass balance of gas in a freeze-dryer.

(Q_{leak}) and that of evacuation by the vacuum pump (Q_{evac}) are balanced. Note that herein, gas leakage refers to both unexpected gas leakage into the drying equipment and operational leakage in the case of equipment with a mechanism that allows inert gas (such as nitrogen) to leak into the equipment for pressure control. The balance equation can be expressed as follows:

Drying chamber:

$$V_1\left(\frac{dP_1}{dt}\right) = \left(\frac{dm_s}{dt}\right)\left(\frac{RT_s}{M_W}\right) - K'_{cc}\left(P_1 - P_2\right) + Q_{leak}. \tag{2.2}$$

Condenser:

$$V_2\left(\frac{dP_2}{dt}\right) = K'_{cc}\left(P_1 - P_2\right) - \left(\frac{dm_c}{dt}\right)\left(\frac{RT_c}{M_W}\right) - Q_{evac}. \tag{2.3}$$

where the gas flow rate from the chamber to the condenser ($K'_{cc}(P_1 - P_2)$) is expressed with the conductance of the path in the dryer, K'_{cc}.

In addition, the water vapor pressure is nonuniform within the equipment; consequently, a large pressure distribution can be generated, particularly in industrial equipment where large amounts of product are loaded. Figure 2.5 shows an example of simulated water vapor flow inside a shelf-type freeze-dryer reported in a

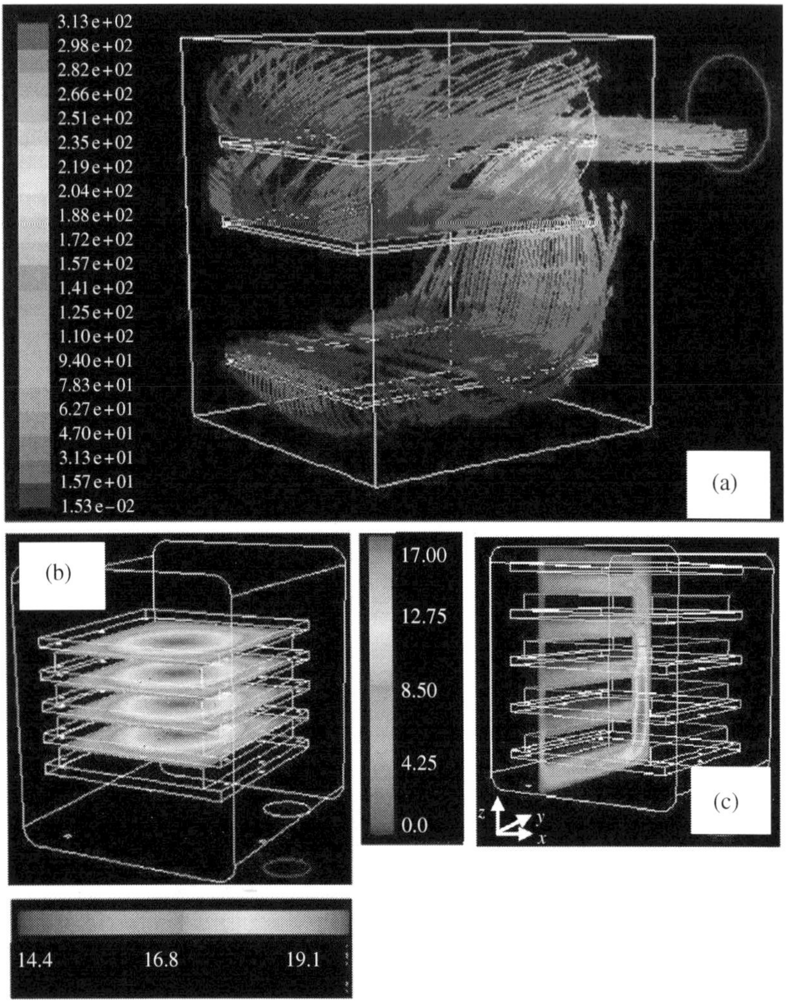

Gas flow in the drying chamber (a). Pressure distribution on the shelf (b). Pressure distribution around the shelf (c)

Figure 2.5 Computational fluid dynamics (CFD) simulation of water vapor flow in a freeze-dryer (Barresi and Marchisio 2018).

previous study (Barresi and Marchisio 2018). As can be seen, a pressure distribution forms inside the equipment on the shelf where the product is placed; the study further reports that the location of the duct connecting the drying chamber to the condenser is an important equipment design parameter.

The flow corresponding to when a compressible fluid attains the speed of sound at a certain cross-section is called choke flow; it is a critical state of compressible flow. The fluid cannot flow above the speed of sound, and when choke flow occurs in a freeze-drying apparatus, the pressure inside the apparatus becomes uncontrollable (Patel et al. 2010). In the equipment, water vapor is generated at a certain rate owing to the progress of sublimation, and it moves through the equipment at a speed corresponding to the rate of sublimation. However, the upper limit of this movement speed is assumed to be reached when more than a certain amount of the product is fed into the system. Once pressure control is lost, the product temperature rises, thereby causing serious product collapse and leading to failure of the drying process. Therefore, in industrial applications, where large amounts of product are loaded to perform drying, the risk of choking for the product feed rate and operating conditions must be predicted and reflected in the operating conditions (Nakagawa and Tamiya 2019; Patel et al. 2010). The cold trap geometry, internal layout of the equipment, and changes in the cold trap performance are also important factors.

2.3 Mathematical Model for Freezing

2.3.1 Thermal Balance during Freezing

Estimating the temperature history of the freezing process under given freezing conditions (such as cooling method and product shape) is useful. In addition to predicting how fast the freezing process will progress under the given temperature conditions, predicting various physicochemical phenomena that occur during the freezing process is also useful. However, experimentally tracking the three-dimensional freezing process inside a product is challenging; in this regard, mathematical simulation, such as the proposed finite element method (described later), is a powerful tool and, thus, is often used (Nakagawa et al. 2007, 2011).

First, a plan view (or three-dimensional view) of the target (solution in a vial) is set up, the interior is divided into a finite number of meshes, and Fourier's heat conduction equation is applied to each mesh. In the calculation, the heat conduction equation is applied to the cooling process until supercooling is released. The boundary conditions are set to reflect the conditions around the vial at the time of freezing, as shown in Figure 2.6. For example, the bottom of the vial is assumed to be cooled by a constant drop in temperature. The heat conduction equation can be written as follows:

$$\rho C_{p\text{-liquid}} \frac{\partial T}{\partial t} = \nabla\left(k \nabla T\right), \tag{2.4}$$

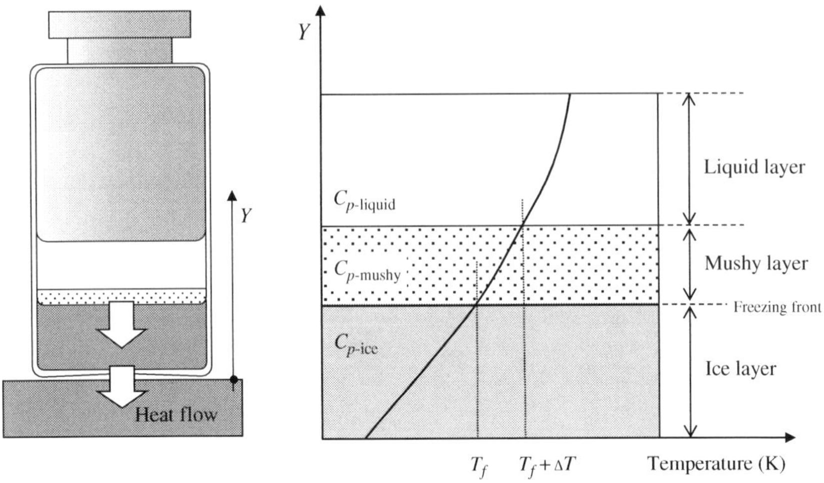

Figure 2.6 Modeling heat transfer of solutions in vials frozen by shelf cooling.

where ρ is the density, C_p is the heat capacity, and k is the thermal conductivity. An area of supercooling begins to form near the bottom of the vial. The timing of the release of supercooling (i.e. the breaking point of supercooling) is unpredictable; therefore, the timing is selected and the temperature distribution at this time is input as the initial condition for the calculation of the freezing process described next. The following heat conduction equation is applied to the freezing process:

$$\rho C_{p\text{-apparent}} \frac{\partial T}{\partial t} = \nabla(k\nabla T) + \dot{Q}_{\mathrm{n}} + \dot{Q}_{\mathrm{c}}, \tag{2.5}$$

where $\dot{Q}_{\mathrm{n}}$ is the enthalpy of solidification released per unit time and per unit volume by the release of supercooling ($\mathrm{W \cdot m^{-3}}$) and $\dot{Q}_{\mathrm{c}}$ is the enthalpy of solidification released per unit time and per unit volume by ice crystal growth after the release of supercooling ($\mathrm{W \cdot m^{-3}}$). The value of $\dot{Q}_{\mathrm{c}}$ is zero until no more supercooling regions exist. Considering that uncooling proceeds at a rate proportional to the degree of supercooling ($T_{\mathrm{f}} - T^*$), the following can be written using the solidification enthalpy ΔH_{f} (Qin et al. 2003):

$$\dot{Q}_n = \Delta H_{\mathrm{f}} k_i \left(T_{\mathrm{f}} - T^*\right), \tag{2.6}$$

where k_i ($\mathrm{kg \cdot m^{-3} \cdot s^{-1} \cdot K^{-1}}$) is the nucleation rate constant. This can be deduced from the ice crystal growth rate v ($\mathrm{m \cdot s^{-1}}$) during the release of supercooling, where v is the rate at which ice crystals grow from a supercooled solution of temperature T° and thickness s (m).

$$k_i = \frac{\nu\rho}{s\left(T_\mathrm{f} - T^\circ\right)}. \tag{2.7}$$

After the release of the supercooling, $\dot{Q}_n$ equals zero, and heat is supplied only for the latent heat corresponding to the amount of ice crystals produced X_{ice}, as follows:

$$\dot{Q}_\mathrm{c} = \Delta H_\mathrm{f} \frac{\partial}{\partial t}\left(\rho X_{\mathrm{ice}}\right). \tag{2.8}$$

This can be written as:

$$\Delta H_\mathrm{f} \frac{\partial}{\partial t}\left(\rho X_{\mathrm{ice}}\right) = -\Delta H_\mathrm{f} \rho \frac{\partial X_{\mathrm{ice}}}{\partial T}\frac{\partial T}{\partial t} = -\Delta H_\mathrm{f} \rho \frac{\Delta X_{\mathrm{ice}}}{\Delta T}\frac{\partial T}{\partial t}. \tag{2.9}$$

The solution in the process of ice crystal growth and freezing as heat flows out of the system is divided into three layers: the unfrozen, mushy (i.e. suspension of ice in water), and frozen layers (Figure 2.6). Each layer is assumed to have a different heat capacity and thermal conductivity. A linear temperature and thermal conductivity gradient of ΔT is assumed to form inside the semi-frozen layer, and the suspended ice fraction in contact with the unfrozen phase is assumed to be zero, whereas that in contact with the frozen phase is assumed to be maximum. Therefore, the apparent heat capacity $C_{p-\mathrm{apparent}}$ $(\mathrm{J \cdot kg^{-1} \cdot K^{-1}})$ and thermal conductivity k_{apparent} $(\mathrm{W \cdot m^{-1} \cdot s^{-1}})$ are applied to each layer, as follows:

$$C_{p-\mathrm{apparent}} = \begin{cases} C_{p-\mathrm{liquid}}, & T_\mathrm{f} + \Delta T < T \\ C_{p-\mathrm{mushy}} = \dfrac{C_{p-\mathrm{liquid}} + C_{p-\mathrm{ice}}}{2} + \dfrac{H_\mathrm{f}\Delta X_{\mathrm{ice}}}{\Delta T}, & T_\mathrm{f} \le T \le T_\mathrm{f} + \Delta T, \\ C_{p-\mathrm{ice}}, & T < T_\mathrm{f} \end{cases} \tag{2.10}$$

$$k_{\mathrm{apparent}} = \begin{cases} k_{\mathrm{liquid}}, & T_\mathrm{f} + \Delta T < T \\ k_{\mathrm{mushy}} = \dfrac{k_{\mathrm{liquid}} - k_{\mathrm{ice}}}{\Delta T}\left(T - T_\mathrm{f}\right) + k_{\mathrm{ice}}, & T_\mathrm{f} \le T \le T_\mathrm{f} + \Delta T. \\ k_{\mathrm{ice}}, & T < T_\mathrm{f} \end{cases} \tag{2.11}$$

2.3.2 Examples of Simulation Results

By applying the abovementioned equations to a finite-element method calculation to obtain a time-dependent solution, the temperature history inside the solution during the freezing process can be calculated. Figures 2.7 and 2.8 show an example of the calculation results (calculated using COMSOL4.0). Evidently,

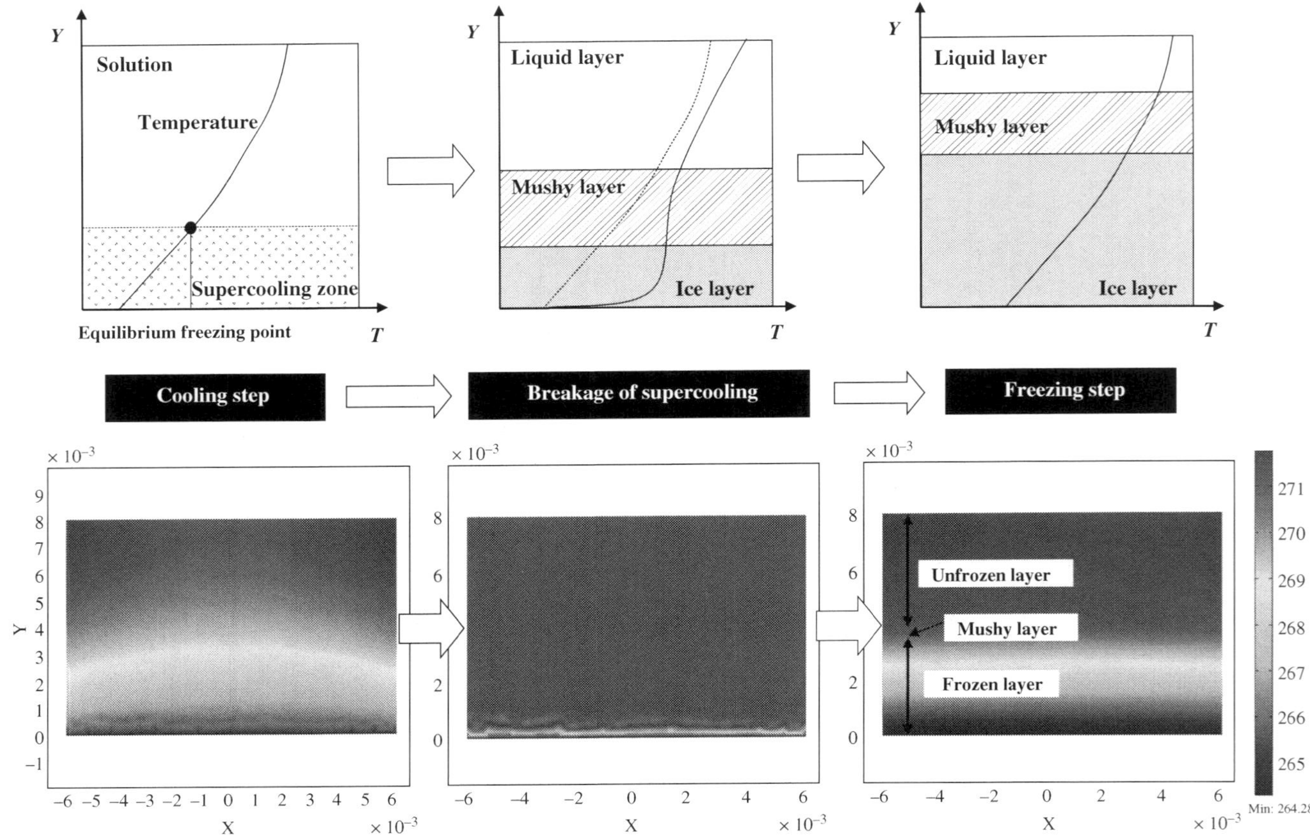

Figure 2.7 Example of simulation results of solution freezing.

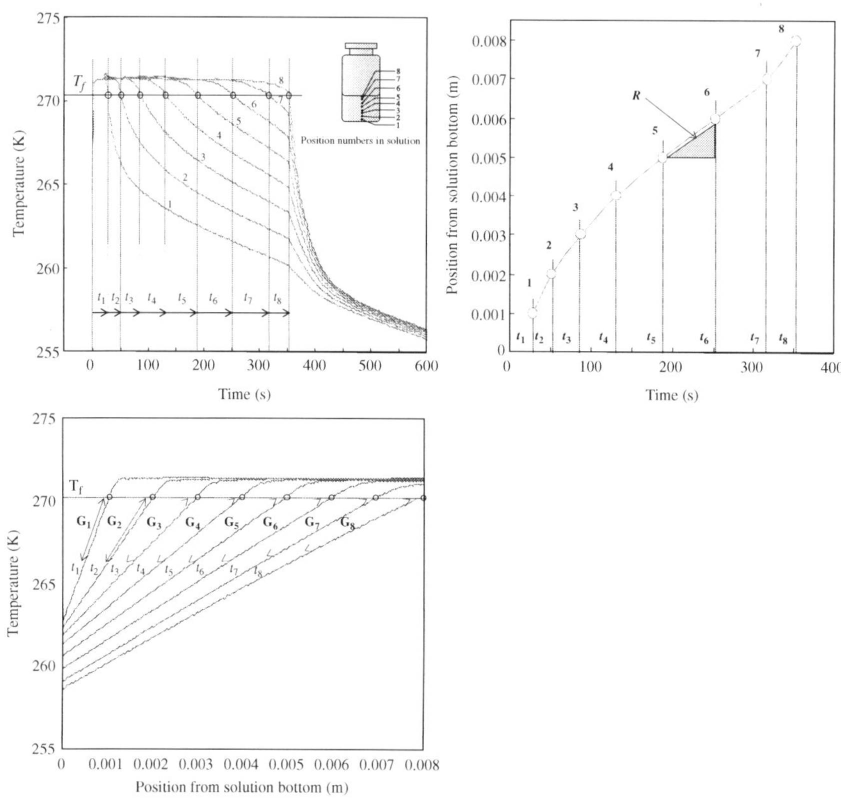

Figure 2.8 Simulated temperature history, temperature distribution, and progress of freezing front during freezing process. *Source:* Nakagawa et al. 2007/Reproduced with permission from John Wiley & Sons.

even if the solution in the vial is cooled at a constant rate, its freezing rate changes over time primarily because the distance between the cooling surface and freezing surface changes; additionally, the temperature gradient inside the solution is also constantly changing. Depending on these factors, a distribution of ice crystal size is generated inside the frozen solution. In addition, the formation of small ice crystals is observed in the region immediately after the release of supercooling; this region can be estimated from the simulation results.

Using the rate of progress of the frozen interface R and the temperature gradient in the frozen layer G obtained from the simulation results, the ice crystal size L^* that forms in the frozen layer can be estimated using the following empirical relationship.

$$L^* = aR^{-0.5}G^{-0.5}, \tag{2.12}$$

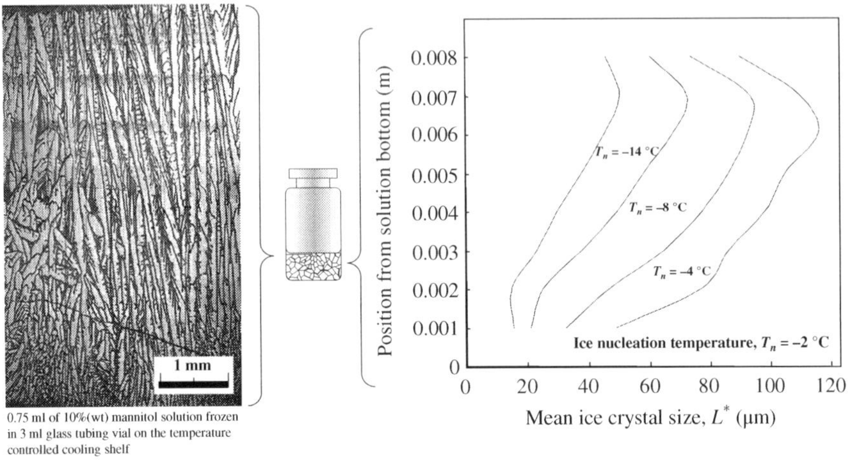

Figure 2.9 Optical microscope image of the ice microstructure and ice crystal size distribution estimated by simulation. *Source:* Kyuya Nakagawa et al. 2007/Reproduced with permission from American Institute of Chemical Engineers.

where proportionality coefficient a is a fitting parameter obtained to match the ice crystal size obtained from the experiment. Figure 2.9 shows the ice crystal patterns obtained from microscopic observations along with the ice crystal size distribution inside the frozen solution deduced from the simulation results. A comparison was performed for different starting temperatures of ice nucleation (i.e. supercooling release temperatures). Evidently from Figure 2.9, the difference in the supercooling temperature has a significant impact on the microstructure of the ice crystals. The ice crystal size calculated from the simulation results indicates that the effect of the degree of supercooling on the ice crystal size is significant. Even when freezing occurs at a constant cooling rate, the ice crystal size is predicted to be smaller at the bottom of the solution and larger at the top. This trend is similar for different freezing conditions; however, the average ice crystal size in the interior is predicted to be larger when ice nucleation begins at higher temperatures.

The microstructure pattern formed by ice crystals is important because it determines the resistance to mass transfer of the water vapor during sublimation. The smaller the ice crystals, the smaller the pore size in the dried layer and the greater the resistance when water vapor moves through this layer. Therefore, the larger the ice crystals, the higher the drying rate (Searles et al. 2001). In addition to the size, the orientation of ice crystals (orientation of pores in the drying layer) is another factor that affects the mass transfer resistance. A complex and intricate network structure increases mass transfer resistance. Using simulations, the extent to which the average ice crystal size varies under various freezing conditions was estimated herein (Figure 2.10).

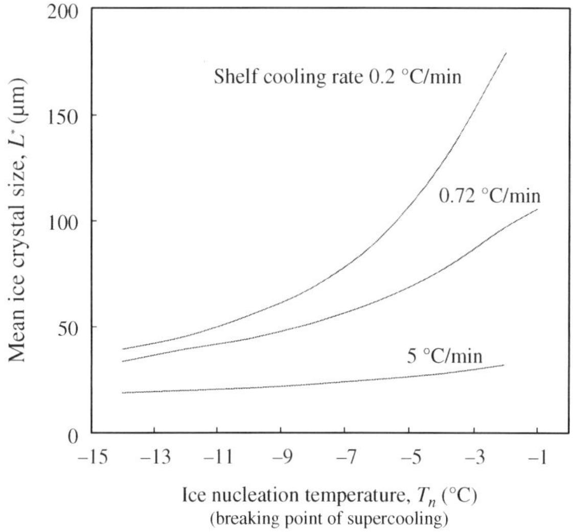

Figure 2.10 Variations of mean ice crystal size depending on the cooling rate and breaking point of super cooling.

Figure 2.11a shows the experimentally estimated permeability (effective diffusion coefficient) values obtained by performing freeze-dried runs to measure the gravimetric change during the selected time (Nakagawa et al. 2007). Evidently, the lower the freezing rate and the smaller the supercooling degree and the larger the ice crystal size, and vice versa. The difference in ice crystal size can cause doubling of the permeability in the drying layer, which predicts that when slow freezing conditions are applied, variations in the breaking point of supercooling cause large variations in the drying rate. By contrast, when a high freezing rate is applied, the variations in ice crystal size due to differences in the supercooling degree are predicted to be small, and if the freezing speed is sufficiently high, the variation in mass transfer resistance due to supercooling is small.

Evaluating the anisotropic microstructures of ice crystals is challenging. Considering the subsequent drying process, an evaluation method that can be linked to the mass transfer resistance of water vapor must be incorporated. A simple and tractable model for describing the effective diffusion coefficient within a pore structure is given by the following equation:

$$D_{\mathrm{A}} = \frac{\varepsilon}{\tau} D_{\mathrm{K}}. \tag{2.13}$$

The model assumes that the effective diffusion coefficient in the pore D_{A} can be written as the pore tortuosity τ and porosity ε (where D_{K} is the Knudsen

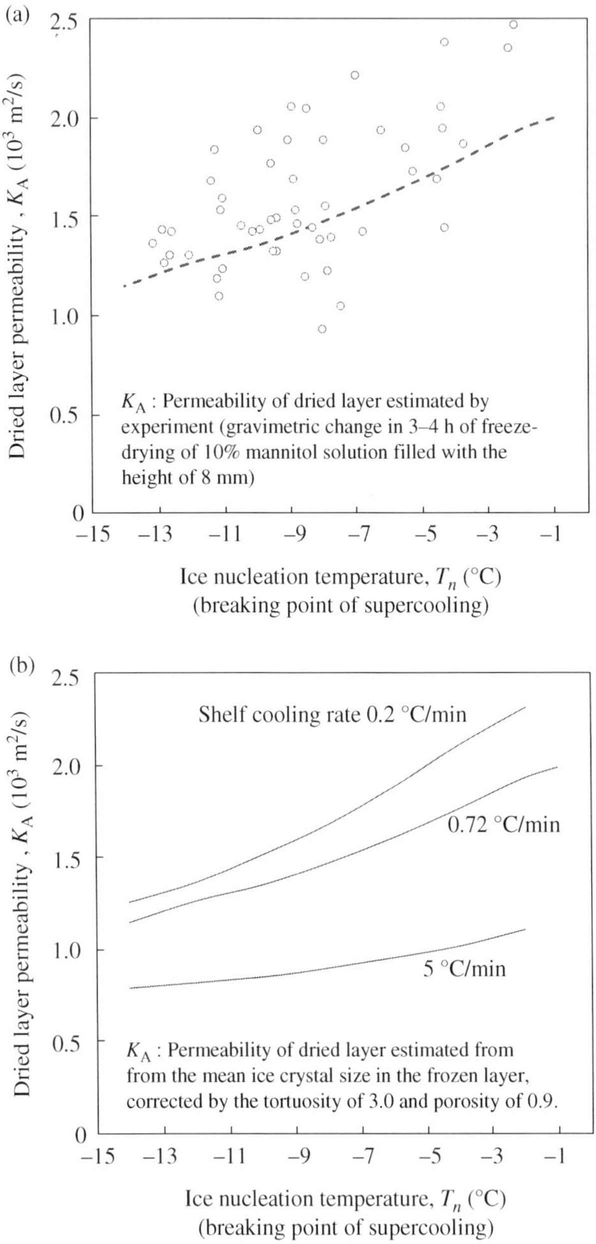

Figure 2.11 Dried layer permeability values estimated from the ice crystal sizes; (a) experimentally obtained; (b) estimated by simulation.

diffusion coefficient), using an average pore size d_p that corresponds with the mean ice crystal size,

$$D_K = \frac{1}{3}\sqrt{\frac{8RT}{\pi M_w}}d_p. \tag{2.14}$$

These equations allow the estimation of the permeability values of the dried layer for given ice crystal sizes, as shown in Figure 2.11b.

The estimation method presented here in applies well to solution or suspension systems. Actual food products often contain numerous foreign substances and are highly viscous (e.g. emulsion and hydrogel), which strongly affects ice crystal formation. In addition, for products of biological origin (particularly livestock and fishery products), numerous studies have reported that ice crystals formed during freezing have a significant impact on tissue structure (Li and Sun 2002; Zhu et al. 2019). The formation of ice crystals in food products depends on the complex interactions between various factors, such as the cooling rate, phase transition, locational differences, and components. Surprisingly, a recent report suggested that freshness also influences the formation of ice crystals in food products (Kominami et al. 2014; Nakazawa and Okazaki 2020; Nakazawa et al. 2020), and further research is needed in this field. However, as most of these factors are reflected in the movement of the freezing front, the freezing front must be tracked to understand the formation of ice crystals (Miyawaki 2018; Miyawaki et al. 1992). Therefore, the simulation can be helpful in predicting ice crystal microstructure formation, though complex factors derived from the characteristics of the food must be considered.

2.4 Mathematical Expression of the Progress of Freeze-Drying

2.4.1 Primary Drying

The primary drying process is the main process wherein ice crystals in the product are removed via sublimation. Primary drying always proceeds at the frozen interface and a drying layer is formed, which corresponds with the degree of progress of drying (sublimation). Unlike drying by evaporation, because water (ice crystals) does not move inside the frozen material, a drying layer corresponding to the progress of sublimation is always formed. The drying layer has a porous structure as a replica of ice crystals. The water vapor generated by sublimation moves through this drying layer and into the vacuum chamber with a viscous flow or diffusion mechanism, and finally reaches the cold trap. The principle of movement of the viscous flow can be expressed by Darcy's law described in the previous chapter, and diffusion is considered to follow the mechanism of Knudsen diffusion. Therefore, the movement of water vapor inside the dried layer is driven by the difference between the water vapor pressure at the sublimation surface and that inside the vacuum chamber. The water vapor pressure at the sublimation surface relative to the sublimation surface temperature is determined by the sublimation rate. The porous structure of the dried layer provides resistance to mass transfer, which is strongly dependent on the porous structure (porosity, pore size, and three-dimensional structure).

The rate of primary drying is determined by the balance between the heat and mass transfer rates. The heat reaching the sublimation surface is used as the sublimation enthalpy, and the higher the sublimation rate, the higher the rate of heat consumption for sublimation. Therefore, the sublimation rate and sublimation surface temperature are determined by the balance between the heat transfer rate to the sublimation surface and the mass transfer rate through the drying layer. Because the thickness of the drying layer varies according to the degree of drying progress, and the mass transfer resistance and thermal resistance also change, the sublimation rate and sublimation surface temperature change accordingly.

When freeze-drying a product on a temperature-controlled shelf, a drying layer forms on the top side of the product as the drying progresses (Figure 2.12a). The thickness of the dried layer is proportional to the degree of drying progress. The heat

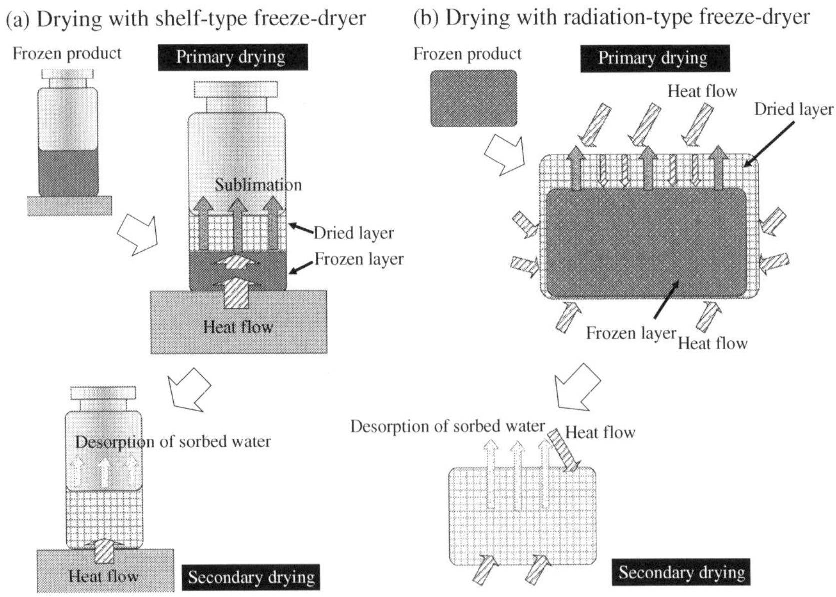

Figure 2.12 Progress of freeze-drying.

supplied to the sublimation surface is primarily by the conductive heat through the frozen layer. Therefore, as the drying progresses, the resistance to mass transfer through the dried layer increases, but the resistance to heat transfer through the frozen layer decreases, which is advantageous in terms of heat transfer efficiency.

Conversely, when drying by radiative heat, as shown in Figure 2.12b, the material surface is heated by radiation and heat is transferred from the material surface to the sublimation surface by conduction through the dried layer. The resistance to both heat and mass transfer increases as drying progresses. In addition, the effective thermal conductivity of the dried layer is low because it is a porous medium, which also contributes to the increase in thermal resistance. Therefore, this type of freeze-dryer is not superior to the shelf-type. However, owing to the simplicity of equipment components and ease of maintenance, they are widely used as industrial equipment for food products. In cases where products are placed directly on shelves and drying progresses in three dimensions, the product shape and product layout also have a significant impact on the drying process because the heat-receiving surface of the radiation and the thickness distribution of the drying layer affect the drying progress.

Figure 2.13 shows the principle of water vapor movement in the dried layer. The gas components in the system consist of water vapor and inert gas. The mass transfer fluxes (N_W and N_{in} [kg/m^{2}s]) of water vapor and inert gas, respectively, can be written as:

$$N_w = -D_w \frac{\partial C_w}{\partial t} + x_w \left(N_w + N_{in} \right), \tag{2.15}$$

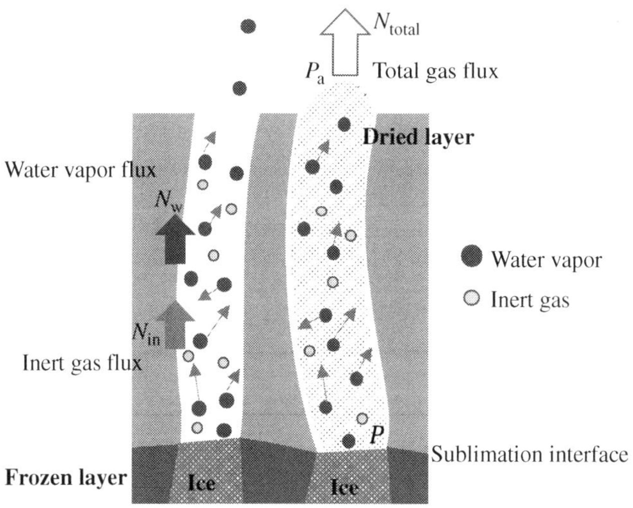

Figure 2.13 Mass transfer within the drying layer during the primary drying process.

$$N_{\text{in}} = -D_{\text{w}} \frac{\partial C_{\text{in}}}{\partial t} + x_{\text{in}} \left(N_{\text{w}} + N_{\text{in}} \right), \tag{2.16}$$

where D_{w} (m^2/s) is the diffusion coefficient of water vapor in the dried layer and C_{w} (kg/m^3) and C_{in} (kg/m^3) are the water vapor and inert gas concentrations in the dried layer respectively. According to Darcy's law, the total flux of these gases ($N_{\text{total}} = N_{\text{w}} + N_{\text{in}}$) is considered to be driven by the pressure difference between the sublimation surface and the outside of the dry layer.

$$N_{\text{total}} = N_{\text{w}} + N_{\text{in}} = v_{\text{w}} \left(C_{\text{w}} + C_{\text{in}} \right) = -\left(\frac{\kappa \bar{P}}{\mu R \bar{T}} \right) \frac{\partial P}{\partial x}, \tag{2.17}$$

where v_{w} (m/s) is the gas velocity in the dried layer, κ (m^2/s) is the gas permeability in the dried layer, μ (Pa s) is the gas viscosity, and $\bar{P}$ and $\bar{T}$ are the mean pressure and temperature in the dried layer respectively.

Thus, the following equation is obtained:

$$N_{\text{w}} = -\left(\frac{\kappa \bar{P}}{\mu} + D_{\text{w}} \right) \frac{M_{w}}{RT} \frac{\partial P}{\partial x}, \tag{2.18}$$

which expresses the principle of mass transfer, wherein the pressure gradient in the drying layer is the driving force and the mass transfer flux is determined by the coefficient $(\kappa \bar{P}/\mu + D_{\text{w}})$, essentially reflecting the diffusive and viscous flow of molecules (Sadikoglu and Liapis 1997). In the former and later chapters, this coefficient is simply referred to as permeability (K_{A}).

2.4.2 Secondary Drying

Secondary drying is the process of removing the water contained within the freeze-concentrated phase by evaporation (which does not involve sublimation). When a solution containing more than one component is frozen, a freeze-concentrated phase is formed and water that does not form ice crystals is distributed inside the freeze-concentrated phase. In eutectic systems, the water and solute components form a solid phase (multicomponent crystalline phase), where as in glass systems, the water and solute components form a glass phase (matrix in amorphous state, glass transition point is observed). Water in the glass matrix strongly restricts its diffusivity. After the ice crystals are removed by primary drying, the water in the freeze-concentrated phase can be reduced to the moisture content level indicated by the glass transition line. Therefore, the moisture content after secondary drying is determined by the final temperature of the product in accordance with its glass transition characteristics. In the actual freeze-drying process, even if primary drying is in progress inside the product, secondary drying may occur if the temperature of the drying layer is above the glass transition temperature. As addressed in the previous section, numerous food items exhibit very low glass transition temperatures; therefore, secondary drying can easily occur simultaneously in products where primary drying is in progress. In addition, when drying with a radiation-type freeze-dryer, the dried layer is often actively heated. Thus, when primary drying is completed, secondary drying is also completed.

The removal of water from the matrix during the secondary drying process is an evaporative phenomenon. However, because water movement is inhibited, the water vapor pressure exhibited by this matrix is apparently lower than that of bulk water during normal observation periods. Therefore, the rate of progress of secondary drying can be expressed by the Arrhenius-type rate equation (Sadikoglu and Liapis 1997), as follows:

$$\frac{\mathrm{d}C_{\mathrm{sw}}}{\mathrm{d}t} = -k_{\mathrm{d}} C_{\mathrm{sw}}, \tag{2.19}$$

$$k_{\mathrm{d}} = k_{\mathrm{d0}} \exp\left(-\frac{E}{RT}\right), \tag{2.20}$$

where C_{sw} is the density of unfreezable water in the dried layer ($\mathrm{kg} \cdot \mathrm{m}^{-3}$), E the activation energy of the desorption of unfreezable water ($\mathrm{J} \cdot \mathrm{mol}^{-1}$), k_{d} the secondary drying dehydration rate constant (s^{-1}), R the gas constant, and k_{d0} the frequency factor (s^{-1}). The values for the frequency factor and desorption energy are given as $k_{\mathrm{d0}} = 3.6298 \times 10^{7}\ \mathrm{s}^{-1}$ and $E = 3185.72\ \mathrm{kJ} \cdot \mathrm{kg}^{-1}$ for freeze-drying skim milk (Liapis and Bruttini 1994). Although these values vary across materials (depending on the extent to which the diffusivity of water is restricted), they are generally considered to be of this order of magnitude.

2.4.3 Practical Approach for Modeling of Freeze-Drying

Freeze-drying of solid foods is often performed by placing the product directly on a tray whose temperature cannot be controlled. Therefore, the progress of drying is three-dimensional depending on the shape of the product and the heat transfer pathway, thus requiring ingenuity in the application of simplified models. Liquid food products are dispensed into individual trays for drying. When drying is performed with conductive heat transfer from the shelves as the primary heat source, it may be regarded as a one-dimensional process if the heat flow from the sides of the product is small. While when it is performed with radiative heat as the primary heat source, it should be viewed as a three-dimensional process.

Numerous foods in the frozen state have a low glass transition temperature (T_{g}'), and in several cases, the product temperature during the primary drying process

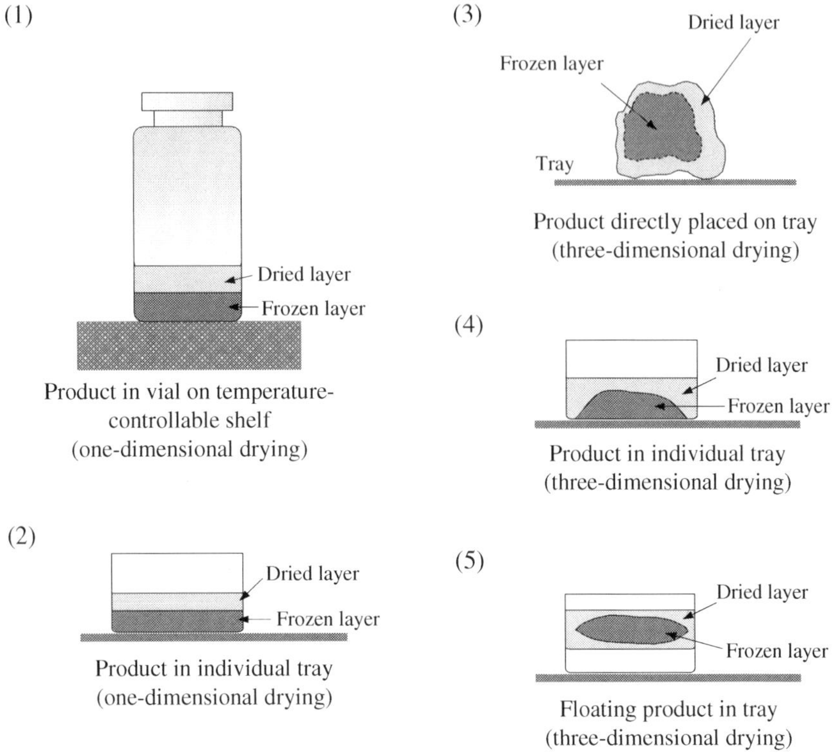

Figure 2.14 Classification of progress of freeze-drying.

exceeds T_g'. Consequently, the product in the tray may float, shrink, or foam during the primary drying step. In these cases, changes in the heat transfer pathway and interfacial surface area for heat and mass transfer occur as drying progresses. Therefore, modeling that rigorously incorporates these phenomena is challenging. The mathematical approach should be considered with a practical objective. Figure 2.14 classifies the progress of freeze-drying.

2.5 One-Dimensional Model (R_p–K_v Model)

The R_P–K_V model is one of the most practical freeze-drying models that is compatible to the process operated with shelf-type freeze-dryer (Giordano et al. 2011; Rambhatla et al. 2004; Velardi and Barresi 2008). It simplifies the various parameters involved in heat and mass transfer, and allows the estimation of the one-dimensional freeze-drying process with limited numbers of parameters (Figure 2.15). Given that

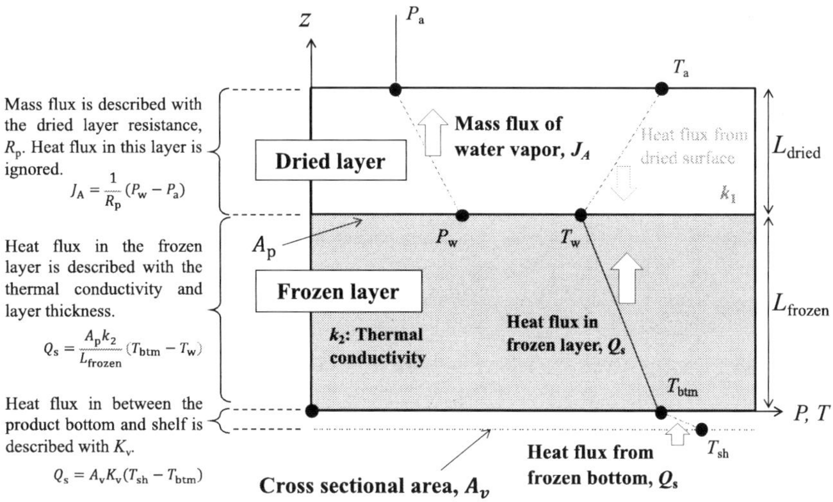

Figure 2.15 Schematic of R_P–K_v model.

this model was developed for a pharmaceutical vial freeze-drying process, care must be taken regarding its suitability when applied to food freeze-drying.

First, the mass transfer equation is expressed, as follows:

$$J_A = \left(-\frac{1}{A}\frac{dm}{dt}\right) = \frac{1}{R_p}\left(P_w - P_a\right), \tag{2.21}$$

where R_p is the mass transfer resistance of the dried layer, which increases with increasing L_{dried}. It increases linearly when the permeability in the dried layer is assumed to be uniform. If the value of J_A and the sublimation surface temperature (T_w) can be obtained experimentally, R_p can be determined and plotted as a function of the drying process ($L_{dried}/(L_{dried} + L_{frozen})$). The obtained R_p (m/s) can be approximated by the following equation, where A_1 and A_2 are the fitting parameters:

$$R_p = R_{p0} + \frac{A_1 L_{dried}}{1 + A_2 L_{dried}}. \tag{2.22}$$

The effective thermal conductivity of the frozen layer (k_2) is close to that of ice at approximately 1.2–1.5 $W \cdot m^{-1} \cdot K^{-1}$. The effective thermal conductivity of the dry layer (k_1) is less than one tenth of this value; thus, the value of this term is very small. Assuming that the heat transfer from the dry layer is negligible, the heat flow reaching the sublimation surface ($J \cdot s^{-1}$) is expressed as follows:

$$Q_s = \frac{A_p k_2}{L_{frozen}}\left(T_{btm} - T_w\right), \tag{2.23}$$

where A_p is the cross-sectional area through which heat flows in the frozen layer. The equation that considers the heat transfer coefficient (K_v) at the bottom of the product where the product and shelf board are in contact is as follows:

$$Q_s = A_v K_v \left(T_{sh} - T_{btm}\right). \tag{2.24}$$

This equation expresses the heat flow ($J \cdot s^{-1}$) entering from the shelf and flowing to the bottom of the product, where A_v is the cross-sectional area through which the heat flows. Assuming that drying is in pseudo-steady state, this heat flux is equal to the heat flux reaching the sublimation surface from the bottom of the product through the freezing layer. Now, assuming $A_p \cong A_v$, Eqs. (2.23) and (2.24) yield:

$$Q_s = \frac{A_p\left(T_{sh} - T_w\right)}{1/K_v + L_{frozen}/k_2}. \tag{2.25}$$

Note that K_v can be regarded as the overall heat transfer coefficient between the shelf and bottom of the frozen product; nonetheless, the effect of neglected heat sources, such as heat flux through the dried layer and/or radiative heat, can also be considered in this parameter.

Assuming a pseudo-steady state of heat and mass transfer in the drying process (i.e. assuming that all the heat delivered to the sublimation interface is used for sublimation), we obtain:

$$Q_s = \Delta H\left(-\frac{dm}{dt}\right). \tag{2.26}$$

Therefore, the following relationship is obtained by applying Eqs. (2.21) and (2.25) to Eq. (2.26).

$$\left(-\frac{dm}{dt}\right) = \frac{A_p}{R_p}\left(P_w - P_a\right) = \frac{A_p}{\Delta H}\frac{\left(T_{sh} - T_w\right)}{\left(1/K_v + L_{frozen}/k_2\right)}. \tag{2.27}$$

Next, Eqs. (2.24) and (2.26) yield:

$$\Delta H\left(-\frac{dm}{dt}\right) = A_p K_v \left(T_{sh} - T_{btm}\right). \tag{2.28}$$

Integrating this over the entire drying time yields:

$$K_v = \frac{\Delta H m_{sub}}{A_p \int_{t=t_1}^{t=t_2}\left(T_{sh} - T_{btm}\right)dt}, \tag{2.29}$$

where m_{sub} is the amount of water sublimated between time $t = t_1$ and t_2. If the product bottom temperature (T_{btm}), shelf temperature (T_{sh}), and amount of sublimation (m_{sub}) can be obtained experimentally during freeze-drying, K_v ($W \cdot m^{-2} \cdot K^{-1}$) can be determined using this formula. In general, K_v depends on P_a and can be fitted using the following equation:

$$K_V = K_C + \frac{K_p P_a}{1 + K_D P_a}, \quad (2.30)$$

where K_C, K_p, and K_D are the fitting parameters. These parameters can be deduced from the experimental data obtained beforehand and then applied to the simulation.

2.6 Three-Dimensional Model (K_{cc} Model)

2.6.1 Case Classification of Freeze-Drying under Radiative Heat

This section presents a mathematical model of freeze-drying that uses radiative heat as the primary heat source. The drying unit typically has multiple shelves; therefore, the model incorporates radiation heaters above and below for a single product, as shown in Figure 2.16. The progression of drying here is formulated separately for Cases A and B.

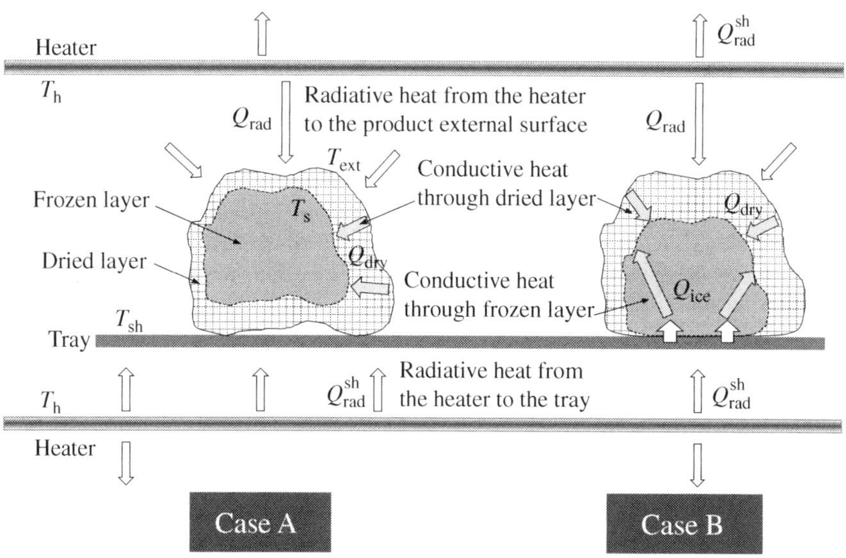

Figure 2.16 Schematic of freeze-drying proceeding with radiative heat.

Case A:

The drying progresses uniformly from the outer surface of the product and the frozen layer recedes into the interior of the product. The heat input from the tray through the drying layer by conduction of electric heat is very small, and only the radiant heat from the heater can be regarded as the heat source. The drying of products placed directly on trays corresponds to this case. This case is also considered to be the case where products float, even when drying is carried out in individual trays.

Case B:

In this case, drying progresses from the outer surface of a portion of the product and the frozen layer recedes into the interior of the product. Moreover, herein, the heat transfer from the tray (through the dried or frozen layer) is large, and the heat flow path to the sublimation surface should consider both the freezing and drying layers. This case is considered to be the case where the contact area between the product and tray is relatively large, or where drying proceeds within separated small pack trays. The division into Cases A and B is intended to simplify the mathematical treatment of this case. If the drying mechanism is strictly considered, the drying process can be assumed to proceed by the Case B mechanism in the early stages of drying and then gradually shift to Case A. If the timing of this transition can be modeled, it can be incorporated into later calculations.

2.6.2 Radiative Heat to Product

First, the amount of radiative heat transfer, which is the primary heat source for the product, is described. Assuming that there are *i* sources of radiation and that the sum of each is the total radiative heat transfer (Q_{rad}), we obtain:

$$Q_{rad} = \sum_i A_i \varepsilon_i \sigma \left(T_i^4 - T_{ext}^4 \right), \tag{2.31}$$

where T_{ext} is the external surface temperature of the product, T_i the temperature of the *i*th radiative source, A_i the area receiving heat from the *i*th radiation source, ϵ_i the emissivity of the *i*th radiation source, and σ the Stefan–Boltzmann coefficient. In this section, only the radiant heat source from heater temperature T_h is considered, as follows:

$$Q_{rad} = A_1 \varepsilon_1 \sigma \left(T_h^4 - T_{ext}^4 \right). \tag{2.32}$$

Simplifying this equation, we obtain:

$$Q_{rad} = H_{R1} \left(T_h - T_{ext} \right), \tag{2.33}$$

where

$$H_{R1} = A_1\varepsilon_1\sigma\left(T_h^2 + T_{ext}^2\right)\left(T_h + T_{ext}\right), \tag{2.34}$$

where H_{R1} (W/K) is a parameter, that is, the heat transfer coefficient multiplied by the surface area involved in radiation. For simplicity, this is treated as a constant value. This coefficient can be experimentally obtained by conducting ice sublimation tests. The heat-receiving area included in the equation depends on the product shape, product layout in the equipment, and location of the heater; thus, experimental values must be obtained according to actual drying processes, as is practical. The emissivity also depends on the material and color of the heater and product. Therefore, the H_{R1} value too must be obtained experimentally, as is practical.

2.6.3 Radiative Heat to Tray

The radiative heat transfer to the product tray that is always in contact with the products must be estimated. Conductive heat from the tray to the product could be a large heat source for the product when the tray is in contact with a frozen product with a large surface area (e.g. Case B). Conversely, this implies that the conductive heat exchange between the tray and product can be negligibly small when the tray is in contact with the dried layer of the product and the contact surface is sufficiently small (i.e. such as in Case A).

The product tray receives heat from the radiation heater as well as the product; simultaneously, the conductive heat is transferred to the product at a lower temperature. The amount of heat transferred by radiation to the product tray is expressed as follows:

$$Q_{rad}^{sh} = H_{R2}\left(T_h - T_{sh}\right), \tag{2.35}$$

where

$$H_{R2} = A_2\varepsilon_2\sigma\left(T_h^2 + T_{sh}^2\right)\left(T_h + T_{sh}\right). \tag{2.36}$$

Under steady-state conditions, the radiative heat transfer to the tray (Q_{rad}^{sh}), heat transfer from the product tray to the product (Q_{sh}), and heat transfer to the sublimation surface through the frozen layer (Q_{ice}) are considered equal.

$$H_{R2}\left(T_h - T_{sh}\right) = H_{btm}\left(T_{btm} - T_{sh}\right) = H_{cf}\left(T_s - T_{btm}\right). \tag{2.37}$$

Therefore, the product tray temperature can be expressed as (the derivation of Q_{sh} and Q_{ice} are expressed in a later section):

$$T_{sh} = \frac{T_h + \alpha T_s}{1 + \alpha},$$

where

$$\alpha = \frac{H_{\text{cf}} H_{\text{btm}}}{H_{\text{R2}}\left(H_{\text{cf}} + H_{\text{btm}}\right)}. \tag{2.38}$$

2.6.4 Conductive Heat Transfer in the Frozen and Dried Layers

To calculate the flux of heat and mass transfer within the drying layer, the thicknesses of the frozen and dried layers must be estimated. When drying proceeds in three dimensions, the outer surface area and sublimation surface area always differ, and the distribution of the thickness of the drying layer depends on the shape of the product. To simplify this situation and calculate the representative layer thicknesses from this three-dimensional shape, the following approximation is made: the outer surface area of the product (A_{ext}), sublimation surface area (A_{s}), and product bottom surface area (A_{btm}) are rearranged in spherical coordinates, as shown in Figure 2.17. The difference in the radii of the spheres obtained by this rearrangement can be regarded as the representative layer thickness (hollow sphere model).

The heat conduction equation for a hollow sphere at steady state is expressed by:

$$\frac{d}{dr}\left(4\pi r^2 k \frac{dT}{dr}\right) = 0, \tag{2.39}$$

where k is the thermal conductivity.

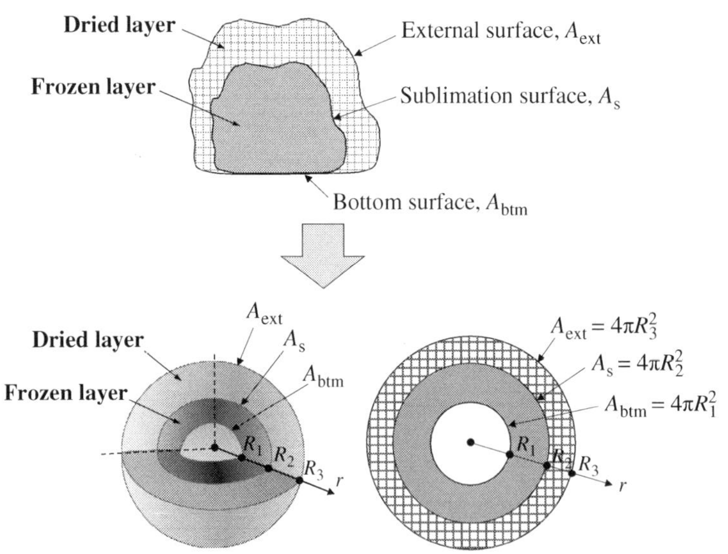

Figure 2.17 Approximation of mean layer thicknesses by the hollow sphere model.

Substituting the boundary conditions (sphere inner surface temperature T_1, outer surface temperature T_2), the rate of heat flow Q (J/s) at the steady state yields:

$$Q = \frac{2k}{\sqrt{1/\pi A_1} - \sqrt{1/\pi A_2}} \left(T_2 - T_1\right), \tag{2.40}$$

where A_1 and A_2 are the surface areas of the sphere at the coordinates $R = R_1$ and R_2 respectively.

Using the equation obtained here, the rate of conductive heat transfer from the outer surface of the product to the sublimation surface through the dried layer (Q_{dry}) can be estimated using:

$$Q_{dry} = H_{cd}\left(T_{ext} - T_s\right), \tag{2.41}$$

where

$$H_{cd} = \frac{2k_{dry}}{\left(1/\sqrt{\pi A_s}\right) - \left(1/\sqrt{\pi A_{ext}}\right)}.$$

Similarly, the rate of conductive heat transfer from the bottom of the product to the sublimation surface through the freezing layer (Q_{ice}) can be estimated using:

$$Q_{ice} = H_{cf}\left(T_s - T_{btm}\right), \tag{2.42}$$

where

$$H_{cf} = \frac{2k_{ice}}{\left(1/\sqrt{\pi A_{btm}}\right) - \left(1/\sqrt{\pi A_s}\right)}.$$

In Case B in Figure 2.16, when the conducted heat transfer from the product tray to the product (Q_{sh}) should be considered, the following heat term must be included.

$$Q_{sh} = H_{btm}\left(T_{btm} - T_{sh}\right), \tag{2.43}$$

where

$$H_{btm} = A_{btm} h_{btm},$$

where h_{btm} is the overall heat transfer coefficient at the bottom of the product. Solid products placed directly on trays often dry immediately from the bottom of the product because of the high heat input. However, when a drying layer is formed at the bottom of the product, the rate of heat flow is significantly reduced owing to the low thermal conductivity of the drying layer. In this case (such as Case A in Figure 2.16), the heat transfer between the product tray and product may be negligible.

In Case A, the heat reaching the sublimation surface (Q_S) is the sum of the heat transferred through the frozen and dried layers:

$$Q_s = Q_{dry} = \frac{T_h - T_s}{\frac{1}{H_{R1}} + \frac{1}{H_{cd}}}. \tag{2.44}$$

In Case B, it is:

$$Q_s = Q_{dry} - Q_{ice} = \frac{T_h - T_s}{\frac{1}{H_{R1}} + \frac{1}{H_{cd}}} - \frac{T_s - T_{sh}}{\frac{1}{H_{btm}} + \frac{1}{H_{cf}}}. \tag{2.45}$$

2.6.5 Three-Dimensional Model (K_{cc} Model) – Water Vapor Transfer

As shown in Figure 2.18, the water vapor that sublimates on the sublimation surface passes through the drying layer, drying chamber, piping, condenser chamber, and ices on the cold-trap surface. Assuming that these processes proceed with the water vapor pressure difference as the driving force and that the permeability in the drying layer is K_A (m^2/s), the conductance between the drying chamber and condenser is K'_{cc} (m^3/s), the mass transfer coefficient at the cold trap surface is k_c (m/s), and the surface area of the cold trap is A_c (m^2), then it can be expressed as:

$$\left(-\frac{dm}{dt}\right) = \frac{M_W}{RT_{av}} \frac{P_s - P_c}{\frac{\left(1/\sqrt{\pi A_s}\right) - \left(1/\sqrt{\pi A_{ext}}\right)}{2K_A} + \frac{1}{K'_{cc}} + \frac{1}{A_c k_c}}. \tag{2.46}$$

where T_{av} is the average temperature in the transfer path of water vapor. The representative thickness of the dried layer was estimated by applying the hollow sphere model, as explained in the previous section. This equation can be simplified by introducing the overall conductance of the path in the equipment from the product surface to the cold trap surface is K_{cc} (m^3/s):

$$\left(-\frac{dm}{dt}\right) = \frac{M_W}{RT_{av}} \frac{P_s - P_c}{\frac{\left(1/\sqrt{\pi A_s}\right) - \left(1/\sqrt{\pi A_{ext}}\right)}{2K_A} + \frac{1}{K_{cc}}}. \tag{2.47}$$

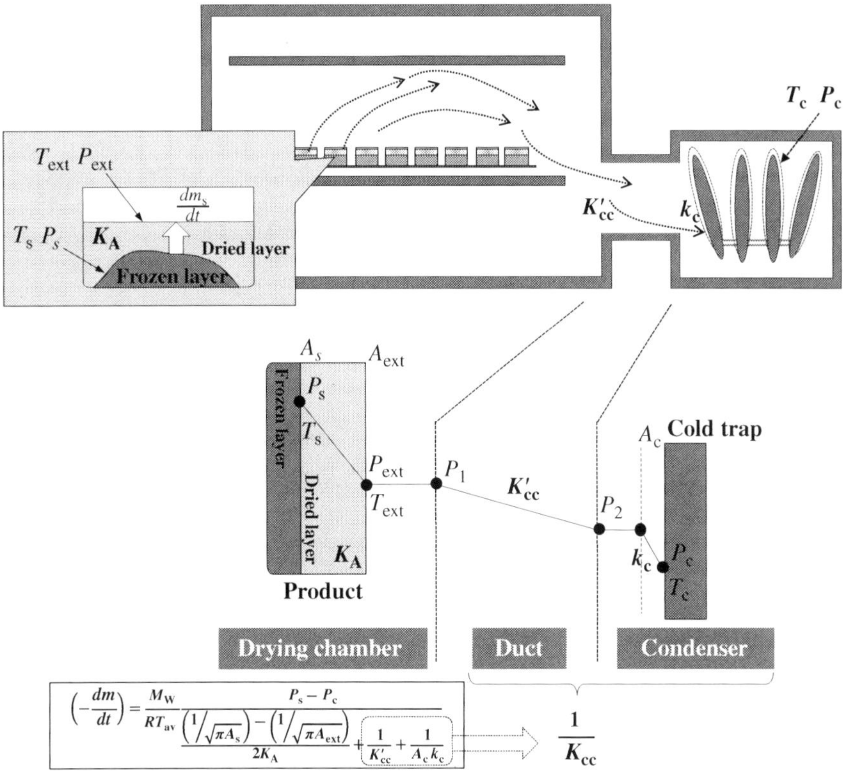

Figure 2.18 Schematic of K_{cc} model.

When the water vapor velocity in the chamber is not sufficiently fast or the trapping capacity (trapping speed) of the cold trap cannot keep up with the rate of water vapor generation, a large difference in the water vapor pressure is produced in the equipment. In such cases, K_{cc} has an impact on the overall sublimation rate. Because the water vapor pressure difference in the device is strongly related to the occurrence of choked flow (Patel et al. 2010), the value of K_{cc} is strongly related to device performance. K_{cc} is an important parameter for scale-up and practical operation (Nakagawa and Tamiya 2019; Nakagawa et al. 2020b); thus, evaluating the fluid flow in the device using the value of K_{cc} is desirable. However, this is an issue for future research.

Assuming a pseudo-steady state of heat and mass transfer during the drying process, we obtain:

$$Q_s = \Delta H\left(-\frac{dm}{dt}\right). \tag{2.26}$$

Therefore, for Case A:

$$\frac{T_{\mathrm{h}}-T_{\mathrm{s}}}{\frac{1}{H_{\mathrm{R1}}}+\frac{1}{H_{\mathrm{cd}}}}=\frac{\Delta HM_{\mathrm{W}}}{RT_{\mathrm{av}}}\frac{P_{\mathrm{s}}-P_{\mathrm{c}}}{\frac{\left(1/\sqrt{\pi A_{\mathrm{s}}}\right)-\left(1/\sqrt{\pi A_{\mathrm{ext}}}\right)}{2K_{\mathrm{A}}}+\frac{1}{K_{\mathrm{cc}}}}, \tag{2.48}$$

whereas for Case B:

$$\frac{T_{\mathrm{h}}-T_{\mathrm{s}}}{\frac{1}{H_{\mathrm{R1}}}+\frac{1}{H_{\mathrm{cd}}}}-\frac{T_{\mathrm{s}}-T_{\mathrm{sh}}}{\frac{1}{H_{\mathrm{btm}}}+\frac{1}{H_{\mathrm{cf}}}}=\frac{\Delta HM_{\mathrm{W}}}{RT_{\mathrm{av}}}\frac{P_{\mathrm{s}}-P_{\mathrm{c}}}{\frac{\left(1/\sqrt{\pi A_{\mathrm{s}}}\right)-\left(1/\sqrt{\pi A_{\mathrm{ext}}}\right)}{2K_{\mathrm{A}}}+\frac{1}{K_{\mathrm{cc}}}}, \tag{2.49}$$

where

$$T_{\mathrm{av}}=\frac{T_{\mathrm{s}}+T_{\mathrm{c}}}{2}. \tag{2.50}$$

From this, T_{s} can be obtained at a selected drying time, and the drying rate at that time can be deduced from the obtained T_{S}. The time required to dry a certain volume fraction can be determined from the drying rate, which can be used to simulate the drying kinetics.

2.6.6 Parameter Estimation by Ice Sublimation Test

To perform a simulation based on the mathematical model described in the previous sections, parameters H_{R1}, H_{R2}, H_{btm}, and K_{cc} are required for the calculation. These parameters can be obtained from the ice sublimation test results using the equation presented in the previous sections, which excludes the contribution of the dried layer. First, a certain amount of ice was sublimated using a dryer in which the test was performed, and the sample temperature (T_{p}), sublimation rate $\left(-\frac{dm}{dt}\right)$, and product tray temperature (T_{sh}) were measured during the process.

Because H_{R1}, H_{R2}, and H_{btm} are all heat transfer coefficients that multiply the corresponding heat transfer area, the test must be conducted with an outer surface area similar to that of the actual product to obtain reliable values. If a linear relationship can be assumed between these parameters and the corresponding heat transfer areas, the obtained parameters can be converted using the area ratio between the ice used in the test and the actual products. Here, H_{R1}, H_{R2}, and H_{btm} obtained from ice sublimation tests are denoted by H_{R1}°, H_{R2}°, and H_{btm}° respectively. The heat input to the ice (Q_{s}°) should be balanced by the sublimation rate:

$$Q_{\mathrm{s}}^{\circ}=Q_{\mathrm{rad}}-Q_{\mathrm{sh}}=H_{\mathrm{R1}}^{\circ}\left(T_{\mathrm{h}}-T_{\mathrm{p}}\right)+H_{\mathrm{btm}}^{\circ}\left(T_{\mathrm{p}}-T_{\mathrm{sh}}\right)=\Delta H\left(-\frac{dm}{dt}\right). \tag{2.51}$$

For several experimentally determined T_p, $\frac{dm}{dt}$, T_{sh}, and T_h, H°_{R1}, H°_{R2}, and H°_{btm} that satisfy the above equations were estimated using a nonlinear optimization tool, which is convenient. As noted earlier, if a linear relationship can be assumed between these parameters and the corresponding heat transfer areas, they can be converted with the coefficient η_i, as follows:

$$\eta_i H_i^\circ = H_i. \tag{2.52}$$

For example, if a nice sublimation test is conducted with a volume that fills a shelf of area 1 m^2, the value of η should be set to 1.0. If an equivalent volume of liquid product is loaded, a loading volume that fills a shelf of area 2 m^2 can be set to 2.0 as an approximation; however, care must be taken to check how well this linearity is ensured.

Finally, K_{cc} can be obtained from the following equation using the experimental values of T_p and $\frac{dm}{dt}$, as follows:

$$\left(-\frac{dm}{dt}\right) = \frac{K_{cc} M_W}{RT_{av}}\left(P_p - P_c\right). \tag{2.53}$$

Note that the value of K_{cc} is a coefficient that reflects the resistance to water vapor transfer in the apparatus; thus, it varies depending on the operating conditions of the device. Therefore, preferably, a value obtained from an ice sublimation test conducted with close operative conditions that are closer to the target of the simulation must be estimated.

2.6.7 Estimation of Essential Surface Area

When the freezing layer recedes into the product as it receives radiant heat from the outer surface of the product, the sublimation surface area decreases as drying progresses. However, the sublimation surface is not necessarily flat and may increase owing to interfacial instability (Toei et al. 1975). Because modeling the dynamic changes in the sublimation surface area, which depend on the product shape and microstructure, is challenging, the product outer surface area (A_{ext}), sublimation surface area (A_s), and product bottom surface area (A_{btm}) were experimentally obtained herein.

First, A_{ext}, A_s, and A_{btm} were assumed to be expressed as functions of the drying process. If the product shape does not change significantly after drying, A_{ext} was assumed as constant. The ratio of A_s and A_{btm} to their initial values was expressed as a function of the volume fraction of the frozen layer, as follows:

$$f\left(V_{ff}\right) = A_s\left(V_{ff}\right) / A_s\left(0\right), \tag{2.54}$$

$$g\left(V_{ff}\right) = A_{btm}\left(V_{ff}\right) / A_{btm}\left(0\right). \tag{2.55}$$

These functions were obtained experimentally. Examples are shown in Figure 2.19. This is an example of freeze-drying a cubic apple block (a cube 10 mm on a side) and liquid food product in a plastic tray. First, the product was freeze-dried and the semi-dried samples were collected at different drying times. The sample was immediately cut and the cross-section was observed. The weight of the samples was measured and the volume fraction of the frozen sections was determined. As shown in Figure 2.19, $f(V_{ff})$ and $g(V_{ff})$ are plotted as functions of $V_{ff}/V_{ff}(0)$; these plots were fitted with the appropriate function. Examples of the correlation equations are shown in the figure.

These surface area values are rearranged into spherical coordinates and used to calculate the thicknesses of the frozen and dried layers. Therefore, the relationship between the external surface area, sublimation surface area, and bottom surface area of the product was assumed in this order. Essentially, depending on the shape of the product, this approximation of the thicknesses of the frozen and dried layers using spherical coordinates may not be appropriate.

2.6.8 Simulation Result Example

An example of the simulation run results is shown in Figure 2.20 (the values of the parameters used are listed in Table 2.1). This is a comparison between the experiment and simulation for the freeze-drying of liquid food in plastic trays (Nakagawa and Ochiai 2017). Evidently, the simulation predicted the drying kinetics and the product temperatures well. The temperature profile of the sublimation surface obtained by the simulation was the temperature at the moving boundary, which differed from the value indicated by the sensor measured at a fixed point. The simulation enabled us to estimate the temperatures, such as the sublimation surface and product surface, which are difficult to measure using sensors. If the temperature of the sublimation surface can be predicted in advance by simulation, determining the drying conditions that prevent collapse and/or bubbling is relatively easy. Additionally, obtaining a design space that shows the interaction between the drying performance and operative parameters, as shown in Figure 2.21, is also easy. However, practical freeze-drying in the industry often involves more complex systems (e.g. soup foods with vegetables). Furthermore, the food industry has numerous cases where in the drying process is performed while allowing a certain degree of collapse and bubbling; however, the model presented here does not consider these phenomena. To make the simulation results a practical evaluation tool for practical processing, the mathematical expression and involved parameters must be further tuned to represent the target products and equipment more appropriately.

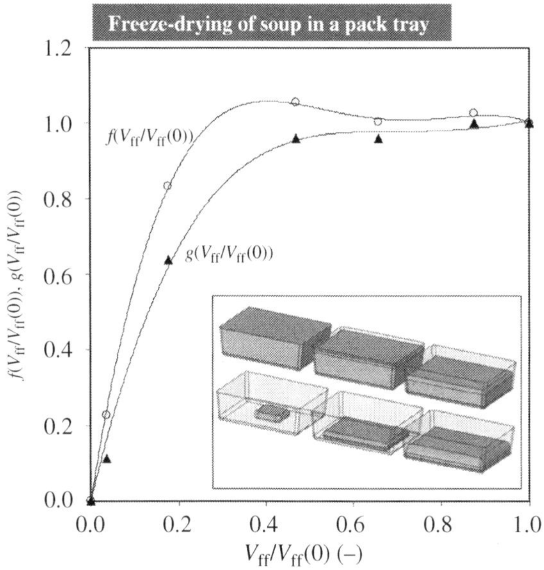

$$f\left(V_{ff}/V_{ff(0)}\right) = 1.3\left(V_{ff}/V_{ff(0)}\right)^3 - 2.2\left(V_{ff}/V_{ff(0)}\right)^2 + 1.9\left(V_{ff}/V_{ff(0)}\right)$$

$$g\left(V_{ff}/V_{ff(0)}\right) = -1.5\left(V_{ff}/V_{ff(0)}\right)^3 + 2.0\left(V_{ff}/V_{ff(0)}\right)^2 + 0.5\left(V_{ff}/V_{ff(0)}\right)$$

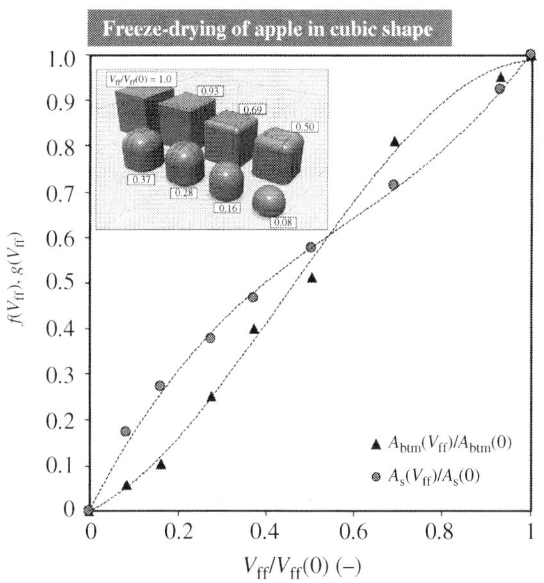

$$f\left(V_{ff}/V_{ff(0)}\right) = 1.3\left(V_{ff}/V_{ff(0)}\right)^3 - 2.2\left(V_{ff}/V_{ff(0)}\right)^2 + 1.9\left(V_{ff}/V_{ff(0)}\right)$$

$$g\left(V_{ff}/V_{ff(0)}\right) = -1.5\left(V_{ff}/V_{ff(0)}\right)^3 + 2.0\left(V_{ff}/V_{ff(0)}\right)^2 + 0.5\left(V_{ff}/V_{ff(0)}\right)$$

Figure 2.19 Experimentally obtained sublimation and bottom surface area during drying and their correlation to the frozen volumes (Nakagawa and Ochiai, 2015, 2017).

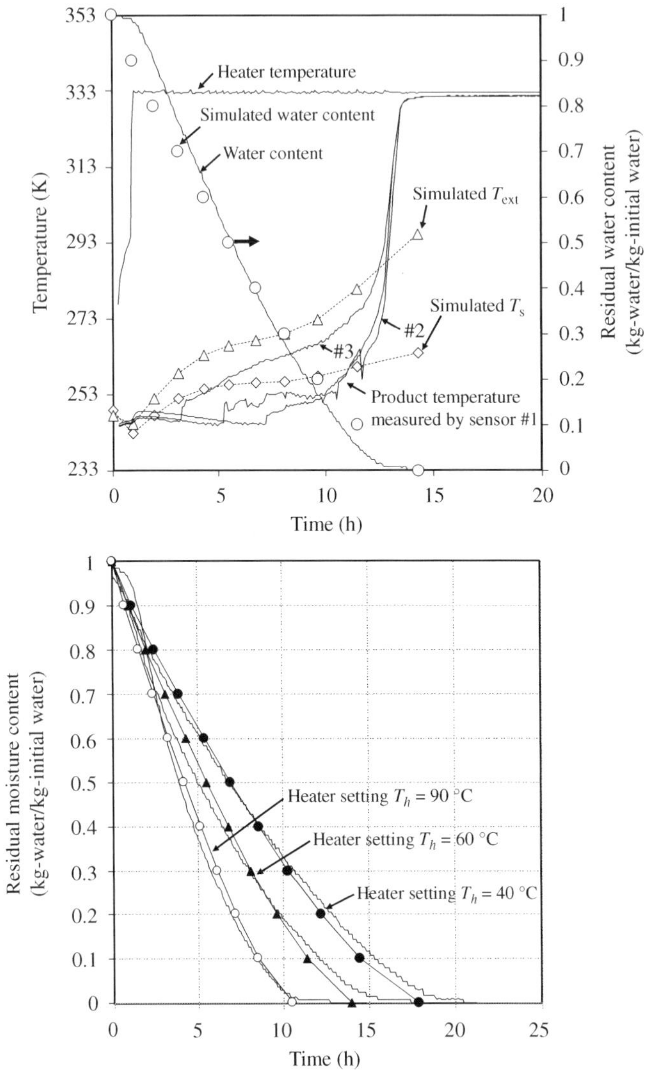

Figure 2.20 Comparison of experiment and simulation (Nakagawa and Ochiai 2017).

2.7 Conclusion

The modeling approaches of food freeze-drying process were presented in this chapter. As stated, modeling approaches have significant implications for the rational design of processes and operations, and its application to the food industry requires the development of robust and practical models. The authors cannot say that the mathematical models presented in this chapter perfectly meet these requirements. First, the most important issue in food production is not only the

Table 2.1 Values applied to the simulation shown in Figures 2.20 and 2.21.

A_{btm} (0) (m^2)	0.063
A_{ext} (0) (m^2)	0.081
H_{R1} (W/K)	0.28
H_{R2} (W/K)	2.6
h_{btm} (W/m^2K)	2.0
K_{cc} (m^3/s)	0.08
K_A (m^2/s)	4.3×10^{-3}
k_{dry} (W/mK)	0.09
k_{ice} (W/K)	1.4
ΔH_f (J/kg)	2,800,000
M_w (kg/mol)	0.018
R (J/K/mol)	8.314
T_{ch} (K)	298.15
T_c (K)	223.15
Total volume of product (m^3)	1.2×10^{-3}
Density (kg/m^3)	790
ε (−)	0.82
τ (−)	1.8

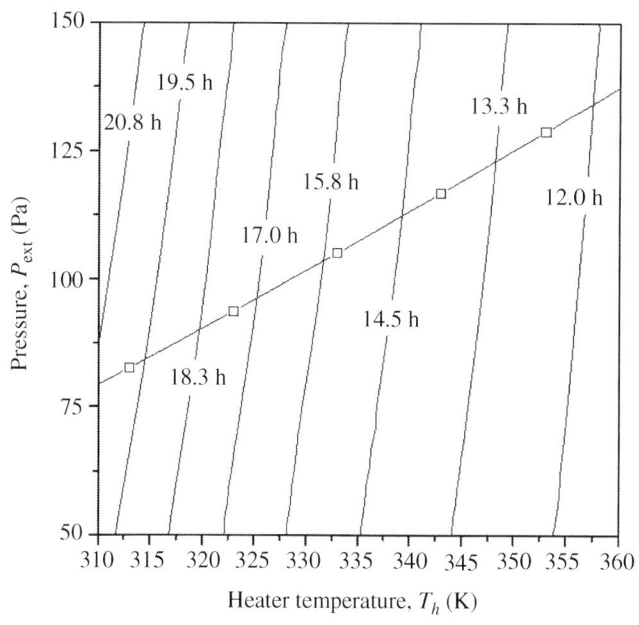

Figure 2.21 Contour line of the total drying time as functions of heater temperature and chamber pressure (Nakagawa and Ochiai 2017).

kinetics of drying, but mainly the control of the end-product quality such as product shelf-life, rehydration ability, taste, texture, flavors, and appearance. Therefore, it is ideal that a mathematical approach can be used to achieve even quality control. Achieving this to a greater degree in food systems, where multiple characteristics combine to determine product quality, will not be easy, but will certainly be an exciting challenge in the future. This approach will be especially significant in freeze-drying, which is well suited for high-quality food production. The author believes that the following are important issues to be addressed in the future.

- ✓ Prediction of the water activity as results of drying operations.
- ✓ Classification and typification of foods in terms of drying.
- ✓ Providing mathematical models (and/or parameters) that can be applied to typified food groups.
- ✓ Understanding the dynamics of structure formation and quality changes in dried products.
- ✓ Quantification of food qualities; estimating the quality from operation variables.
- ✓ Advancement of drying equipment in terms of energy consumption and quality control.

References

Alves-Filho, O. (2002). Combined innovative heat pump drying technologies and new cold extrusion techniques for production of instant foods. *Drying Technology* 20 (8): 1541–1557. https://doi.org/10.1081/DRT-120014051.

Bai, Y., Rahman, M.S., Perera, C.O. et al. (2001). State diagram of apple slices: glass transition and freezing curves. *Food Research International* 34 (2): 89–95. https://doi.org/10.1016/S0963-9969(00)00128-9.

Barresi, A.A. and Marchisio, D.L. (2018). Computational fluid dynamics data for improving freeze-dryers design. *Data in Brief* 19: 1181–1213. https://doi.org/10.1016/j.dib.2018.05.141.

Chou, S.K. and Chua, K.J. (2001). New hybrid drying technologies for heat sensitive foodstuffs. *Trends in Food Science & Technology* 12 (10): 359–369. http://dx.doi.org/10.1016/S0924-2244(01)00102-9.

Claussen, I.C., Andresen, T., Eikevik, T.M. et al. (2007a). Atmospheric freeze drying – modeling and simulation of a tunnel dryer. *Drying Technology* 25 (12): 1959–1965. https://doi.org/10.1080/07373930701727275.

Claussen, I.C., Ustad, T.S., Strmmen, I. et al. (2007b). Atmospheric freeze drying – a review. *Drying Technology* 25 (6): 947–957. https://doi.org/10.1080/07373930701394845.

Colucci, D., Fissore, D., Mulet, A. et al. (2017). On the investigation into the kinetics of the ultrasound-assisted atmospheric freeze drying of eggplant. *Drying Technology* 35 (15): 1818–1831. https://doi.org/10.1080/07373937.2016.1277738.

Colucci, D., Fissore, D., Rossello, C. et al. (2018). On the effect of ultrasound-assisted atmospheric freeze-drying on the antioxidant properties of eggplant. *Food Research International* 106: 580–588. https://doi.org/10.1016/j.foodres.2018.01.022.

Di Matteo, P., Donsi, G., and Ferrari, G. (2003). The role of heat and mass transfer phenomena in atmospheric freeze-drying of foods in a fluidised bed. *Journal of Food Engineering* 59 (2–3): 267–275.

Garcia-Perez, J.V., Carcel, J.A., Riera, E. et al. (2012). Intensification of low-temperature drying by using ultrasound. *Drying Technology* 30 (11–12): 1199–1208.

Giordano, A., Barresi, A.A., and Fissore, D. (2011). On the use of mathematical models to build the design space for the primary drying phase of a pharmaceutical lyophilization process. *Journal of Pharmaceutical Sciences* 100 (1): 311–324. https://doi.org/10.1002/jps.22264.

Heldman, D. and Hohner, G. (1974). An analysis of atmospheric freeze drying. *Journal of Food Science* 39 (1): 147–155.

Kominami, Y., Watanabe, M., and Suzuki, T. (2014). Effects of postmortem physiological changes on ice crystal morphology in the fish muscle tissue. *Trans JSRAE* 31 (2): 47–56.

Li, B. and Sun, D.-W. (2002). Novel methods for rapid freezing and thawing of foods–a review. *Journal of Food Engineering* 54 (3): 175–182.

Liapis, A.I. and Bruttini, R. (1994). A theory for the primary and secondary drying stages of the freeze-drying of pharmaceutical crystalline and amorphous solutes: comparison between experimental data and theory. *Separations Technology* 4 (3): 144–155. http://dx.doi.org/10.1016/0956-9618(94)80017-0.

Meryman, H.T. (1959). Sublimation freeze-drying without vacuum. *Science* 130 (3376): 628–629. http://dx.doi.org/10.1126/science.130.3376.628.

Miyawaki, O. (2018). Water and freezing in food. *Food Science and Technology Research* 24 (1): 1–21.

Miyawaki, O., Abe, T., and Yano, T. (1992). Freezing and ice structure formed in protein gels. *Bioscience, Biotechnology, and Biochemistry* 56 (6): 953–957.

Mumenthaler, M. and Leuenberger, H. (1991). Atmospheric spray-freeze drying: a suitable alternative in freeze-drying technology. *International Journal of Pharmaceutics* 72 (2): 97–110.

Nakagawa, K., Ochiai, T. (2015). A Mathematical Model of Multi-Dimensional Freeze-Drying for Food Products. *Journal of Food Engineering* 161: 55–67.

Nakagawa, K. and Ochiai, T. (2017). Design space estimation in the freeze-drying of instant soup product. *Japan Journal of Food Engineering* 18 (2): 115–123.

Nakagawa, K. and Tamiya, S. (2019). Influence of inner vapor transfer property of a freeze dryer on the design space of drying process. *AICHE Journal* 65 (5): e16571.

Nakagawa, K., Hottot, A., Vessot, S. et al. (2007). Modeling of freezing step during freeze-drying of drugs in vials. *AICHE Journal* 53 (5): 1362–1372. http://dx.doi.org/10.1002/aic.11147.

Nakagawa, K., Surassmo, S., Min, S.-G. et al. (2011). Dispersibility of freeze-dried poly(epsilon-caprolactone) nanocapsules stabilized by gelatin and the effect of freezing. *Journal of Food Engineering* 102 (2): 177–188. https://doi.org/10.1016/j.jfoodeng.2010.08.017.

Nakagawa, K., Tamiya, S., Do, G. et al. (2018a). Observation of glassy state relaxation during annealing of frozen sugar solutions by X-ray computed tomography. *European Journal of Pharmaceutics and Biopharmaceutics* 127: 279–287. https://doi.org/10.1016/j.ejpb.2018.02.027.

Nakagawa, K., Tamiya, S., Sakamoto, S. et al. (2018b). Observation of microstructure formation during freeze-drying of dextrin solution by in-situ X-ray computed tomography. *Frontiers in Chemistry* 6 (418): https://doi.org/10.3389/fchem.2018.00418.

Nakagawa, K., Horie, A., and Nakabayashi, M. (2020a). Modeling atmospheric freeze-drying of food products operated above glass transition temperature. *Chemical Engineering Research and Design* 163: 12–20. https://doi.org/10.1016/j.cherd.2020.08.017.

Nakagawa, K., Tamiya, S., and Ochiai, T. (2020b). Evaluation of degree of collapse and its relationship with retention of organic volatiles in freeze-dried dextrin matrices. *Industrial & Engineering Chemistry Research* 59 (40): 18298–18306.

Nakagawa, K., Horie, A., Nakabayashi, M. et al. (2021). Influence of processing conditions of atmospheric freeze-drying/low-temperature drying on the drying kinetics of sliced fruits and their vitamin C retention. *Journal of Agriculture and Food Research* 6: 100231.

Nakazawa, N. and Okazaki, E. (2020). Recent research on factors influencing the quality of frozen seafood. *Fisheries Science* 86 (2): 231–244.

Nakazawa, N., Wada, R., Fukushima, H. et al. (2020). Effect of long-term storage, ultra-low temperature, and freshness on the quality characteristics of frozen tuna meat. *International Journal of Refrigeration* 112: 270–280.

Patel, S.M., Chaudhuri, S., and Pikal, M.J. (2010). Choked flow and importance of Mach I in freeze-drying process design. *Chemical Engineering Science* 65 (21): 5716–5727. https://doi.org/10.1016/j.ces.2010.07.024.

Perera, C.O. and Rahman, M.S. (1997). Heat pump dehumidifier drying of food. *Trends in Food Science & Technology* 8 (3): 75–79.

Pikal, M.J. and Shah, S. (1990). The collapse temperature in freeze drying: dependence on measurement methodology and rate of water removal from the glassy phase. *International Journal of Pharmaceutics* 62 (2): 165–186. https://doi.org/10.1016/0378-5173(90)90231-R.

Qin, F.G., Zhao, J.C., Russell, A.B. et al. (2003). Simulation and experiment of the unsteady heat transport in the onset time of nucleation and crystallization of ice from the subcooled solution. *International Journal of Heat and Mass Transfer* 46 (17): 3221–3231.

Rambhatla, S., Ramot, R., Bhugra, C. et al. (2004). Heat and mass transfer scale-up issues during freeze drying: II. Control and characterization of the degree of supercooling. *AAPS PharmSciTech* 5 (4): 54–62. https://doi.org/10.1208/pt050458.

Ratti, C. (2001). Hot air and freeze-drying of high-value foods: a review. *Journal of Food Engineering* 49 (4): 311–319. http://dx.doi.org/10.1016/S0260-8774(00)00228-4.

Sadikoglu, H. and Liapis, A.I. (1997). Mathematical modelling of the primary and secondary drying stages of bulk solution freeze-drying in trays: parameter estimation and model discrimination by comparison of theoretical results with experimental data. *Drying Technology* 15 (3–4): 791–810. http://www.scopus.com/inward/record.url?eid=2-s2.0-0031102433&partnerID=40&md5=f7ca8af340231d848fcca763f185cb6d.

Searles, J.A., Carpenter, J.F., and Randolph, T.W. (2001). The ice nucleation temperature determines the primary drying rate of lyophilization for samples frozen on a temperature-controlled shelf. *Journal of Pharmaceutical Sciences* 90 (7): 860–871. https://doi.org/10.1002/jps.1039.

Slade, L., Levine, H., and Reid, D.S. (1991). Beyond water activity: recent advances based on an alternative approach to the assessment of food quality and safety. *Critical Reviews in Food Science and Nutrition* 30 (2–3): 115–360. https://doi.org/10.1080/10408399109527543.

Sobral, P.J.A., Telis, V.R.N., Habitante, A.M.Q.B. et al. (2001). Phase diagram for freeze-dried persimmon. *Thermochimica Acta* 376 (1): 83–89. http://dx.doi.org/10.1016/S0040-6031(01)00533-0.

Telis, V.R.N. and Sobral, P.J.A. (2001). Glass transitions and state diagram for freeze-dried pineapple. *LWT - Food Science and Technology* 34 (4): 199–205. http://dx.doi.org/10.1006/fstl.2000.0685.

Telis, V.R.N. and Sobral, P.J.A. (2002). Glass transitions for freeze-dried and air-dried tomato. *Food Research International* 35 (5): 435–443. http://dx.doi.org/10.1016/S0963-9969(01)00138-7.

Toei, R., Okazaki, M., and Asaeda, M. (1975). The stability of plane sublimation and a model of zone sublimation in freeze-drying of porous bodies. *Journal of Chemical Engineering of Japan* 8 (4): 282–288. https://doi.org/10.1252/jcej.8.282.

Velardi, S.A. and Barresi, A.A. (2008). Development of simplified models for the freeze-drying process and investigation of the optimal operating conditions. *Chemical Engineering Research and Design* 86 (1): 9–22.

Wang, Z., Gong, J., Wang, Q., & Qiao, X. (2021). Emergency management science and technology: An international transdisciplinary platform. *Emergency Management Science and Technology*, 1(1), 1–3. https://doi.org/10.48130/EMST-2021-0001.

Wolff, E. and Gibert, H. (1988). Développements technologiques nouveaux en lyophilisation. *Journal of Food Engineering* 8 (2): 91–108. https://doi.org/10.1016/0260-8774(88)90057-X.

Wolff, E. and Gibert, H. (1990). Atmospheric freeze-drying part 1: design, experimental investigation and energy-saving advantages. *Drying Technology* 8 (2): 385–404.

Zhu, Z., Zhou, Q., and Sun, D.-W. (2019). Measuring and controlling ice crystallization in frozen foods: a review of recent developments. *Trends in Food Science & Technology* 90: 13–25.

3

Freeze Drying of Fruits and Vegetables

Yogesh Kumar[1] *and Rajat Suhag*[2]

[1] *Department of Agricultural and Food Sciences, University of Bologna, Cesena, FC, Italy*
[2] *Faculty of Agricultural, Environmental and Food Sciences, Free University of Bolzano, Piazza Università, Bolzano, Italy*

3.1 Introduction

In general, drying refers to the process of producing a solid product by removing water by heating (Mujumdar 2020). There are multiple strategies for drying food. Since fruits and vegetables have a limited shelf life, drying them can increase their shelf life while keeping the quality of the product. Freeze drying, also known as lyophilization, is a popular method for producing high-quality food powders and solids. Given that it functions at low temperatures and under high vacuum, it is a preferred technique for drying foods that include substances that are thermally sensitive and prone to oxidation. Application of freeze drying to a variety of fruits and vegetables has already been reported in the literature (Bhatta et al. 2020).

The assessment of product quality is a strategy that is frequently employed in drying experiments. This is explained by the necessity of determining whether the dried product that is produced has an acceptable quality in addition to factors like process efficiency and modeling. This implies that in order to increase consumer acceptance, physical, biochemical, and their prospective food applications must be investigated (Richter Reis et al. 2022). This chapter therefore outlines the physical and biochemical changes brought by freeze drying to various fruits and vegetables. Furthermore, it discusses the use of freeze drying for the encapsulation of fruit and vegetable extracts, as well as freeze-dried fruits and vegetables in food applications.

Freeze Drying of Food Products: Fundamentals, Processes and Applications, First Edition.
Edited by Roji Balaji Waghmare, Manoj Kumar, and Parmjit Singh Panesar.

3.2 Physical Characteristics of Freeze-Dried Fruits and Vegetables

3.2.1 Color

Natural color preservation in dried food products is critical because a consumer's first impression is often based on how something looks. Color, along with size, shape, and other factors, contributes to appearance and serves as an important suggestive criterion for quality control. In the food sector, color modification has compelled improving the visual appeal by adding coloring agents, encouraging high customer acceptance of products, especially dried fruits and vegetables. One of the most difficult aspects of drying processes is performing the operation in such a way that the end product has an appealing color (Calín-Sánchez et al. 2020).

The color of fruits and vegetables during drying is generally influenced by reactions such as enzymatic and non-enzymatic browning (Richter Reis et al. 2022). Enzymatic browning is the oxidation of phenolic compounds in fruits and vegetables by the action of the polyphenol oxidase enzyme, which initiates the formation of brown pigments (melanins). Contrarily, non-enzymatic browning is caused by many non-enzymatic reactions such as Maillard, caramelization, and ascorbic acid oxidation. The presence of large amounts of reducing sugars like glucose and fructose is connected to color changes in carbohydrate-rich dried fruits and vegetables. When exposed to air at high temperatures, these compounds undergo the Maillard reaction by reacting with amino groups from proteins. Maillard reaction time can be reduced by quickly drying the product to 15–20% moisture. Furthermore, the loss of functionality of heat-sensitive compounds such as anthocyanins and carotenes during drying causes color changes. For these reasons, drying methods that require less time and temperature are preferred, reducing browning and preserving the color of fruits and vegetables (Sagar and Suresh Kumar 2010).

When describing the color of dried fruits and vegetables, the CIE (L^* a^* b^*) color scale is commonly used (Al Maiman et al. 2021; Kumar et al. 2021a).The three parameters in the CIE scale: L^* indicates perceptual lightness (0 = black, 50 = middle gray, and 100 = white), a^* is red/green coordinate (negative = green, 0 = neutral, and positive = red) and b^* is yellow/blue coordinate (negative = blue, 0 = neutral, and positive = yellow). Another commonly used parameter is the total color difference (ΔE), which is calculated as shown in Eq. (3.1).

$$\Delta E = \sqrt{\Delta L^{*2} + \Delta a^{*2} + \Delta b^{*2}} \tag{3.1}$$

In comparison to other drying methods, freeze drying of red pepper (Pinar cultivar) showed the least total color change, according to Pinar et al. (2021). Freeze drying of red pepper decreased the lightness (L^*), increased the yellowness (b^*), and had no statistically significant change on the redness (a^*). Reduced lightness could be caused by the Maillard reaction, which produces dark pigments. Differences in redness value should be considered for red pepper to assess the color change that was preserved during freeze drying, making it the preferred method of drying. Furthermore, freeze-dried pumpkin, carrot (Suwannarak and Phanumong 2021), and mango (Link et al. 2017) presented a similar behavior of increased L^* and decreased a^* and b^* values in comparison to fresh samples. In terms of ΔE, pumpkin had a value of 47.75 followed by mango (28.89) and carrot (22.26) (Link et al. 2017; Suwannarak and Phanumong 2021), indicating higher color differences in pumpkin and least in carrot due to freeze drying.

Freeze drying process increased the lightness, greenness and yellowness of kiwi fruits (Akar and Barutçu Mazı 2019; Domin et al. 2020). Freeze drying conditions used during processing may also influence the color of the final product. Increase in the freeze-drying pressure from 12 to 100 Pa reduced the lightness value (from 65.3 to 58.3), while increased the yellowness (from 22.3 to 28.3) and greenness (from −3.2 to −6.8) of kiwi fruit. Interestingly, the ΔE decreased with increase in the freeze-drying pressure, indicting less color changes in kiwi fruits (Domin et al. 2020). Whereas, freeze drying time (120–1440 minutes) had no statistically significant effect on the lightness and greenness of kiwi fruits. This indicates that freeze drying of kiwi maybe helpful in retention of chlorophyll. Yellowness of kiwi fruits increased with increase in the freeze-drying time from 120 to 480 minutes, further increase in drying time to 720 minutes reduced the yellowness but further increase to 1440 minutes had no significant changes (Akar and Barutçu Mazı 2019).

Varying the water activity resulted in a change in the color characteristics of freeze-dried persimmon slices (González et al. 2020). Lightness did not differ significantly from control up to a water activity of 0.430 and decreased with further increase in water activity because of the water content uptake. ΔE values were greater than 3 for all the samples, indicating an appreciable total color change which can be detected by human eye, and the ΔE value increased with increase in water activity. Furthermore, greater color differences were observed from the water value of 0.430 (González et al. 2020), showing that it is a critical value at which the enzymatic reactions can start (Telis and Martínez-Navarrete 2010).

3.2.2 Shrinkage

The dried substance undergoes physical changes during the drying process. One of the essential yet undesirable changes is shrinkage, which occurs as a result of the product's volume being reduced by the evaporation of a considerable amount of water. The volume change is accompanied by a change in shape, and the amount of evaporated water influences the hardness of the material, causing the dry material to fracture and distort, lowering the ultimate quality of the product (Ciurzyńska et al. 2022). Furthermore, drying rate has been shown to influence the shrinkage, the product shrink and become denser at low drying rate whereas, high drying rates may limit the shrinkage to some extent because of case hardening (Richter Reis et al. 2022). To compute shrinkage, the solid displacement method and the toluene displacement method can be used with Eq. (3.2):

$$S = \frac{V_o - V_f}{V_o} \times 100 \quad (3.2)$$

where, S is shrinkage in %, V_0 and V_f are volume before and after drying, respectively.

In a recent study, Feng et al. (2022) reported that freeze-dried vegetables were able to nearly perfectly maintain cubic shape in comparison to fruits (Figure 3.1). In particular, the starchy rhizomes were better able to keep the original shape similar to just frozen and inhibited shrinkage. Furthermore, fruits with a high soluble sugar content showed significant volume shrinkage. For instance, cantaloupe, pineapple, mango and apple with soluble sugar content of 600, 623.85, 594.11 and 690.13 mg/g dry basis, respectively had a shrinkage ratio of 67.99%, 59.97%, 55.29%, and 43.35%, respectively (Feng et al. 2022). Polysaccharides' anti-shrinkage function could be explained by two considerations. First, as concentration increases, the number of inter-/intramolecular cross-linking sites

Figure 3.1 Cubes of different freeze-dried fruit and vegetable. *Source:* Shuhan Feng et al. 2022/Reproduced with permission from Elsevier.

between cell wall polysaccharides may increase (Buchtová and Budtova 2016). Second, polysaccharides may raise the temperature of gelatinization of the overall network, bringing it closer or even above the temperature of sublimation, which is useful in preventing collapse (Roos 2010). Whereas, during freeze drying, if the temperature of sublimation exceeds the temperature of melting, it may decrease the viscosity of the amorphous solute mixture leading to the collapse of the structure (Pikal and Shah 1990). Therefore, these changes may be dependent on the conditions of freeze-drying process and the chemical composition of the fruit and vegetable matrix. However, in the above described study of Feng et al. (2022) similar freeze-drying parameters were applied on all the fruits and vegetables therefore, the difference in the shrinkage was majorly due to the differences in the chemical compositions of the fruits and vegetables.

Additionally, freeze-dried carrots exhibited less shrinkage than convective and microwave drying methods, with average shrinkage of 60.33%, 68.95%, and 64.23%, respectively. Freeze drying was also coupled with convective and microwave drying technologies. Interestingly, shrinkage was reduced when carrot was freeze-dried before convective and microwave drying than when carrot was solely freeze-dried. Whereas freeze drying after convective drying resulted in less shrinkage than freeze-dried carrot, freeze drying after microwave drying resulted in somewhat higher shrinkage than freeze-dried carrot (Ciurzyńska et al. 2022). This demonstrates that freeze drying can generate superior results when combined with other drying technologies, but the sequence of combination may yield different results, which must be tuned and taken into account.

3.2.3 Rehydration Capacity

Rehydration is the process by which a dried material absorbs water when exposed to a large volume of water. Designing, developing, and maximizing convenience and ready-to-eat foods – the primary use of dried foods – requires maintaining a quick rehydration and the preservation of the food's organoleptic features, particularly in the case of vegetables and fruits. Rehydration capacity can be determined by Eq. (3.3).

$$\mathrm{RC} = \frac{M_r - M_d}{M_f - M_d} \times 100 \tag{3.3}$$

where, M_r, M_d, and M_f are the mass of rehydrated, dried, and fresh samples, respectively in (g).

The ability of dried fruits and vegetables to rehydrate has also been regarded as a quality indicator. According to studies, freeze drying is preferable over other drying methods for rehydrating kiwi fruit pieces (Akar and Barutçu Mazı 2019) and red pepper (Pinar et al. 2021). Rehydration is made easier with an intact porous structure, such as the one found in freeze-dried products (Akar and Barutçu Mazı 2019). Furthermore, according to Feng et al. (2022), rehydration capacity of freeze-dried fruits and vegetables ranged from 2.86% to 7.39%, with ash gourd standing out with a high rehydration capacity of 17.64%. The mono/disaccharide concentration of dried fruits and vegetables was connected to their varied rehydrating capacity. Rehydrating capacity correlated negatively with soluble sugar content (−0.61) and total sugar content (−0.5). This negative correlation was explained through two considerations (i) in the first stage of drying, the dissolution of sugar on the surface of fruits and vegetables may form a film, thereby making it difficult for water to be absorbed by the porous structure through capillary flow (Agnieszka and Andrzej 2010) and (ii) high sugar content promotes shrinkage (described in details in Section 2.2), causing loss of porous structure and reduces the ability of dried fruits and vegetables to rehydrate (Lewicki 1998). The freeze-drying pressure is a significant component that also has an impact on the hydration capacity. The ability of kiwi fruits to rehydrate was decreased as the freeze- drying pressure was increased (Domin et al. 2020). During freeze dying, a dense collapsed structure is created under higher pressures, which causes the porous capillaries to constrict and lose their ability to absorb water.

In terms of rehydration kinetics, freeze-dried tomatoes were found to have a rehydration rate that was four times faster than that of hot air-dried tomatoes and six times higher than that of infrared-dried tomatoes. The ability of freeze-dried tomatoes to rehydrate was also influenced by the temperature of the rehydrating medium. Higher rehydrating capacities and higher final equilibrium moisture levels were seen as the temperature of rehydrating medium was increased, for instance, a rehydrating capacity of 52% was found at 50 °C, compared to 37% at 20 °C. The fact that the rehydrated samples fell short of having the same amount of moisture as fresh samples, however, shows that the drying process is irreversible (Lopez-Quiroga et al. 2020).

3.2.4 Microstructure

Understanding the microstructure of dried fruits and vegetables can give us crucial information about their overall quality. In order to aid the study of food microstructure, especially that of fresh and dried products, substantial focus has recently been paid to several imaging techniques like scanning electron microscopy (Akar and Barutçu Mazı 2019; Ciurzyńska et al. 2022; Link et al. 2017), transmission electron microscope (Zhang et al. 2018), light microscope (González et al. 2020), and X-ray microtomography scanning (Lopez-Quiroga et al. 2020; Vergeldt et al. 2014).

Scanning electron microscopic imaging of freeze-dried carrots showed large number of pores of small diameter compared to convective- and microwave-dried carrots, creating a delicate porous structure with minimal structural damage

(Ciurzyńska et al. 2022). Similarly, a spongy structure with a large number of small pores with preserved original structure was observed for freeze-dried mango (Link et al. 2017) and kiwi fruits (Akar and Barutçu Mazı 2019). In addition, a recent thorough investigation on the freeze drying of 12 fruits and vegetables by Feng et al. (2022) found that, with the exception of pear and apple, the average pore size of the majority of freeze-dried samples matched the tissue morphology of the fresh samples (Figure 3.2). A compact microstructure with evidence of cell shape was typically generated by small-celled materials (such as carrot, pumpkin, and pineapple) with few or minimal intercellular gaps. Without the original cell

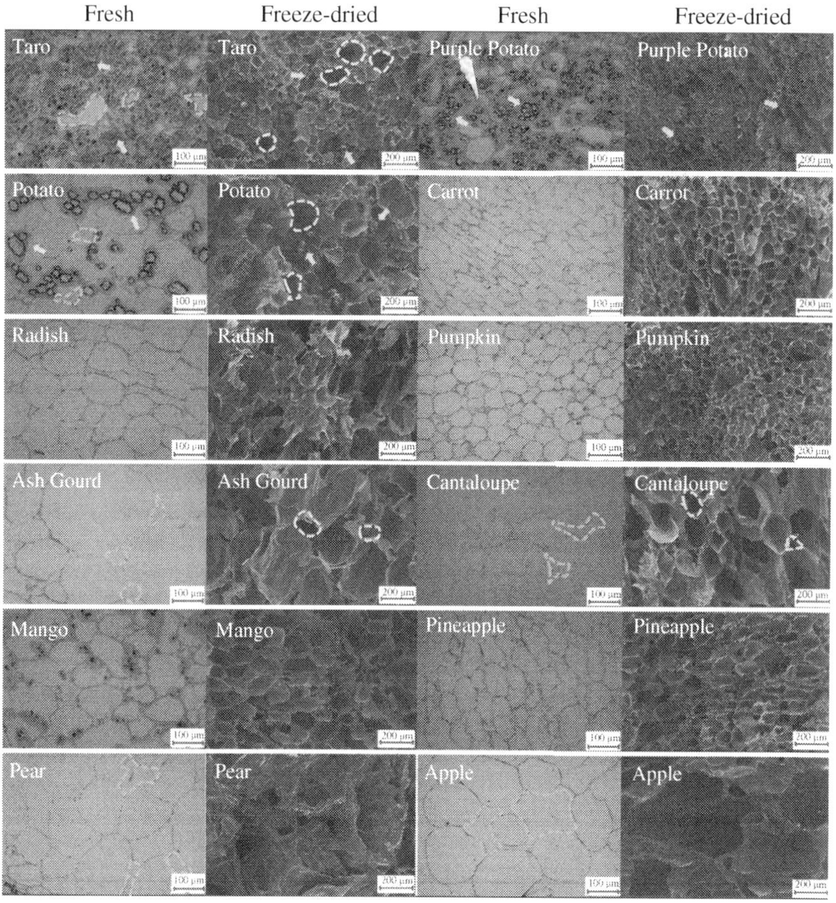

Figure 3.2 The representative microstructure of 12 fresh and freeze-dried fruits and vegetables. Image of fresh materials using light microscopy (×10); image of dried materials using scanning electron micrograph (×100). Red arrow: starch granules; Red dashed lines: large intercellular space; Orange dashed lines: freeze-dried pores derived from the intercellular spaces. *Source:* Shuhan Feng et al. 2022/Reproduced with permission from Elsevier.

shape, the large-celled materials (like apple and pear) with vast intercellular spaces created a spongy microstructure. This could be accounted for by the formation of ice crystals during freezing, which is significantly influenced by the structure of the fresh material's cells (Li et al. 2018). The extracellular space would be frozen first due to the slow freezing rate, which would increase the concentration of extracellular solutes and result in cell dehydration. Large extracellular ice crystals are created during this process, which deforms cells (Schudel et al. 2021). Additionally, the susceptibility of various fruits and vegetables to freezing varies, with apples having been demonstrated to be one of the highly sensitive fruit (Vallespir et al. 2019). The physical qualities of freeze-dried fruits and vegetables, which include less shrinkage, greater rehydration, and better or retained color attributes, are also backed by these structural characterizations.

3.3 Biochemical Characteristics of Freeze-dried Fruits and Vegetables

3.3.1 Antioxidant Activity

To comprehend a food's potential dietary benefit and/or role in the food industry, it is essential to identify the antioxidant components and antioxidant activity in the food of interest. Although a wide range of foods can contain antioxidants, the richness of fruits and vegetables makes them stand out (da Silva et al. 2021; Kiokias and Oreopoulou 2021). Additionally, customers are placing a growing amount of demand on the market for foods that have undergone minimal processing and do not contain artificial ingredients. However, because fruits and vegetables are perishable, availability of fresh fruits and vegetables is a concern that must be addressed. Therefore, it is possible to choose dried fruits and vegetables to provide healthy foods that are high in antioxidant components and, at the same time, are of natural origin. As a result, there has been a lot of work put into drying of fruits and vegetables. In drying studies, the purpose of antioxidant assays is to determine the effect of drying conditions and/or drying techniques on antioxidant components. To prevent the loss of antioxidants, drying processes with lower drying temperatures and faster drying are preferable, and freeze drying may provide such benefits.

Vacuum freeze-dried dropped immature *Citrus reticulata* Blanco fruits showed a higher antioxidant activity by azino-bis [3-ethylbenzthiazoline-6-sulfonic acid] (ABTS), 2,2-diphenyl-1-picrylhydrazyl radical (DPPH) and ferric-reducing ability of plasma (FRAP) assays. Antioxidant activity values in ABTS assay ranged from 12.21 to 13.55 mM/L Trolox and 6.68 to 8.15 mM/L Trolox and in DPPH assay from 15.27 to 16.72 mM/L Trolox and 11.11 to 14.48 mM/L Trolox for vacuum freeze-dried and hot air-dried samples, respectively. Similarly, FRAP assay showed values from 7.31 to 9.07 mM/L Trolox and 5.29 to 6.33 mM/L Trolox for vacuum freeze-dried and hot air-dried samples, respectively (Kumar et al. 2021b). Also, for wild pomegranate flavedo freeze drying exhibited the maximum antioxidant properties of 91%, 63.58%, and 340.35 μM Fe^{2+} 100 g^{-1} by DPPH, metal chelating activity and FRAP assays, respectively. Higher antioxidant activity of freeze-dried flavedo was regarded to the better stability of antioxidant compounds at low temperature during the freeze-drying process compared to mechanical cabinet drying method (Hamid et al. 2020).

Li et al. (2020) dried Huyou (*Citrus changshanensis*) fruit using freeze drying, microwave vacuum drying (MVD) and their different combinations (i) freeze drying followed by MVD (FD-MVD), (ii) MVD followed by freeze drying (MVD-FD), and (iii) MVD followed by freeze drying and again MVD (MVD-FD-MVD). All the drying methods alone or in combinations reduced the antioxidant capacity in comparison to the fresh fruit. In both DPPH and FRAP assays, the order of antioxidant capacity was maximum in fresh fruit followed by FD-MVD > freeze drying > MVD-FD-MVD > MVD-FD > MVD.

Ethanolic extracts of freeze-dried placenta and pericarp from both the Shanghay and Socrates cultivars of pepper fruits, as well as placenta from the Red Knight cultivar, showed comparable or improved antioxidant activity when compared to fresh samples. In contrast, ethanolic extracts of freeze-dried Capel hot placenta and pericarp of Capel hot and Red Knight pepper cultivars showed lower antioxidant activity than fresh samples. These findings imply that freeze drying may improve antioxidant extraction in some circumstances (Materska 2014). Similarly, in comparison to fresh fruits, ethanolic extracts of freeze-dried mango, papaya, muskmelon, and watermelon fruits showed similar antioxidant activity assessed using DPPH and FRAP tests (Shofian et al. 2011). Furthermore, freeze drying of kiwi fruits showed slight reduction in the antioxidant activity compared to fresh fruits but, increasing the pressure of freeze drying from 12 to 100 Pa had no significant effect on the antioxidant activity (Domin et al. 2020).

Sarkar et al. (2020) dried mango leather using different drying methods and reported the DPPH scavenging activity in the order: freeze drying ≥ microwave drying > hot air drying > sun drying. ABTS and FRAP assays, on the other hand, indicated a similar order with freeze drying > microwave drying > sun drying > hot air drying. Tan et al. (2021) found that drying three tomato cultivars (18,131, 18,146, and 1862) via air drying and oven drying produced variable antioxidant activity as determined by the DPPH and ABTS assays. The DPPH assay revealed no significant differences between the 18,131 cultivars after freeze drying and oven drying compared to fresh samples, whereas cultivars 18,146 and 1862 showed a decrease in the DPPH value after freeze drying and oven drying but no significant difference between the drying methods. Furthermore, the ABTS assay revealed a decrease in antioxidant activity for all three cultivars after freeze and oven drying when compared to fresh samples, but interestingly, all three cultivars showed similar ABTS values for both drying methods even when the ABTS values of fresh cultivars were significantly different.

The results of multiple studies show that freeze drying, as opposed to other drying techniques that require subjecting fruits and vegetables to higher temperatures for longer periods of time, has enormous potential to retain or preserve better the antioxidant activity of fruits and vegetables. Furthermore, the antioxidant activity test employed impacts the antioxidant measurements; thus, more than one assay should be performed to examine the effect of different drying or processing procedures on the antioxidant activity of foods for a better understanding.

3.3.2 Polyphenols

The largest class of plant secondary metabolites is known as polyphenols, and the structure of these compounds can range from simplest phenols with hydroxyl groups attached to the aromatic ring to extremely complicated polymeric compounds like tannins and lignans (Zeb 2020). Fruits and vegetables contain a lot of polyphenols, which can be found alone or in conjunction with other types of phytochemicals (Carrillo et al. 2019). Phenolic compounds are the primary determinants of antioxidant properties in fruits and vegetables, and hence form the foundation of antioxidants.

The total phenolic content of mango leather dried using different methods ranged from 4.1 to 7.2 mg/g dry basis, with freeze-dried mango leather having the highest value followed by microwave drying > hot aid drying > sun drying. Similar order was reported for the total flavonoid content of mango leather dried using freeze drying > microwave drying > hot air drying > sun drying. This was further supported by the best polyphenol composition of freeze-dried mango leather with higher amounts of phenols than the leathers dried with other techniques. Gallic acid was the major phenolic acid in mango leather and freeze-dried sample had a 3-, 22-, and 2-time gallic acid content compared to microwave-, sun-, and hot air-dried samples, respectively. Sinapic acid the second major phenolic acid found in the mango leather was found to be 80, 10.6, and 195 times higher in freeze-dried samples than microwave-, sun-, and hot air-dried, respectively. Furthermore, major flavonoids quercetin followed by rutin were maximum for freeze-dried mango leather (Sarkar et al. 2020).

The total phenolic content of three tomato cultivars (18,131, 18,146, and 1862) was decreased by freeze drying and oven drying. However, for cultivars 18,131 and 1862, there was no significant difference between freeze-dried and oven-dried samples, however for cultivar 18,146, freeze-dried samples had a higher total phenolic content than oven-dried samples. The main phenolic acid identified in all three tomato cultivars was ferulic acid, which was shown to be more abundant in freeze-dried samples than in oven-dried ones for all three cultivars despite both methods of drying reducing its amount relative to fresh tomatoes. Surprisingly, the main flavonoid rutin increased considerably after freezing and oven drying. Possible reason reported for this increase was the breakdown of cell structure during drying, and release of bound phenolics from the matrix (Tan et al. 2021).

Freeze drying was found to be the best method to retain the polyphenols of papaya fruits (Minuye et al. 2021). Total phenolic content and total flavonoid content of freeze-dried papaya fruits were 41.47 mg/g gallic acid equivalent and 1.89 mg/g quercetin equivalent, respectively. Furthermore, the second best total phenolic content was for oven-dried papaya fruits with a value of 14.78 mg/g gallic acid equivalent showing approximately 64% reduction compared to freeze-dried papaya. Other drying methods like solar glass house, solar open, solar tray, and refractance window drying all had lower total phenolic and flavonoid content. Similarly, for total carotenoids and β-carotene contents freeze drying showed far higher retention in papaya fruits.

Pico et al. (2019) produced banana flour using oven- and freeze-drying methods, and showed that the drying methods had no significant effect on the extractable and hydrolysable phenolic content of dried banana flour. Phenolic profile analysis included major phenolic acids (gallic and caffeic acid), flavan-3-ols (epicatechin and catechin), and flavonols (quercetin-3-O-gluoside and myricetin) in banana flour. Epicatechin was the abundant phenolic compound with content 1.70 and 1.93 mg/100 g for oven- and freeze-dried banana flours, respectively. Similarly, several other fruits and vegetables like maqui (*Aristotelia chilensis*) berries (Quispe-Fuentes et al. 2018), different cultivars of African eggplant (Mbondo et al. 2018), and blueberries, cherries, strawberry, and cranberries (Nemzer et al. 2018) have been freeze-dried (Table 3.1).

3.3.3 Ascorbic Acid

During handling, processing, and storage of fruits and vegetables vitamins are lost. The ability of vitamins to preserve their properties under chemical, physical, and/or thermal stress is referred to as stability. Vitamin C, also known as ascorbic acid, is a water-soluble vitamin that is abundant in fresh fruits and vegetables. Ascorbic acid is most vulnerable to destruction when fruit and vegetables are subjected to drying process involving high temperature and duration.

After freeze drying to 15% and 5% moisture content, the ascorbic acid content of capsicum increased from 68 mg/100 g to 78 and 110.2 mg/100 g, respectively. Furthermore, after 50 days of storage at 5 °C of freeze-dried capsicum (moisture content: 15%), the ascorbic acid content decreased by approximately 36%, whereas

Table 3.1 Polyphenolic content of dried fruit and vegetables.

Fruit or vegetable	Drying methods	Total phenolic/flavonoid content	Major polyphenols	References
Mango	Freeze drying	TPC = 7.2 mg/g db TFC = 3.27 mg/g db	***Phenolic acids:*** Gallic acid Sinapic acid ***Flavonoids:*** Quercetin Rutin	Sarkar et al. (2020)
	Microwave drying	TPC = 6.3 mg/g db TFC = 2.75 mg/g db		
	Hot air drying	TPC = 4.7 mg/g db TFC = 2.31 mg/g db		
	Sun drying	TPC = 4.1 mg/g db TFC = 2.21 mg/g db		
Tomato (Cultivar 18,131)	Freeze drying and oven drying	Similar TPC for both freeze- and oven-dried samples	***Phenolic acids:*** Ferulic acid 4-hydroxycinnamic acid ***Flavonoids:*** Rutin Quercetin Naringenin	Tan et al. (2021)
Tomato (Cultivar 18,146)		Significantly higher TPC for freeze-dried samples		
Tomato (Cultivar 1862)		Similar TPC for both freeze and oven-dried samples		
Papaya	Oven drying	TPC = 14.78 mg/g GAE TFC = 1.33 mg/g QE TC = 52.46 μg/g β-Carotene = 6.40 μg/g	Not reported	Minuye et al. (2021)
	Solar glass house	TPC = 12.01 mg/g GAE TFC = 1.95 mg/g QE TC = 71.62 μg/g β-Carotene = 15.02 μg/g		

	Solar open	TPC = 6. mg/g GAE TFC = 0.80 mg/g QE TC = 45.10 μg/g β-Carotene = 5.16 μg/g		
	Solar tray	TPC = 11.05 mg/g GAE TFC = 1.06 mg/g QE TC = 49.97 μg/g β-Carotene = 8.24 μg/g		
	Refractance window	TPC = 13.40 mg/g GAE TFC = 1.38 mg/g QE TC = 84.55 μg/g β-Carotene = 22.55 μg/g		
	Freeze drying	TPC = 41.57 mg/g GAE TFC = 1.89 mg/g QE TC = 103.74 μg/g β-Carotene = 33.66 μg/g		
Banana	Oven drying	Extractable phenolic content = 337.79 mg/100 g GAE Hydrolysable phenolic content = 1061.22 mg/100 g GAE	***Phenolic acids*** Gallic acid Caffeic acid ***Flavan-3-ols*** Epicatechin	Pico et al. (2019)
	Freeze drying	Extractable phenolic content = 335.66 mg/100 g GAE Hydrolysable phenolic content = 1082.9 mg/100 g GAE	Catechin ***Flavonols*** Quercetin-3-O-gluoside Myricetin	
Maqui berries	Freeze drying	TPC = 30.65 mg GAE/g DM TFC = 30.41 mg QE/g DM	Ellagic acid Gallic acid Protocatechuic acid	Quispe-Fuentes et al. (2018)

(Continued)

Table 3.1 (Continued)

Fruit or vegetable	Drying methods	Total phenolic/flavonoid content	Major polyphenols	References
	Convective drying	TPC = 23.40 mg GAE/g DM TFC = 24.55 mg QE/g DM		
	Sun drying	TPC = 31.93 mg GAE/g DM TFC = 12.38 mg QE/g DM		
	Infrared drying	TPC = 27.87 mg GAE/g DM TFC = 15.72 mg QE/g DM		
	Vacuum drying	TPC = 30.31 mg GAE/g DM TFC = 22.39 mg QE/g DM		
African eggplant (Cultivar Sangawili)	Freeze drying	TPC = 595.80 mg/100 g GAE	Not reported	Mbondo et al. (2018)
	Sun drying	TPC = 560.58 mg/100 g GAE		
African eggplant (Cultivar Manyire green)	Freeze drying	TPC = 714.01 mg/100 g GAE		
	Sun drying	TPC = 422.57 mg/100 g GAE		
African eggplant (Cultivar S00047A)	Freeze drying	TPC = 494.60 mg/100 g GAE		
	Sun drying	TPC = 547.08 mg/100 g GAE		
African eggplant (Cultivar AB2)	Freeze drying	TPC = 657.27 mg/100 g GAE		
	Sun drying	TPC = 628.32 mg/100 g GAE		
African eggplant (Cultivar Aubergine blanche)	Freeze drying	TPC = 574.31 mg/100 g GAE		
	Sun drying	TPC = 447.32 mg/100 g GAE		
Blueberry	IQF	TPC = 1.68% TFC = 1.70%	Not reported	Nemzer et al. (2018)
	Freeze drying	TPC = 2.56% TFC = 2.31%		
	Hot air drying	TPC = 1.92% TFC = 1.92%		

	Refractance window drying	TPC = 2.26% TFC = 2.14%
Cherry	IQF	TPC = 1.10% TFC = 1.02%
	Freeze drying	TPC = 1.64% TFC = 1.37%
	Hot air drying	TPC = 1.33% TFC = 1.12%
	Refractance window drying	TPC = 1.46% TFC = 1.29%
Cranberry	IQF	TPC = 2.47% TFC = 1.36%
	Freeze drying	TPC = 2.24% TFC = 2.07%
	Hot air drying	TPC = 1.69% TFC = 1.46%
	Refractance window drying	TPC = 2.19% TFC = 1.90%
Strawberry	IQF	TPC = 1.68% TFC = 1.70%
	Freeze drying	TPC = 2.56% TFC = 2.31%
	Hot air drying	TPC = 1.92% TFC = 1.92%
	Refractance window drying	TPC = 2.26% TFC = 2.14%

TPC = total phenolic content; TFC = total flavonoid content; TC = total carotenoids; GAE = gallic acid equivalent; QE = quercetin equivalent; IQF = individual quick freezing; db = dry basis.

capsicum with moisture content of 5% decreased by nearly 50% after 155 days under the same conditions. Moreover, degradation rate of ascorbic acid increased with increase in storage temperature and decreased with decrease in the moisture content (Rahman et al. 2015).

When compared to other drying processes, freeze-dried red pepper cultivars (Bozok and Pinar) had the highest ascorbic acid concentration. The ascorbic content of freeze-dried Bozok and Pinar red pepper cultivars was 2.39 and 2.92 mg/g, respectively. The second highest ascorbic acid level was found following microwave drying at 600 W for the Bozok cultivar, and open sun drying for the Pinar cultivar (Pinar et al. 2021).

Berries have a wide range of ascorbic acid concentration (Table 3.2). The ascorbic acid concentration of freeze-dried berries was highest in strawberries, with 1023.8 mg/100 g, followed by cranberries (150.3 mg/100 g), blueberries (35.77 mg/100 g), and cherries (28.3 mg/100 g). In comparison to hot air and refractance window drying, freeze-dried cherry, cranberry, and strawberry had higher ascorbic acid content. Both the freeze-dried and refractance window-dried blueberries had similar ascorbic acid contents (Nemzer et al. 2018).

Table 3.2 Ascorbic acid content of different freeze-dried fruits and vegetables.

Fruits and vegetables	Drying method	Ascorbic acid content	References
Capsicum	Freeze drying	Fresh = 68 mg/100 g Freeze dried (15% moisture content) = 78 mg/100 g Freeze dried (5% moisture content) = 110.2 mg/100 g	Rahman et al. (2015)
Red pepper (Cultivar Bozok)	Freeze drying	2.39 mg/g	Pinar et al. (2021)
Red pepper (Cultivar Pinar)		2.92 mg/g	
Blueberry	Freeze drying	35.77 mg/100 g	Nemzer et al. (2018)
	Hot air drying	18.74 mg/100 g	
	Refractance window drying	38.09 mg/100 g	
Cherry	Freeze drying	28.30 mg/100 g	
	Hot air drying	26.30 mg/100 g	
	Refractance window drying	22.60 mg/100 g	

Table 3.2 (Continued)

Fruits and vegetables	Drying method	Ascorbic acid content	References
Cranberry	Freeze drying	150.30 mg/100 g	
	Hot air drying	36.60 mg/100 g	
	Refractance window drying	107.4 mg/100 g	
Strawberry	Freeze drying	1023.80 mg/100 g	
	Hot air drying	42.40 mg/100 g	
	Refractance window drying	216.40 mg/100 g	
Kiwi	Freeze drying	175.30 mg/100 g	Akar and Barutçu Mazı (2019)
	Hot air drying	56.01 mg/100 g	
	Hot air–microwave-assisted vacuum drying	149.83 mg/100 g	
	Vacuum drying	61.27 g/100 g	

Akar and Barutçu Mazı (2019) reported that the ascorbic acid content of kiwi fruits decreased from 249.17 mg/100 g to 56.01, 61.27, 149.83, and 175.30 mg/100 g after hot air, vacuum, hot air–microwave-assisted vacuum, and freeze drying, respectively. Showing a maximum retention after freeze drying. According to the ascorbic acid breakdown mechanism described, under aerobic conditions, ascorbic acid oxidizes to dihydroxy-ascorbic acid, which might then undergo oxidation to generate 2,3-deketogluonic acid irreversibly (Peleg et al. 2018). Ascorbic acid degradation was steady during the first six hours of the freeze drying of kiwi fruits and then was nearly constant thereafter. The first six hours were better fitted by zero and first order kinetic models, but after that the models showed deviation (Akar and Barutçu Mazı 2019).

3.4 Freeze Drying for Encapsulation

Encapsulation is a phenomenon where a core material is enclosed by a food-grade wall material in order to prevent chemical and physical reactions and maintain the biological, functional, and physicochemical characteristics of the core material (Singh et al. 2022). Freeze drying is a prominent encapsulation technique for sensitive chemicals/compounds. This technique includes freezing the solution by submitting the feed to extremely low temperatures (−50 and −80 °C), followed by low pressure to sublime the ice crystals. The whole process requires a lot of energy and lasts for around 24 to 48 hours. This process is expensive due to the usage of a vacuum pump (Rezvankhah et al. 2020). The main findings related to the use of freeze drying as an encapsulation technique are described in Table 3.3.

Table 3.3 Application of freeze drying as encapsulation technique.

Type of fruits and vegetable	Wall materials	Main findings	References
Acerola	Maltodextrin and gum arabic	• Encapsulation effectiveness was more than 50% for polyphenolic compounds and total flavonoids. • Better physico-chemical properties compared to puree	Rezende et al. (2018)
Blackberry pomas	Maltodextrin with 10 and 20 dextrose equivalents (DE)	• Better anthocyanin retention • Better results for hygroscopicity, moisture content, acidity, water activity using maltodextrin 10DE	Yamashita et al. (2017)
Star fruit (Averrhoa carambola) pomace	Maltodextrin (≤20DE)	• Higher retention of phenolic compounds (97.2%) as compared to spray dried (79.07%)	Saikia et al. (2015)
Model fruit juice	Gum arabic and maltodextrin	• Greater gallic acid levels	Ramírez et al. (2015)
Beetroot juice	Soy protein isolate	• Encapsulation efficiency varies from 47.79% to 92.47% according to variables	Tumbas Šaponjac et al. (2020)
Black carrot juice	Maltodextrin 20DE, gum arabic and tapioca starch	• Highest concentration of total anthocyanin for maltodextrin 20DE and followed by gum arabic and tapioca starch • Higher antioxidant activity, Water solubility index encapsulation efficiency (98.51%) as compared to spray-dried encapsulated sample	Murali et al. (2015)

Table 3.3 (Continued)

Type of fruits and vegetable	Wall materials	Main findings	References
Cherry juice	Maltodextrin DE10 and arabic gum	• Higher concentration of total phenolic and monomeric anthocyanins as compared to fresh juice • Higher retention of anthocyanins during storage • No significant change in color parameters	Sanchez et al. (2015)
Red wine (Cabernet Sauvignon grapes)	Maltodextrin	• Higher concentration of total polyphenols (3.7 time to wine) • No significant in the concentration of total polyphenols during storage	Sanchez et al. (2013)
Cloudberry	Maltodextrins	• Improved the storage stability of phenolics • Improved antioxidant activity during storage	Laine et al. (2008)
Grape skin and seeds (Bordered grape)	Maltodextrin	• Lower anthocyanin retention values than spray-dried samples	De Souza et al. (2015)
Grape skin (Cabernet Sauvignon)	Maltodextrin Gum arabic	• High encapsulation efficiency and anthocyanin retention up to 89.6% and 91.5%, respectively.	Stoll et al. (2016)
Jaboticaba pomace	Maltodextrin, Pectin Soy protein isolate	• Stabilized the anthocyanins • Extend the shelf life of anthocyanins	Pereira Souza et al. (2017)
Olive leaf	Maltodextrin Trehalose dehydrates	• The wall materials and the C/W ratio had a significant effect on the encapsulation efficiency and the antioxidant, thermal properties	González-Ortega et al. (2020)
Orange peel oil	Maltodextrin Gelatin	• High encapsulation efficiency (75.75%) and encapsulation yield (90.19%) • Encapsulated oil showed antioxidant and antibacterial properties	de Araújo et al. (2020)

Rezende et al. (2018) used freeze drying to encapsulate bioactive compounds extracted from acerola pulp using maltodextrin (MD) and gum arabic as a wall material. The results indicated that for polyphenolic compounds and total flavonoids, the microencapsulation efficiency was more than 50% with good physico-chemical properties. Furthermore, Yamashita et al. (2017) microencapsulated the anthocyanin rich blackberry pomas by freeze drying using MD with 10 and 20 dextrose equivalents (DE) as a wall material. According to the findings, MD DE10 (76%) retained anthocyanins more effectively in microcapsules compared to MD DE20 (68%). The powder formulated with MD DE10 performed much better in all physicochemical studies (hygroscopicity, moisture content, acidity, water activity, and color indices) than the powder formulated with MD DE20. Freeze-dried encapsulated star fruit pomace had higher retention of phenolic compounds (97.2%) as compared to spray-dried encapsulated star fruit pomace (79.07%) (Saikia et al. 2015). Ramírez et al. (2015) optimized the encapsulation of model fruit juice in gum arabic and MD using spray- or freeze-drying, with the concentration of gallic acid as a response. According to response surface methodology optimization, the freeze-dried encapsulates produced with a wall blend ratio near to 100% gum arabic and a core concentration from 10% to 20% exhibited higher gallic acid levels.

Tumbas Šaponjac et al. (2020) optimized the beetroot juice encapsulation by freeze drying using soy protein isolate as a wall material. The results indicated that encapsulation efficiency of beet root based on phenolic content varies from 47.79% ($X1 = 100$, $X2 = 2$ and $X3 = 15$ minutes) to 92.47% ($X1 = 150$, $X2 = 0$ and $X3 = 15$ minutes) according to combination of independent variable such as the ratio of wall : core (X1), dilution of beetroot juice (X2) and mixing time (X3). Moreover, Murali et al. (2015) encapsulated the black carrot juice in different wall materials such as MD 20DE, gum arabic and tapioca starch using freeze and spray drying. The results showed that wall/carrier material had a significant ($p = 0.01$) effect on the total anthocyanin content. The highest concentration of total anthocyanin was found in the powder having MD 20DE (1650.00 mg/100 g dry matter in freeze dried and 1461.23 mg/100 g dry matter in case of spray dried at 150 °C) as a wall material and followed by gum arabic and tapioca starch. Freeze-dried encapsulated powder had highest anthocyanin content as compared to spray-dried in all three types of wall materials. Freeze drying encapsulated sample having MD 20DE as a core material had the highest antioxidant activity (329.40 mmol Trolox/g dry matter), water solubility index (96.13%), and encapsulation efficiency (98.51%) among all the sample and with respective spray-dried encapsulated samples. Sanchez et al. (2015) studied the monomeric anthocyanins and color stability of developed freeze-dried encapsulated cherry juice having MD DE10 and Arabic gum (80 : 20) at a ratio of 25% of encapsulating agents and compared with the fresh and concentrate cherry juice stored at 38 °C. The results indicated that encapsulated powder had the concentration of total phenolic and monomeric anthocyanins higher than fresh juice and less than juice concentrate. The retention of anthocyanin in juice and juice concentrate (decreased to 11% after 33 days of storage in both) during storage dropped rapidly, although it is significantly more stable in encapsulated powder (remained at around 90% after 33 days).After 33 days, the L^* and b^* values in juice and juice concentrate decreased while the a^* value increased, resulting in greater tone (h^0) and brownish juice, but no significant change was seen in encapsulated powder. Furthermore, encapsulated red wine having MD as a wall material showed higher concentration of total polyphenols (3.7 time to wine) and not significant change after 15 days storage at 38 °C. The encapsulated powder contains less than 1% ethanol and can be used to other foods as a source of polyphenols (Sanchez et al. 2013). Microencapsulation of cloudberry (Rubus chamaemorus) improved the storage stability of phenolic and antioxidant activity improved during storage (Laine et al. 2008).

Recently, González-Ortega et al. (2020) analyzed the encapsulation efficiency, antioxidant capacity, and physical properties of freeze-dried encapsulated olive leaves extract (OLE) having MD and trehalose (TR) as a wall material. The results showed that sample containing 30% solid (X1), and 100 MD : TR ratio (X2) and 0.05 OLE : matrix ratio (X3) showed the highest total phenolic concentration (TPC) encapsulated efficiency (99.33%) and highest TPC yield (75.22%) obtained at formulation X1 = 30, X2 = 0 and X3 = 0.05 and for other combination it was above 70%. Glass transition in samples formed of MD–TR combinations ranged from 111 to 175 °C, and it was higher than that observed in samples made of TR as a single component (100 °C). Microencapsulated powders showed positive but low a^* and positive b^* values in the range +11 to 22, which indicated a light yellow to light brownish color. Overall, the process variables (X1, X2 and X3) have an impact on the encapsulation effectiveness of total phenolic compounds and color characteristics of the encapsulated powder. Furthermore, de Araújo et al. (2020) studied the antimicrobial and antioxidant activities of microencapsulated sweet orange essential oil using MD and MD/gelatin (GEL) blend. Wall materials were used in three formulations: X1- only MD; X2-containing MD and GEL at ratio of 2 : 1 and X3-containing MD and GEL at ratio of 1 : 1. The results indicated that all three formations have high yield (above 80%) for the microencapsulation, particularly the yield for X1, X2, and X3 were 81.14%, 90.19%, and 89.09%, respectively. Whereas encapsulation efficiency did not differ among the wall material and encapsulation efficiency was around 75%. All three formations effectively inhibit the growth of *S. aureus* and *E. coli* with bacterial growth inhibition zones diameters between 13.0 and 18.5 mm. Sample X3 had the highest value of the total phenolic content (10.53 mg GAE/g), while X1 and X2 showed the similar concentration of total phenolic (6.43 mg GAE/g). In addition, sample X1 had highest value of antioxidant activity followed by X2 and X3.

3.5 Applications of Freeze-dried Fruits and Vegetables

Fruits and vegetables are good source of vitamins, minerals, and amino acids (Liu et al. 2022). Freeze-dried fruits and vegetables and powder were used as antioxidants and functional ingredients. The freeze-dried blueberries, raspberries, blackberries, pomegranates pomace, and beetroots powders were used as the sources of phenolic compounds to enrich different types of chocolates such as DCH (dark chocolate with 53% cocoa content sweetened with sucrose), DCH98S (dark chocolate with 98% cocoa content sweetened with sucrose), DCH98Est (dark chocolate with 98% cocoa content sweetened with erythritol with stevia), and MCH (milk chocolate with 40% cocoa content sweetened with sucrose). Overall color parameters (L^*, a^*, and b^*) values is increased in freeze-dried fruit powder-added chocolates as compared to control sample. Beetroots freeze-dried powder-added chocolates had lower score of organoleptic assessment score as compared to control and other freeze-dried fruits powder added chocolates. The MCH chocolate that had been enhanced with blueberry powder had the largest percentage increase in the overall polyphenol content (around 80%) (Żyżelewicz et al. 2021). Furthermore, authors fortified (10% of the flour by weight) breads containing oil (Ranawana et al. 2016b) and not containing oil (Ranawana et al. 2016a) as an ingredient with freeze-dried vegetables such as carrot, tomato, beetroot, or broccoli. The addition of freeze-dried vegetables enhanced the nutritional and antioxidant characteristics of both breads. Breads prepared with addition of oil and freeze-dried vegetable powders showed similar textural and storage properties compared to the control, but the bread containing freeze-dried vegetable powder prepared without oil were poor in texture and storage properties as compared to the oil-containing breads (Figures 3.3 and 3.4).

Figure 3.3 Physical appearance and size comparison of the baked oil containing breads. From left to right: plain, carrot, tomato, beetroot and broccoli. *Source:* Viren Ranawana et al. 2016/Reproduced from MDPI/CC BY 4.0.

Figure 3.4 Oil-free bread samples showing transverse structure and loaf size. From left to right: plain bread, carrot bread, tomato bread, beetroot bread, broccoli bread. *Source:* Viren Ranawana et al. 2016/Reproduced from MDPI/CC BY 4.0.

All four types of freeze-dried vegetables powder significantly reduced lipid oxidation in oil-containing fresh bread, although oil-free breads were linked with lower lipid oxidation both in their fresh form and following gastro-intestinal digestion due to the absence of oil. In both types of bread, beetroot, carrot, and broccoli reduced protein oxidation, but they had no impact during gastrointestinal digestion. Overall, bread prepared with beetroot and broccoli powder showed higher storage life as compared to other breads. Furthermore, addition of freeze-dried orange powder (3.6%, 6.2%, 5%, 10% and 17.5%) (Benjamin et al. 2007) and freeze-dried maqui berry powder (0.5%, 1%, 1.5% and 2.5%) (Lee 2014) in cake improved the specific volume, organoleptic, and textural properties.

Beverage (1 g lyophilized powder + 30 mL of distilled water) drinks were prepared using integral exploitation freeze-dried carrot, pumpkin, lentil sprouts, raspberry, strawberry, and apple: MO1 (30% carrot, 30% pumpkin, and 40% lentil sprouts), MO2 (MO1 + 10% raspberry), MO3 (MO1 + 10% strawberry), MO4 (MO1 + 10% apple), MO5 (MO1 + 5% raspberry + 5% strawberry), MO6 (MO1 + 5% raspberry + 5% apple), MO7 (MO1 + 5% strawberry and 5% apple), and MO8 (MO1 + 4% apple, 3% raspberry, and 3% strawberry). Beverages enriched with lyophilized fruit powder showed higher antioxidant activity, phenols, reducing power and vitamin C as compared to control (MO1) sample. Particularly, beverage enriched with raspberry (MO2), strawberry (MO3), and combination of both (MO5) showed the higher total phenolic content, antioxidant activities, and color value among all. The samples enhanced with strawberry powder (MO3) showed the highest increase in vitamin C content approx. 186% over the control (Bochnak-Niedźwiecka and Świeca 2020). In addition, Demina et al. (2021) prepared the milkshake (strawberry-raspberry, apple-beet milkshake, and carrot-pumpkin milkshake) milkshake using freeze-dried fruit and vegetable powder (Figure 3.5). Results indicate that, strawberry–raspberry milkshake had a significant increase in vitamin content and for a carrot–pumpkin milkshake, the vitamin C content

Figure 3.5 Appearance of milkshakes prepared using freeze-dried fruit and vegetable powder (a) strawberry-raspberry, (b) apple-beet, and (c) carrot-pumpkin. *Source:* Ekaterina Demina et al. 2021/Reproduced from EDP Sciences/CC BY 4.0.

increased by 2.5 times, the amount of carotene by 100 times, the amount of vitamin E by six times, potassium by 15%, sodium by 29%, and iron doubled. The following parameters have increased due to the complete chemical makeup of vegetable, fruit, and berry powders. Finally, according to organoleptic evaluation results prepared milkshake is acceptable to consume.

Furthermore, Duthie et al. (2013) studied the antioxidant effect of freeze-dried vegetable powders on the lipid and protein oxidative stability of cooked turkey meat patties. The freeze-dried powders of beetroot, broccoli, carrot, celery, green pea, onion, red pepper, swede, spinach, tomato, and yellow pea were added (10 g) to the basic burger (10 g of vitamin E stripped corn oil and 145 g of lean raw turkey meat). The results indicated that the addition of freeze-dried vegetable powder increased the concentration of carbohydrate, fiber, total phenols, vitamin C, α-tocopherol, γ-tocopherol, and lutein. In particular the patties enrich by green and yellow pea powder had highest content of carbohydrate (5.69/100 g) and broccoli powder enrich patties showed the highest content of fiber (4.08/100 g) among all the samples. The concentrations of vitamin C and α-tocopherol increased by 10 and 4 times, respectively, after the addition of tomato and red pepper powders. The freeze-dried powders improved the oxidative stability (tomato > green pea > red pepper > onion > yellow pea > spinach) by 20–30%. Moreover, Raikos et al. (2015) evaluated the impact of storage period at 4 °C on the oxidative stability of the dispersed phase of mayonnaise by adding freeze-dried vegetable powder (5% w/w). The results indicated that the type of vegetable used to reformulate mayonnaise was crucial in preventing oxidation and followed the order beetroot > carrot = onion with respect to antioxidant capability. Broccoli induced a pro-oxidant effect and the rate of oxidation after four weeks storage was 42 times higher compared with the control. Mayonnaise's oxidative stability was greatly enhanced by the addition of beetroot, whether it was fresh or freeze-dried.

Różyło et al. (2019) added freeze-dried elderberry and chokeberry (in the amounts of 1%, 2%, 3%, 4%, and 5%) powder as natural colorants in gluten-free wafer sheets. The results indicated that addition of lyophilized powder decreased the lightness (L^*) and yellowness (b^*) values, while redness (a^*) increased with the addition of chokeberry and decreased with the addition of elderberry. The color difference (ΔE) of wafer sheets increased with increased in the concentration of freeze-dried elderberry and chokeberry powder. In particular, freeze elderberry powder containing wafer sheets had higher ΔE value as compared to chokeberry wafer sheets. It could be because the pigments in elderberry are more stable than those in chokeberry, and the Maillard reaction during baking may be stronger in the case of elderberry powder (Delgado-Andrade et al. 2010). Besides, addition of freeze-dried elderberry and chokeberry powders significantly increase the minerals content, including iron, potassium, calcium, magnesium, and sodium. Overall gluten free wafers obtained by adding 4% and 5% of freeze-dried elderberry and chokeberry powders were most acceptable according to sensory evaluation results.

3.6 Advantages and Disadvantages of Freeze Drying

The prominence of freeze-drying is based on some well-known advantages and few disadvantages as compared to competing methods.

Advantages:

- The freeze-dried samples have higher stability at room temperature compared to dried using other methods and higher rehydration capacity (Ciurzyńska and Lenart 2011).
- The retention of morphological, biochemical, and immunological properties of fruits and vegetables (Oyinloye and Yoon 2020).
- Freeze-dried fruits and vegetables are significantly crisper than hot air drying (Pan et al. 2008).
- Encapsulation of bioactive compounds (from fruit and vegetable by-products) by freeze drying had high encapsulation efficiency, better anthocyanin retention, and extend their shelf-life with better physical characteristics (Marcillo-Parra et al. 2021).
- Freeze-dried encapsulated powder showed higher encapsulation yield, better antioxidant, and antibacterial properties as compared to other encapsulation techniques (Marcillo-Parra et al. 2021).
- Longer shelf life of freeze-dried products due to removal of water by 95–99.5% (Dziki 2020).
- Under vacuum conditions, oxidizable compounds are highly protected, with low loss in volatiles and heat-sensitive nutritional and aromatic compounds (Dziki 2020).
- Food preservation technologies, such as freeze-drying, can play a crucial role in ensuring food security and safety during emergencies, natural disasters, or other

crises. It may explore the implications of such technologies in terms of sustainable emergency food management, ensuring adequate nutrition, and mitigating food shortages during challenging times. The importance of various emergency events and their management has recently been highlighted as well (Wang et al., 2021).

Disadvantages:
- Expensive to setup the unit at small level scale.
- Freeze drying is an expensive operation for producing a dehydrated product due to high energy consumption and high operating and maintenance costs.
- At high vacuum pressure volatile compounds can by removed.
- Drying parameters needs to be optimized for different food products.

3.7 Conclusion

Freeze-drying is a popular method for dehydrating various fruits and vegetables. Despite the longer processing time and higher cost, it is recommended for a higher quality final product. Freeze-drying can cause some losses in vitamins (ascorbic acid) and other valuable bioactive compounds. In comparison to other dehydration techniques, freeze drying is the most effective at maintaining the physical properties and thermally sensitive components of fruits and vegetables. Additionally, freeze-dried fruit and vegetable quality characteristics including increased rehydration and improved or preserved color are advantageous for food product applications and increase the acceptability of dried fruit and vegetable by consumers.

References

Agnieszka, C. and Andrzej, L. (2010). Rehydration and sorption properties of osmotically pretreated freeze-dried strawberries. *Journal of Food Engineering* https://doi.org/10.1016/j.jfoodeng.2009.10.022.

Akar, G. and Barutçu Mazı, I. (2019). Color change, ascorbic acid degradation kinetics, and rehydration behavior of kiwifruit as affected by different drying methods. *Journal of Food Process Engineering* https://doi.org/10.1111/jfpe.13011.

Al Maiman, S.A., Albadr, N.A., Almusallam, I.A. et al. (2021). The potential of exploiting economical solar dryer in food preservation: storability, physicochemical properties, and antioxidant capacity of solar-dried tomato (Solanum lycopersicum) fruits. *Food* 10 (4): 734. https://doi.org/10.3390/foods10040734.

Benjamin, A.C., Akingbala, J.O., and Baccus-Taylor, G.S.H. (2007). Effect of drying and storage on flavour quality of orange (Citrus cinensis (Linn) Osbeck) peel for cupcakes. *Journal of Food, Agriculture and Environment* https://doi.org/10.1234/4.2007.975.

Bhatta, S., Janezic, T.S., and Ratti, C. (2020). Freeze-drying of plant-based foods. *Food* https://doi.org/10.3390/foods9010087.

Bochnak-Niedźwiecka, J. and Świeca, M. (2020). Quality of new functional powdered beverages enriched with lyophilized fruits-potentially bioaccessible antioxidant properties, nutritional value, and consumer analysis. *Applied Sciences (Switzerland)* https://doi.org/10.3390/app10113668.

Buchtová, N. and Budtova, T. (2016). Cellulose aero-, cryo- and xerogels: towards understanding of morphology control. *Cellulose* 23 (4): 2585–2595. https://doi.org/10.1007/s10570-016-0960-8.

Calín-Sánchez, Á., Lipan, L., Cano-Lamadrid, M. et al. (2020). Comparison of traditional and novel drying techniques and its effect on quality of fruits, vegetables and aromatic herbs. *Foods* 9 (9): 1261. https://doi.org/10.3390/foods9091261.

Carrillo, J.Á., Pilar Zafrilla, M., and Marhuenda, J. (2019). Cognitive function and consumption of fruit and vegetable polyphenols in a young population: is there a relationship? *Food* https://doi.org/10.3390/foods8100507.

Ciurzyńska, A., Janowicz, M., Karwacka, M. et al. (2022). The effect of hybrid drying methods on the quality of dried carrot. *Applied Sciences* 12 (20): 10588. https://doi.org/10.3390/app122010588.

Ciurzyńska, A. and Lenart, A. (2011). Freeze-drying—application in food processing and biotechnology—a review. *Polish Journal of Food and Nutrition Sciences* https://doi.org/10.2478/v10222-011-0017-5.

da Silva, L.C., Viganó, J., de Souza Mesquita, L.M. et al. (2021). Recent advances and trends in extraction techniques to recover polyphenols compounds from apple by-products. *Food Chemistry: X* 12: 100133. https://doi.org/10.1016/j.fochx.2021.100133.

de Araújo, J.S.F., de Souza, E.L., Oliveira, J.R. et al. (2020). Microencapsulation of sweet orange essential oil (Citrus aurantium var. dulcis) by liophylization using maltodextrin and maltodextrin/gelatin mixtures: preparation, characterization, antimicrobial and antioxidant activities. *International Journal of Biological Macromolecules* 143: 991–999. https://doi.org/10.1016/j.ijbiomac.2019.09.160.

De Souza, V.B., Thomazini, M., Balieiro, J.C.D.C., and Fávaro-Trindade, C.S. (2015). Effect of spray drying on the physicochemical properties and color stability of the powdered pigment obtained from vinification byproducts of the Bordo grape (Vitis labrusca). *Food and Bioproducts Processing* https://doi.org/10.1016/j.fbp.2013.11.001.

Delgado-Andrade, C., Conde-Aguilera, J.A., Haro, A. et al. (2010). A combined procedure to evaluate the global antioxidant response of bread. *Journal of Cereal Science* https://doi.org/10.1016/j.jcs.2010.05.013.

Demina, E., Simonenkova, A., Luneva, O. et al. (2021). Usage of freeze-dried vegetable and fruit-berry powders in milkshake technology. In: *E3S Web of Conferences*. https://doi.org/10.1051/e3sconf/202127903022.

Domin, M., Dziki, D., Kłapsia, S. et al. (2020). Influence of the Freeze-drying Conditions on the Physicochemical Properties and Grinding Characteristics of Kiwi. *International Journal of Food Engineering* https://doi.org/10.1515/ijfe-2018-0315.

Duthie, G., Campbell, F., Bestwick, C. et al. (2013). Antioxidant effectiveness of vegetable powders on the lipid and protein oxidative stability of cooked Turkey meat patties: implications for health. *Nutrients* https://doi.org/10.3390/nu5041241.

Dziki, D. (2020). Recent trends in pretreatment of food before freeze-drying. *Processes* https://doi.org/10.3390/pr8121661.

Feng, S., Bi, J., Yi, J. et al. (2022). Cell wall polysaccharides and mono-/disaccharides as chemical determinants for the texture and hygroscopicity of freeze-dried fruit and vegetable cubes. *Food Chemistry* 395: 133574. https://doi.org/10.1016/j.foodchem.2022.133574.

González-Ortega, R., Faieta, M., Di Mattia, C.D. et al. (2020). Microencapsulation of olive leaf extract by freeze-drying: effect of carrier composition on process efficiency and technological properties of the powders. *Journal of Food Engineering* https://doi.org/10.1016/j.jfoodeng.2020.110089.

González, C.M., Llorca, E., Quiles, A. et al. (2020). Water sorption and glass transition in freeze-dried persimmon slices. Effect on physical properties and bioactive compounds. *LWT - Food Science and Technology* https://doi.org/10.1016/j.lwt.2020.109633.

Hamid, Thakur, N.S., Thakur, A., and Kumar, P. (2020). Effect of different drying modes on phenolics and antioxidant potential of different parts of wild pomegranate fruits. *Scientia Horticulturae* https://doi.org/10.1016/j.scienta.2020.109656.

Kiokias, S. and Oreopoulou, V. (2021). A review of the health protective effects of phenolic acids against a range of severe pathologic conditions (Including coronavirus-based infections). *Molecules* https://doi.org/10.3390/molecules26175405.

Kumar, A., Begum, A., Hoque, M. et al. (2021a). Textural degradation, drying and rehydration behaviour of ohmically treated pineapple cubes. *LWT - Food Science and Technology* 142: 110988. https://doi.org/10.1016/j.lwt.2021.110988.

Kumar, D., Ladaniya, M.S., Gurjar, M. et al. (2021b). Quantification of flavonoids, phenols and antioxidant potential from dropped citrus reticulata blanco fruits influenced by drying techniques. *Molecules* https://doi.org/10.3390/molecules26144159.

Laine, P., Kylli, P., Heinonen, M., and Jouppila, K. (2008). Storage stability of microencapsulated cloudberry (Rubus chamaemorus) phenolics. *Journal of Agricultural and Food Chemistry* https://doi.org/10.1021/jf801868h.

Lee, H.J. (2014). Antioxidant activity and properties characteristics of pound cakes prepare using freeze dried maquiberry (Aristotelia chilensis [Mol.]) powder. *The Korean Journal of Food and Nutrition* https://doi.org/10.9799/ksfan.2014.27.6.1067.

Lewicki, P.P. (1998). Effect of pre-drying treatment, drying and rehydration on plant tissue properties: a review. *International Journal of Food Properties 1* (1): 1–22. https://doi.org/10.1080/10942919809524561.

Li, D., Zhu, Z., and Sun, D.W. (2018). Effects of freezing on cell structure of fresh cellular food materials: a review. *Trends in Food Science and Technology* https://doi.org/10.1016/j.tifs.2018.02.019.

Li, L., Zhang, M., Chitrakar, B., and Jiang, H. (2020). Effect of combined drying method on phytochemical components, antioxidant capacity and hygroscopicity of Huyou (Citrus changshanensis) fruit. *LWT - Food Science and Technology* https://doi.org/10.1016/j.lwt.2020.109102.

Link, J.V., Tribuzi, G., and Laurindo, J.B. (2017). Improving quality of dried fruits: a comparison between conductive multi-flash and traditional drying methods. *LWT - Food Science and Technology 84*: 717–725. https://doi.org/10.1016/j.lwt.2017.06.045.

Liu, Y., Zhang, Z., and Hu, L. (2022). High efficient freeze-drying technology in food industry. *Critical Reviews in Food Science and Nutrition* https://doi.org/10.1080/10408398.2020.1865261.

Lopez-Quiroga, E., Prosapio, V., Fryer, P.J. et al. (2020). Model discrimination for drying and rehydration kinetics of freeze-dried tomatoes. *Journal of Food Process Engineering* https://doi.org/10.1111/jfpe.13192.

Marcillo-Parra, V., Tupuna-Yerovi, D.S., González, Z., and Ruales, J. (2021). Encapsulation of bioactive compounds from fruit and vegetable by-products for food application—a review. *Trends in Food Science and Technology* https://doi.org/10.1016/j.tifs.2021.07.009.

Materska, M. (2014). Bioactive phenolics of fresh and freeze-dried sweet and semi-spicy pepper fruits (Capsicum annuum L.). *Journal of Functional Foods* https://doi.org/10.1016/j.jff.2014.02.002.

Mbondo, N.N., Owino, W.O., Ambuko, J., and Sila, D.N. (2018). Effect of drying methods on the retention of bioactive compounds in African eggplant. *Food Science & Nutrition* 6 (4): 814–823. https://doi.org/10.1002/fsn3.623.

Minuye, M., Getachew, P., Laillou, A. et al. (2021). Effects of different drying methods and ascorbic acid pretreatment on carotenoids and polyphenols of papaya fruit in Ethiopia. *Food Science & Nutrition* https://doi.org/10.1002/fsn3.2324.

Mujumdar, A.S. (2020). Drying of solids: principles, classification, and selection of dryers. In: *Handbook of Industrial Drying* (ed. A.S. Mujumdar). CRC Press https://doi.org/10.1201/9780429289774.

Murali, S., Kar, A., Mohapatra, D., and Kalia, P. (2015). Encapsulation of black carrot juice using spray and freeze drying. *Food Science and Technology International* https://doi.org/10.1177/1082013214557843.

Nemzer, B., Vargas, L., Xia, X. et al. (2018). Phytochemical and physical properties of blueberries, tart cherries, strawberries, and cranberries as affected by different drying methods. *Food Chemistry* https://doi.org/10.1016/j.foodchem.2018.04.047.

Oyinloye, T.M. and Yoon, W.B. (2020). Effect of freeze-drying on quality and grinding process of food produce: a review. *Processes* https://doi.org/10.3390/PR8030354.

Pan, Z., Shih, C., McHugh, T.H., and Hirschberg, E. (2008). Study of banana dehydration using sequential infrared radiation heating and freeze-drying. *LWT - Food Science and Technology* https://doi.org/10.1016/j.lwt.2008.01.019.

Peleg, M., Normand, M.D., Dixon, W.R., and Goulette, T.R. (2018). Modeling the degradation kinetics of ascorbic acid. *Critical Reviews in Food Science and Nutrition* https://doi.org/10.1080/10408398.2016.1264360.

Pereira Souza, A.C., Deyse Gurak, P., Marczak, D.F., and L. (2017). Maltodextrin, pectin and soy protein isolate as carrier agents in the encapsulation of anthocyanins-rich extract from jaboticaba pomace. *Food and Bioproducts Processing* https://doi.org/10.1016/j.fbp.2016.12.012.

Pico, J., Xu, K., Guo, M. et al. (2019). Manufacturing the ultimate green banana flour: impact of drying and extrusion on phenolic profile and starch bioaccessibility. *Food Chemistry* https://doi.org/10.1016/j.foodchem.2019.124990.

Pikal, M.J. and Shah, S. (1990). The collapse temperature in freeze drying: dependence on measurement methodology and rate of water removal from the glassy phase. *International Journal of Pharmaceutics* https://doi.org/10.1016/0378-5173(90)90231-R.

Pinar, H., Çetin, N., Ciftci, B. et al. (2021). Biochemical composition, drying kinetics and chromatic parameters of red pepper as affected by cultivars and drying methods. *Journal of Food Composition and Analysis* 102: 103976. https://doi.org/10.1016/j.jfca.2021.103976.

Quispe-Fuentes, I., Vega-Gálvez, A., and Aranda, M. (2018). Evaluation of phenolic profiles and antioxidant capacity of maqui (Aristotelia chilensis) berries and their relationships to drying methods. *Journal of the Science of Food and Agriculture* https://doi.org/10.1002/jsfa.8938.

Rahman, M.S., Al-Rizeiqi, M.H., Guizani, N. et al. (2015). Stability of vitamin C in fresh and freeze-dried capsicum stored at different temperatures. *Journal of Food Science and Technology* https://doi.org/10.1007/s13197-013-1173-x.

Raikos, V., Neacsu, M., Morrice, P., and Duthie, G. (2015). Anti- and pro-oxidative effect of fresh and freeze-dried vegetables during storage of mayonnaise. *Journal of Food Science and Technology* https://doi.org/10.1007/s13197-015-1897-x.

Ramírez, M.J., Giraldo, G.I., and Orrego, C.E. (2015). Modeling and stability of polyphenol in spray-dried and freeze-dried fruit encapsulates. *Powder Technology* https://doi.org/10.1016/j.powtec.2015.02.060.

Ranawana, V., Campbell, F., Bestwick, C. et al. (2016a). Breads fortified with freeze-dried vegetables: quality and nutritional attributes. Part II: breads not containing oil as an ingredient. *Food* 5 (4): 62. https://doi.org/10.3390/foods5030062.

Ranawana, V., Raikos, V., Campbell, F. et al. (2016b). Breads fortified with freeze-dried vegetables: quality and nutritional attributes. Part 1: breads containing oil as an ingredient. *Food* https://doi.org/10.3390/foods5010019.

Rezende, Y.R.R.S., Nogueira, J.P., and Narain, N. (2018). Microencapsulation of extracts of bioactive compounds obtained from acerola (Malpighia emarginata DC) pulp and residue by spray and freeze drying: chemical, morphological and chemometric characterization. *Food Chemistry* https://doi.org/10.1016/j.foodchem.2018.02.026.

Rezvankhah, A., Emam-Djomeh, Z., and Askari, G. (2020). Encapsulation and delivery of bioactive compounds using spray and freeze-drying techniques: a review. *Drying Technology* 38 (1–2): 235–258. https://doi.org/10.1080/07373937.2019.1653906.

Richter Reis, F., Marques, C., de Moraes, A.C.S., and Masson, M.L. (2022). Trends in quality assessment and drying methods used for fruits and vegetables. *Food Control* 142: 109254. https://doi.org/10.1016/j.foodcont.2022.109254.

Roos, Y.H. (2010). Crystallization, collapse, and glass transition in low-water food systems. In: *Water Properties in Food, Health, Pharmaceutical and Biological Systems: ISOPOW 10* (ed. D.S. Reid, T. Sajjaanantakul, P.J. Lillford, and S. Charoenrein), 335–351. Wiley-Blackwell https://doi.org/10.1002/9780470958193.ch25.

Różyło, R., Wójcik, M., Dziki, D. et al. (2019). Freeze-dried elderberry and chokeberry as natural colorants for gluten-free wafer sheets. *International Agrophysics* https://doi.org/10.31545/INTAGR/109422.

Sagar, V.R. and Suresh Kumar, P. (2010). Recent advances in drying and dehydration of fruits and vegetables: a review. *Journal of Food Science and Technology* https://doi.org/10.1007/s13197-010-0010-8.

Saikia, S., Mahnot, N.K., and Mahanta, C.L. (2015). Optimisation of phenolic extraction from Averrhoa carambola pomace by response surface methodology and its microencapsulation by spray and freeze drying. *Food Chemistry* https://doi.org/10.1016/j.foodchem.2014.08.064.

Sanchez, V., Baeza, R., and Chirife, J. (2015). Comparison of monomeric anthocyanins and colour stability of fresh, concentrate and freeze-dried encapsulated cherry juice stored at 38°C. *Journal of Berry Research* https://doi.org/10.3233/JBR-150106.

Sanchez, V., Baeza, R., Galmarini, M.V. et al. (2013). Freeze-drying encapsulation of red wine polyphenols in an amorphous matrix of maltodextrin. *Food and Bioprocess Technology* https://doi.org/10.1007/s11947-011-0654-z.

Sarkar, T., Salauddin, M., Hazra, S.K., and Chakraborty, R. (2020). Effect of cutting edge drying technology on the physicochemical and bioactive components of mango (Langra variety) leather. *Journal of Agriculture and Food Research* https://doi.org/10.1016/j.jafr.2020.100074.

Schudel, S., Prawiranto, K., and Defraeye, T. (2021). Comparison of freezing and convective dehydrofreezing of vegetables for reducing cell damage. *Journal of Food Engineering* https://doi.org/10.1016/j.jfoodeng.2020.110376.

Shofian, N.M., Hamid, A.A., Osman, A. et al. (2011). Effect of freeze-drying on the antioxidant compounds and antioxidant activity of selected tropical fruits. *International Journal of Molecular Sciences* https://doi.org/10.3390/ijms12074678.

Singh, H., Kumar, Y., and Meghwal, M. (2022). Encapsulated oil powder: processing, properties, and applications. *Journal of Food Process Engineering* 45 (8): e14047. https://doi.org/10.1111/jfpe.14047.

Stoll, L., Costa, T.M.H., Jablonski, A. et al. (2016). Microencapsulation of anthocyanins with different wall materials and its application in active biodegradable films. *Food and Bioprocess Technology* https://doi.org/10.1007/s11947-015-1610-0.

Suwannarak, J. and Phanumong, P. (2021). Preparation and some physicochemical properties of freeze-dried vegetables carving. *Food Research* 5 (6): 290–299. https://doi.org/10.26656/fr.2017.5(6).755.

Tan, S., Ke, Z., Chai, D. et al. (2021). Lycopene, polyphenols and antioxidant activities of three characteristic tomato cultivars subjected to two drying methods. *Food Chemistry* https://doi.org/10.1016/j.foodchem.2020.128062.

Telis, V.R.N. and Martínez-Navarrete, N. (2010). Application of compression test in analysis of mechanical and color changes in grapefruit juice powder as related to glass transition and water activity. *LWT - Food Science and Technology* https://doi.org/10.1016/j.lwt.2009.12.007.

Tumbas Šaponjac, V.T., Čanadanović-Brunet, J.M., Ćetković, G.S. et al. (2020). Optimisation of beetroot juice encapsulation by freeze-drying. *Polish Journal of Food and Nutrition Sciences* https://doi.org/10.31883/pjfns/115153.

Vallespir, F., Rodríguez, Ó., Eim, V.S. et al. (2019). Effects of freezing treatments before convective drying on quality parameters: vegetables with different microstructures. *Journal of Food Engineering* https://doi.org/10.1016/j.jfoodeng.2019.01.006.

Vergeldt, F.J., Van Dalen, G., Duijster, A.J. et al. (2014). Rehydration kinetics of freeze-dried carrots. *Innovative Food Science & Emerging Technologies* https://doi.org/10.1016/j.ifset.2013.12.002.

Wang, Z., Gong, J., Wang, Q., & Qiao, X. (2021). Emergency management science and technology: An international transdisciplinary platform. *Emergency Management Science and Technology*, 1(1), 1–3.

Yamashita, C., Chung, M.M.S., dos Santos, C. et al. (2017). Microencapsulation of an anthocyanin-rich blackberry (Rubus spp.) by-product extract by freeze-drying. *LWT - Food Science and Technology* https://doi.org/10.1016/j.lwt.2017.05.063.

Zeb, A. (2020). Concept, mechanism, and applications of phenolic antioxidants in foods. *Journal of Food Biochemistry* https://doi.org/10.1111/jfbc.13394.

Zhang, Z., Wei, Q., Nie, M. et al. (2018). Microstructure and bioaccessibility of different carotenoid species as affected by hot air drying: study on carrot, sweet potato, yellow bell pepper and broccoli. *LWT - Food Science and Technology* https://doi.org/10.1016/j.lwt.2018.05.061.

Żyżelewicz, D., Oracz, J., Bilicka, M. et al. (2021). Influence of freeze-dried phenolic-rich plant powders on the bioactive compounds profile, antioxidant activity and aroma of different types of chocolates. *Molecules* https://doi.org/10.3390/molecules26227058.

4

Freeze-Drying of Meat and Seafood Products

Rubén Agregán[1], Noemí Echegaray[1], Paulo E.S. Munekata[1], Manoj Kumar[2], Ruben Dominguez[1], Mirian Pateiro[1,3], and Jose M. Lorenzo[1,4]

[1] *Centro Tecnológico de la Carne de Galicia, Avda. Galicia n° 4, Parque Tecnológico de Galicia, San Cibrao das Viñas, Ourense, Spain*
[2] *Chemical and Biochemical Processing Division, ICAR-Central Institute for Research on Cotton Technology, Mumbai, India*
[3] *Research and Development Department, Idaho Milk Products, Jerome, ID, USA*
[4] *Área de Tecnología de los Alimentos, Facultad de Ciencias de Ourense, Universidad de Vigo, Ourense, Spain*

4.1 Introduction

The demand for meat and seafood products is growing more and more in recent times on account of their important nutritional values (i.e. high protein content, presence of essential amino acids, vitamins, and minerals) (Ekezie et al. 2019; Esua et al. 2021; Gómez et al. 2019; Munekata et al. 2020). In fact, current meat consumption exceeds 337 million metric tons per year which is expected to increase by 48 million metric tons by 2027 (FAOSTAT 2020). For its part, the present consumption of seafood is estimated at more than 112 million metric tons per year, forecasting an increase of more than 80 tons by 2027 (FAO 2021). Nonetheless, meat and seafood products are extremely perishable foodstuffs, which represent an excellent medium for the propagation of microorganisms while contributing to potential foodborne illness (Das et al. 2021; Umaraw et al. 2020). At the same time, the deterioration of these foods also affects the nutritional and organoleptic quality of the commercialized nourishments (Elmas et al. 2020). Thus, spoilage of meat and seafood products can be due to microbiological, physical, and chemical reactions. In this context, the use of different methods to preserve both meat and seafood products is essential to supply safe and high-quality nourishments.

One of the most important and widely used conservation methods throughout history has been the traditional drying, where moisture is removed from a product after

Freeze Drying of Food Products: Fundamentals, Processes and Applications, First Edition.
Edited by Roji Balaji Waghmare, Manoj Kumar, and Parmjit Singh Panesar.

exposing it to a heat source, either natural (sun) or artificial (Akpinar 2006; Hashim et al. 2014; Onwude et al. 2016). However, this traditional conservation technique can have many negative effects on meat and seafood products quality traits, such as color, texture, flavor, and nutritional value (Coombs et al. 2017), given that the application of high temperatures or long exposure times can damage the food features. Thus, traditional hot drying can favor consumer rejection, which in turn possesses relevant impacts on the economic aspects of the food industry (Šimat et al. 2021).

Concurrently, driven by the present demand from consumers for high-quality natural products without the addition of synthetic additives, the meat and seafood industry can rely on technologies that are less used in this sector such as freeze-drying. This technique of conservation allows nonperishable food to be obtained with a very high quality (biological, nutritional, and organoleptic features) (Assegehegn et al. 2019) without the need to use additives and/or more aggressive methods such as traditional hot-air drying. Freeze-drying is a process where the water present in the product is sublimated by directly transitioning from solid water to vaporous water and then desorbing the water from the "dry" layer (Adams et al. 2015; Haseley and Oetjen 2018; Nowak and Jakubczyk 2020). This preservation method is broadly used for the stabilization of several high-quality foods (especially nourishments with high added value), biological materials, and pharmaceutical products (e.g. bacteria, mammals' cells, proteins, and vaccines), since it maintains the quality of the dry product because the removal of water reduces or even eliminates the chemical, biochemical, and microbiological processes (Nowak and Jakubczyk 2020; Uscanga et al. 2021). As a result, freeze-drying has significant advantages over other food preservation techniques, including the use of hot air as a traditional drying method, considering that this process preserves to a very high extent the aroma and taste of the original fresh food in addition to keeping its initial appearance almost intact (Jia et al. 2019; Meda and Ratti 2005). At the same time, freeze-drying guards the composition and prevents the loss of nutritional and bioactive substances, particularly suitable for heat-sensitive compounds (e.g. vitamins and polyphenols) (Berk 2013; Cheng-Hai et al. 2008). Moreover, the freeze-drying method permits to obtain foodstuffs that are easily reconstituted with a fast rehydration speed and because of this method it leads to the formation of a highly porous structure of the freeze-dried product (Meda and Ratti 2005). In addition, the freeze-dried food conserves the moisture content at a low level, being an ideal food for meals, tourism, meetings, and entertainment (Liu et al. 2022). Nevertheless, this technological process continues to entail structural changes at the matrix level that directly influence the quality of the final product, though moderately compared to more classic processes. Thus, parameters, such as color, odor, texture, rehydration property, bulk properties, flow property, water activity (a_w), and retention of nutrients and volatile compounds, can be affected (Sablani 2006), requiring optimization. Furthermore, it should be noted that freeze-drying has always been considered the most expensive process for processing a dehydrated product, which can compromise the viability of the fare obtained (Ratti 2001).

4.2 Principles of Freeze-Drying Technology

Food raw materials contain large amount of water, which in the case of meat and seafood products can represent more than 75% of its composition (Cozzolino and Murray 2012; Pereira and Vicente 2013). This water, as in any food, can be divided into bound water or free water depending on whether it is attached to the matrix (other water molecules, proteins, sugars, salts, etc.) by different interaction forces (hydrogen bridges, electrostatic forces, etc.) or not, respectively (Mathlouthi 2001; Wang et al. 2018). Free water freezes completely during a freezing process, while bound water remains in a liquid state. That is why in the freeze-drying process itself all the free water is removed, while only a small part of the bound water is eliminated (Haseley and Oetjen 2018). Therefore, the drying process by means of freeze-drying is a highly complex mechanism that consists of several steps. Specifically, freeze-drying occurs in three levels (Figure 4.1). Initially the product is frozen (generally at atmospheric pressure); or rather the free water in the food matrix is frozen. Subsequently, the primary drying takes place, that is, the proper freeze-drying occurs, since it is here where the frozen water sublimates (turning directly from water in a liquid state to water vapor). This stage is generally carried out at reduced pressure. Finally, in a third interval, secondary drying is performed (known as desorption drying), where the bound water is eliminated, reaching the final moisture of the required product (Nowak and Jakubczyk 2020).

The parameters used during freeze-drying (time, freezing temperatures, freezing rate, primary and secondary drying temperatures, and working pressure) must be adjusted to achieve a product with the appropriate characteristics. However, economic aspects must also be considered, since the freezing and primary drying temperatures are very demanding of energy, making the freeze-drying process highly expensive (Haseley and Oetjen 2018). In fact, the economic aspect is one of the biggest disadvantages of this conservation process. For this reason, its employment in the food industry, including meat and seafood sector, is limited since it implies a cost overrun on the products obtained.

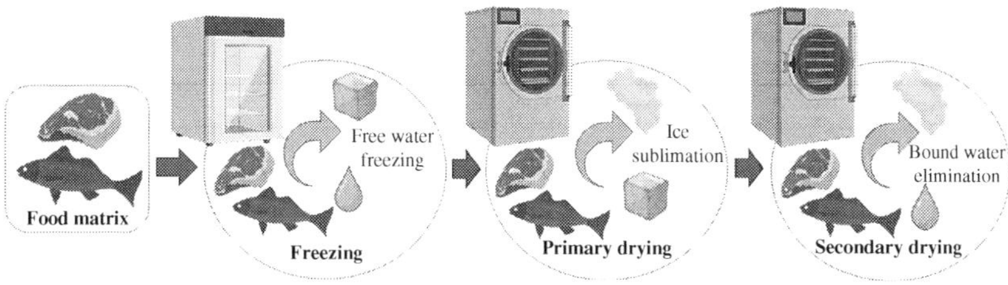

Figure 4.1 Schematic representation of freeze-drying process.

4.3 Effect of Freeze-Drying on the Sensory Attributes of Meat and Seafood Products

Highly protein and fat matrices, such as meat and seafood products, are far from those usually dried by the freeze-drying technique, mostly represented by fruits (strawberry, raspberry, blueberry, blackberry, apple, or banana) and vegetables (asparagus, tomato, broccoli, carrot, pepper, or eggplant). Recent research has revealed the benefits that this drying method has on these types of food matrices, also reporting certain drawbacks to its application. At the same time, fundamental aspects in achieving a sensorially acceptable product have been pointed out.

The influence of freeze-drying technology on the quality factors of meat and seafood products are discussed next and summarized in Table 4.1.

Table 4.1 Influence of freeze-drying process on sensory quality parameters of meat and seafood products.

Quality parameter	Product	Main findings	Reference
Color	Broiler chicken breast	• Increased L^* and b^* values. • Decreased a^* value. • After cooking improved the color of the sample.	Babić et al. (2009)
	Yellow croaker fish	• Decreased L^* parameter. • Increased a^* and b^* values.	Kim et al. (2020)
	Fish oil	• Brighter and less dark powder than that obtained by spraying when microencapsulating.	Pang et al. (2017)
Texture	Chinese style ham	• The precise calculation of the collapse temperature made it possible to work just below this value and enabled a highly porous meat structure with good rehydration properties.	Ma et al. (2018)
	Chicken meat hydrolysate	• A high storage temperature positively affected the solubilization capacity of the dried product. • An excessive increase in the freezing rate caused a decrease in the solubility of the future dry powder.	Söbeli and Kayaardı (2020)
	Yellow croaker fish	• Notable increase in hardness and the energy required to chew the sample.	Kim et al. (2020)
	Silver carp surimi	• Remarkable increase in the emulsifying stability index of myofibrillar proteins. • Decreased activity index of myofibrillar proteins.	Niu et al. (2019)

Table 4.1 (Continued)

Quality parameter	Product	Main findings	Reference
Aroma	Golden pompano fish	• Low temperatures during the drying process preserved the 1-octen-3-ol alcohol, responsible for the typical fishy smell. • Low production of aldehyde compounds, improving the overall aromatic profile. • Ester production was higher compared to other drying methods (twice as many compounds as using microwave).	Zhang et al. (2019)
Nutrients	Yellow croaker fish	• Increased levels of amino acids due to the concentration of dry matter. • Especially important increase in asparagine and serine.	Kim et al. (2020)
	Sea cucumber	• Good preservation of polyunsaturated fats. • Maintenance of higher levels of omega-3 PUFAs in the phospholipid fraction compared to the classical method using hot air.	Liu et al. (2021)

L^*: lightness; a^*: redness; b^*: yellowness; PUFAs: polyunsaturated fatty acids.

4.3.1 The Challenge of Imitating Fresh Meat

The appearance of food, especially meat, is one of the most important and determining quality parameters in its acceptance or rejection by the consumer, as it is usually the first attribute to be evaluated at the point of sale (Tomasevic et al. 2021). Similarly, the nutritional value is another parameter that greatly determines the quality and reputation of food (Karwowska et al. 2021). Meat drying by conventional thermal dehydration methods, such as hot air or sun, is a slow process that affects flavor and color, degrading aromas and darkening the surface of the product (Santchurn et al. 2012). These transformations spoil the appearance of meat. There is also a loss of nutrients (vitamins and amino acids) and a lower rate of rehydration. On the other hand, significant changes have been evidenced in physicochemical properties, such as the porous structure, shrinkage, and bulk density, of conventionally dried meat products (Aksoy et al. 2019; AykınDinçer 2021).

In order to prevent all these adverse effects on meat and meat products, modern drying methods more respectful with the product such as freeze-drying have been studied and developed during the last decades. In theory, when a meat product is subjected to freeze-drying, its sensory and nutritional characteristics are largely preserved, imitating very well the original fresh product once rehydrated. However,

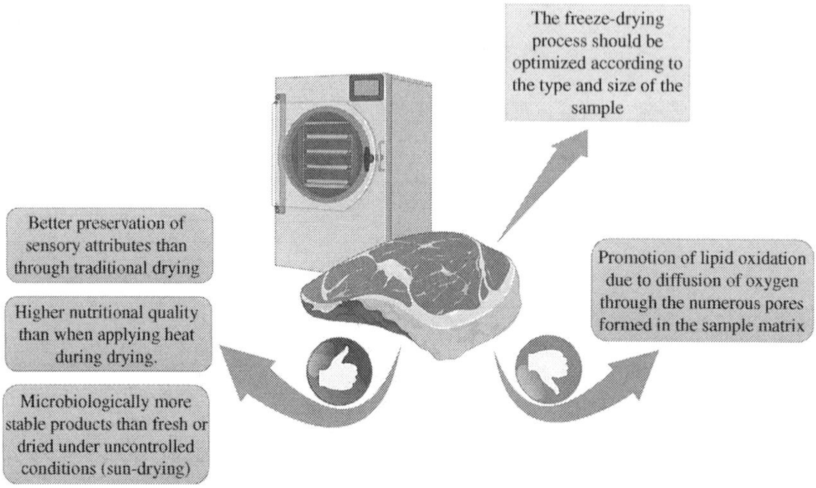

Figure 4.2 Scheme of how the freeze-drying process can affect meat.

recent studies have shown that freeze-drying might cause some significant changes in the meat that could call this claim into question (Figure 4.2). A comprehensive study about the influence of this drying technology on certain quality parameters of chicken breast meat showed interesting results. Thus, the color of the rehydrated meat pieces, regardless of the freeze-drying conditions used (variation of the sample thickness, freezing rate, time of drying phases, and pressure), presented a lightness (L^*) and yellowness (b^*) values much higher than those of the fresh product. On the contrary, the redness (a^*) value of the dry product decreased when applying an operating pressure under 25 Pa. The L^* and b^* values of the freeze-dried chicken breast were better matched to those of the fresh product once both were cooked. On the other hand, it seems that the freezing rate, regardless of the pressure, has no clear influence on the color attribute (Babić et al. 2009).

Despite the benefits that this drying technology offers, a prior study of the operating conditions according to the sample is always necessary, since an incorrect setting of the freeze-drying parameters may lead to an unacceptable sensory product at the same time that it can cause economic losses. When Babić et al. (2009) freeze-dried chicken breasts of 0.7 and 1.3 cm thickness, found that a freezing rate of 3 hours, pressure of 30 Pa, and an operation time of 31 hours or a rate of 6 hours, pressure of 25 Pa, and time of 18 hours (for more details consult the article) were not suitable for the selected volumes and shapes of the sample, giving rise to products that were too hard once they were rehydrated and cooked, which were rejected by the members of a sensory panel. This proves that the freeze-drying process should be optimized depending on the sample and its size.

This optimization takes into account many and varied factors, the modulation of which will be decisive in order to obtain a sensory acceptable product. In addition to the parameters mentioned earlier, the maximum operating temperature is another critical value calculated based on another temperature closely linked to the sample, called collapse temperature, which is generally the maximum allowable temperature for a solute system that does not crystallize but remains amorphous (Pikal and Shah 1990). During freeze-drying, the first drying or ice sublimation is the stage that usually consumes more energy, and to properly optimize a drying process it must be operated at a temperature close to the collapse temperature, which is particular to each substrate (Pikal and Shah 1990). The ice crystals formed once the sample is frozen disappear and the space originally occupied by them remains empty, leaving a porous sponge honeycomb structure. An increase in temperature can cause the loss of structural rigidity in the product and above the collapse temperature the closing of the holes formed by the ice (Ma et al. 2014). Structure collapse in freeze-dried foods has been reported to negatively affect physicochemical quality (Karathanos et al. 1996). Therefore, it is advisable to calculate the collapse temperature to prevent this problem as Ma et al. (2018) made in ham and tenderloin. When these samples reached −31.9 and −30.8 °C, respectively, their structures were found to collapse. Carrying out the freeze-drying process at these temperatures will result in a highly porous structure with excellent rehydration properties. It is especially important to comply with this operational requirement in meat matrices due to the large size of some pieces (Stapley 2008).

Likewise, the speed with which the freeze-drying temperature is reached was found to have a major effect on the quality of meat product. A sensory analysis performed on freeze-dried chicken breasts showed a better texture result when the temperature drop was slow. This could be due to a better rehydration of the dried meat pieces (Babić et al. 2009). Indeed, Söbeli and Kayaardı (2020) reported a decrease in the solubility of freeze-dried boneless and hydrolyzed chicken residues (in powder form) when the freezing rate was increased. According to the authors, this caused a low specific surface area, which is well-known to affect the thermodynamic and aqueous solubility of powdery matrices.

4.3.2 The High Opportunities of Freeze-Drying in Seafood

Freeze-drying method has also been used successfully in seafood products in order to better preserve their sensory and physicochemical qualities than more traditional drying methods. As in the case of meat products, there are still not many studies in this regard. However, the few available have reported considerable potential for the maintenance of quality parameters in fish, though it can also have some disadvantages (Figure 4.3). Due to the absence of high temperatures

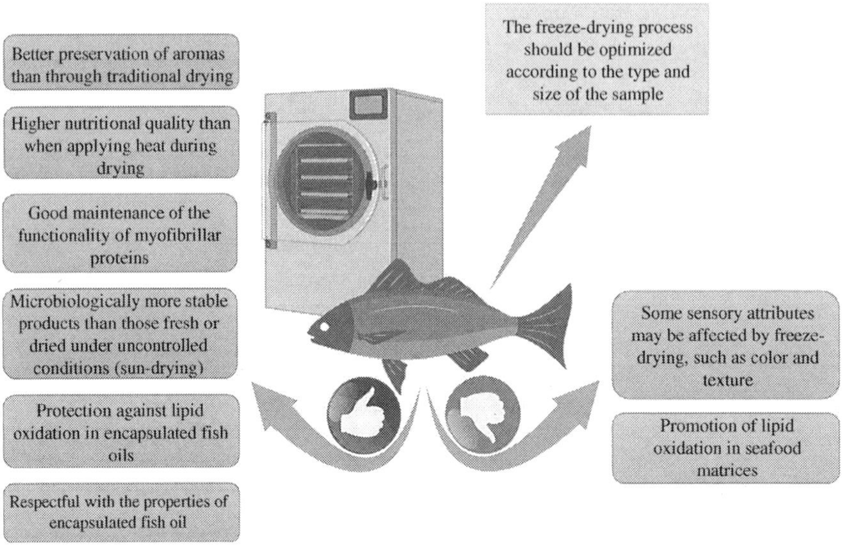

Figure 4.3 Scheme of how the freeze-drying process can affect seafood products.

during the drying process, freeze-drying better preserves the volatile compounds in fish, which can be especially intense, compared to other drying procedures, such as classic hot-air drying and microwave drying, with and without vacuum, being found up to 63 volatiles substances by 54, 45, and 37 of the previous methods, respectively. Freeze-drying preserved 1-octen-3-ol, the alcohol responsible for the typical fishy odor, in golden pompano fillets, one of the most widely traded fish in China and Southeast Asia. Unlike the application of hot-air and microwave drying, freeze-drying contributed less aldehydes to the overall aroma of the fillets (42.72% compared to 60.98% for microwave drying), improving the sensorial quality of the product. On the other hand, the number of esters detected, responsible for the meat flavor characteristics of fish, was higher when the fillets were freeze-dried, preserving twice as many compounds as by microwave drying. This is because hot air may remove volatile compounds from small molecules and decompose those that are heat sensitive (Zhang et al. 2019).

A freeze-drying process has the potential to not only preserve the nutrients in the seafood but also increase their concentration. Kim et al. (2020) noted that the amino acid content almost doubled (from 229.84 to 403.06 mg/100 g sample) in yellow croaker, a fish widely distributed in the Yellow Sea and East China. Drying allowed the concentration of dry matter, contributing with an extra content of amino acids asparagine and serine compared to the other two drying methods tested in the study, hot-air and vacuum drying at low temperature. On the other hand, the color of fish samples was affected during freeze-drying. L^* value was reduced, while a^* and b^* values were increased. A similar result was reported for texture, with a significant increase in hardness, as well as the energy required to chew the sample. Negative aspects that were minimized using the alternative drying methods to freeze-drying are previously mentioned.

Freeze-drying technology in seafood is especially focused on the drying of matrices rich in thermosensitive active compounds, opening a window toward obtaining innovative products from traditional marine species or other less consumed such as the sea cucumber. Seafood-based products are increasingly oriented toward more select markets, including nutraceuticals, functional foods, cosmetics, and pharmaceuticals. Because freeze-drying is a complex and costly process for the industry, its use is mainly addressed to the production of derivatives, such as collagen powder or shark cartilage (Manuel et al. 2011).

The aforementioned sea cucumber is a source of a wide variety of compounds with biological properties, such as collagen, saponins, lectins, polyphenols, peptides, or polyunsaturated fatty acids (PUFAs) (Hossain et al. 2020). Precisely, omega-3 fatty acids, which play an important role in well-being and prevention of cardiovascular disease (CVD) (Agregán et al. 2022), have been found at high levels in the viscera of this marine invertebrate, specially eicosapentaenoic acid (C20:5*n*-3). Freeze-drying of this sea cucumber byproduct has reasonably preserved the original fatty acid content of the fresh product, raising the levels of omega-3 PUFAs in the phospholipid fraction compared to conventional drying with hot air, and demonstrating its potential use in the processing of functional foods and pharmaceuticals (Liu et al. 2021). The absence of high temperatures in freeze-drying processes better protects PUFAs against oxidative rancidity, providing a better quality dry product (Cittadini et al. 2022). Specifically, omega-3 fatty acids are highly unstable due to oxidant factors such as temperature.

Dehydrated fish proteins represent a rich source of highly digestible amino acids that can be used in the formulation of fish and shellfish and in the enrichment of food products, among other applications (Shaviklo 2015). The freeze-drying procedure was found not to excessively damage the functionality of myofibrillar proteins, maintaining a good standard of quality. Niu et al. (2019) reported an increase in the emulsifying stability index of up to twofold after freeze-drying myofibrillar protein from silver carp despite the partial denaturation caused by the process. Gong et al. (2016) confirmed that the protein matrix is endowed with greater molecular flexibility and surface hydrophobicity after being freeze-dried, thus improving the stability of the emulsions subsequently formed from the dry product. On the other hand, freeze-drying was found to significantly affect the activity index of silver carp myofibrillar proteins (from 1.36 to 1.27 m^2/g), though the change was not as marked as the previously increase in emulsifying stability (Niu et al. 2019).

Freeze-drying has recently been studied as a technology for the encapsulation of nutraceuticals, especially those that, as in the case of fish oil, are sensitive to high temperatures during the drying process. Specifically, encapsulation by freeze-drying is capable of producing microcapsules of a higher quality than those produced by the spray-drying method, which, though more used because it is cheaper, has certain disadvantages such low solubility. However, encapsulation by freeze-drying results in pore formation and enhances rehydration of the powdery matrix (Khoshnoudi-Nia et al. 2022). The rehydration capacity of docosahexaenoic acid (C22:6*n*-3) microencapsulated by freeze-drying was in the range of 36–48 seconds, while the same fatty acid microencapsulated by spray drying was between 89 and 102 seconds, more than double (Karthik and Anandharamakrishnan 2013). In another study, Pang et al. (2017) reported a brighter and less dark powder than that obtained by spraying. This indicates a more respectful process with the encapsulated fat by avoiding the browning produced when too high temperatures are used.

4.4 Stability of Freeze-Dried Meat and Seafood Products

It is well-known that the dehydrated foods are more stable than fresh foods. It is the loss of water that produces a reduction in a_w, retarding food spoilage by inhibiting microbial growth, chemical reactions, and enzymatic activity (Ngamwonglumlert and Devahastin 2017). As already discussed in this chapter, the removal of water causes deep changes in the food matrix. For this reason, alternative technologies to the classic hot-air drying, such as freeze-drying, were developed with the intention of preventing any possible chemical, biochemical, or physical modification of the food, attempting to achieve a rehydrated product with similar characteristics to the fresh original. Despite the fact that a highly porous microstructure such as that obtained by freeze-drying has been found to favor the rehydration of the dry product (Ma et al. 2018) leading to the obtaining products very similar to fresh once rehydrated. This could be an inconvenience toward the conservation of the product during the storage since a microstructure with a high number of pores is very hygroscopic, promoting the uptake of water from the medium. This premature moisture absorption reduces the shelf life of the dried product during storage. On the other hand, a microstructure of this type leaves a large surface area exposed to the action of elements, such as light and oxygen, and factors recognized pro-oxidants (Ngamwonglumlert and Devahastin 2017).

The meat matrix is a very perishable element as it is a source of high-quality proteins, fatty acids, vitamins, and minerals (Umaraw et al. 2020). The same occurs with fish, especially with regard to fatty acids, as its high concentration of easily oxidized PUFAs favors the rancidity of the product (Gómez et al. 2019b). The most recent and noteworthy discoveries on the effect of freeze-drying on the stability of meat and seafood products are exposed and discussed in the following paragraphs and summarized in Table 4.2.

Table 4.2 Ability of the freeze-drying process to stabilize meat and seafood products.

Factor affecting conservation	Product	Main findings	Reference
Microbiology	*Longissimus dorsi* muscle of male goat	• ~2.5 log cfu/g reduction in the amount of aerobics after freeze-drying.	Rahman et al. (2005)
	Broiler chicken breast	• TAMB remained below 5 log cfu/g (lower than the acceptable microbial limit of 7 log cfu/g) after 8 mo of storage. • LAB decrease by more than 1.5 log units after 2 mo of storage.	Cantalejo et al. (2016)
	Bombay duck fish	• Decreased TVC and TFC by 0.1 and 0.2 logarithmic units, respectively, with respect to fresh fish.	Nagwekar et al. (2017)
VBN	Yellow croaker fish	• VBN levels increased from ~10 to ~33 mg N/100 g, but bellow the limit established for human intake (35–40 mg N/100 g) and for that obtained by hot-air drying (~66 mg N/100 g).	Kim et al. (2020)
Lipid oxidation	Meat-based baby food	• Promotion of cholesterol oxidation. • The 7-ketocholesterol was the main compound of direct oxidation found.	Evangelisti et al. (2004)
	Minced beef	• Formation of twice as many peroxides as by vacuum drying and ultrasonic-assisted vacuum drying.	Aksoy et al. (2019)
	Atlantic and horse mackerel	• Sudden increase in peroxides after 8 wk of storage (~1000 meq O_2/kg lipid). • Structural changes in proteins. • Lower solubility of myofibrillar proteins in salt solution.	Sarkardei and Howell (2007)
	Omega-3 PUFAs	• The encapsulation process by freeze-drying kept the peroxide levels practically unaltered. • Better protection against primary oxidation than the spray-drying process.	Melgosa et al. (2019)

TAMB: total aerobic mesophilic bacteria; LAB: lactic acid bacteria; TVC: total viable count; TFC: total fungal count; VBN: volatile basic nitrogen; PUFAs: polyunsaturated fatty acids; meq: milliequivalents.

4.4.1 Advantages and Disadvantages in the Preservation of Meat by Freeze-Drying

Freeze-drying is considered one of the safest forms of food preservation, it drastically reduces the a_w when water is removed by sublimation and thus delays the development cycle of spoilage microorganisms. In this way, a highly stable product is obtained. As meat is a highly perishable product, freeze-drying would offer the possibility of storage for long periods of time until consumption. Moreover, meat processed in this way has the additional advantage of being easily handled during shipping and transportation (Cantalejo et al. 2016). During a study on the shelf life of *broiler* chicken breast, freeze-drying and subsequent vacuum packaging managed to keep the pieces of meat suitable for consumption during four months of storage. Total aerobic mesophilic bacteria (TAMB) were maintained below 5 log units (more than 2 log units of margin with respect to what is considered the acceptable microbial limit, 7 log cfu/g) during eight months. Surprisingly, freeze-drying slightly increased the number of lactic acid bacteria (LAB) compared to the initial one. Nevertheless, after only two months of storage, the number was reduced by more than 1.5 log cfu/g and after four months the LAB count was below 1 log unit. On the contrary, the fresh undried sample showed an increase of up to almost 2 log cfu/g in two months of storage, which proves the effectiveness of the freeze-drying process to microbiologically stabilize a chicken breast such as the one used in the study (Cantalejo et al. 2016).

When the ability to reduce the concentration of microorganisms in goat meat by sun-drying was compared to some alternative preservation methods, including freeze-drying, a significant difference was found in microbiologically stabilizing the *longissimus dorsi* muscle matrix. Freeze-drying caused a further reduction in aerobic counts and in the number of *Pseudomonas* spp. and *Staphylococcus* spp. than that achieved by sun-drying. The analysis of the fresh product presented an amount of aerobics of almost 7 log cfu/g, which after drying by sublimation were reduced to 4.5 log cfu/g, while in the sun-dried meat an increase slightly higher than 1 log unit was found (Rahman et al. 2005).

Although freeze-drying was observed to guarantee an increase in the shelf file of meat due to the formation of a hostile environment for microbes, problems related to lipid oxidation observed in different studies could reverse or seriously limit conservation. Evangelisti et al. (2004) reported a promotion of cholesterol oxidation in freeze-dried meat-based baby foods. The 7-ketocholesterol compound was found to be the main product of direct oxidation, which was cataloged as a potential indicator of cholesterol oxidation by the authors. Compared with a similar product, which was only subjected to homogenization without a freeze-drying process, a higher oxidation was noted. The oxidative destabilization seemed to be out of control at times, leading to a marked deterioration of the lipid fraction and aggravated by the oxidative damage induced by subsequent mincing of the matrix, a step that may cause a strong increase in the amount of cholesterol oxidation products. The work of Aksoy et al. (2019) supported these outcomes in an attempt to study the level of oxidation caused by a freeze-drying process in minced beef. The authors reported a pronounced oxidation compared to other unconventional drying methods. The formation of peroxides in freeze-dried meat was found to double that formed by vacuum drying and ultrasound-assisted vacuum drying (10.42 vs. 5.71–6.91 and 5.02–5.63, respectively). The high porosity of the matrices submitted to a sublimation process seems to be behind the high oxidation observed with respect to other drying methods assayed. According to the study carried out by Rahman et al. (2002), the high number of pores formed by freeze-drying compared to other drying techniques, such as hot-air and vacuum drying at 70 °C, causes a high value of fat peroxidation throughout the storage of the product. This is attributed to a high oxygen diffusion caused by the increase in the surface area of the matrix after being subjected to sublimation.

4.4.2 Suitability of Freeze-Drying to Stabilize Seafood Products

As in the case of meat, the seafood matrix was found to be more microbiologically stable after freeze-drying than after traditional sun-drying. When Bombay duck fish was submitted to drying, comparing different techniques (classical vs. modern), in all of them the reduction in moisture content gave rise to an a_w lower than 0.6 (Nagwekar et al. 2017), which in principle ensures a safe and durable product in microbiological terms, since microbial growth stops below this critical value (Perera and Rahman 1997). Nevertheless, the possibility of achieving greater protection against spoilage and/or pathogenic bacteria has led to considerate treatments that do not include exposure of food material to the outside environment. Nagwekar et al. (2017) reported significantly higher total viable counts (TVC) and total fungi counts (TFC) in sun-drying fish than in other less traditional techniques including freeze-drying. This is because outdoor drying conditions are not fixed, and variables, such as temperature, humidity, and air velocity, fluctuate unpredictably. On the other hand, handling and exposure to the surrounding environment is a source of contamination. In fact, instead of decreasing, microbial levels increased from 5.38 log colony-forming units (cfu)/g to 6.43 log cfu/g (TVC) and from 4.99 log cfu/g to 6.34 cfu/g (TFC). However, freeze-drying reduced the concentration of these groups of microorganisms by 0.1 and 0.2 log units, respectively, compared to fresh fish. Despite the reduction in microbial levels achieved by this technique, maintaining temperatures below 0 °C could retain microorganisms during the drying process.

Trimethylamine (TMA) levels found in the above freeze-dried fish exceed that of fresh fish, but remain below those found in sun-dried fish or by other means, such as hot-air drying and the novel solar conduction drying technique. Unfortunately, these TMA values in freeze-dried fish have been reported to be above the acceptable limit for fresh fish (10–15 mg N/100 g) (Nagwekar et al. 2017). Such a high value of TMA could compromise the quality of the product since the concentration of this volatile nitrogenous compound increases throughout the storage as it is indicative of the bacterial deterioration suffered by this type of matrices over time, generating an intense smell reminiscent of ammonia. Nonetheless, the microbial load of the fish and its a_w have been reduced by the freeze-drying process applied, which undoubtedly helps to stabilize the product. Similarly, the production of volatile basic nitrogen (VBN), which includes TMA, is another parameter that measures the degree of deterioration of fish and therefore its acceptability for consumption. Kim et al. (2020) reported that despite the increase in VBN concentration after freeze-drying yellow croaker (from ~10 to ~33 mg N/100 g), the value remained close to but below the limit considered adequate for human intake (35–40 mg N/100 g) (Immaculate et al. 2013), suggesting that freeze-drying may be a good method to maintain freshness and good quality of dried fish compared to hot-air drying, which produced too much VBN (~66 mg N/100 g), exceeding the previously mentioned limit.

Oxidation in fish is an even more serious problem than in meat due to the high content of PUFAs, which go rancid easily, generating compounds with an unpleasant odor and that can be a health hazard as they are potential promoters of severe human pathologies, such as atherosclerosis and cancer, and of process related to cellular aging (Domínguez et al. 2019). In relation to this, hydroperoxides produced in lipid oxidation reactions have been found to produce toxic effects on cells, even at low concentrations (Angeli et al. 2011). Hydroperoxide formation in freeze-dried Atlantic mackerel and horse mackerel was reported to increase sharply after the first three weeks of storage at 22 °C, peaking at around 1000 milliequivalents (meq) O_2/kg lipid for the two fish species at the eight weeks. Freeze-drying was found to affect the structure and composition of fish lipids by producing prominent long-term changes. Alterations involving —CH groups were identified. These modifications were more intense in the freeze-dried Atlantic mackerel oil than in the freeze-dried horse mackerel oil, possibly due to the higher percentage of eicosapentaenoic acid (C20:5*n*-3, EPA) (6.23% vs. 2.02%) in the former of the mentioned fish. In addition, structural changes in the proteins of both species were also noted, which led to a lower solubility of the myofibrillar proteins in salt solution. The formation of intramolecular crosslinks and hydrophobic interactions were reported (Sarkardei and Howell 2007).

Despite that freeze-drying is a drying technique conceived to attempt to protect food matrix and preserve its organoleptic characteristics, the damage caused by the drying process might be important, especially in long-term storage. For this reason, the addition of antioxidants could be considered when processing seafood in this way (Sarkardei and Howell 2007).

Freeze-drying has recently been studied for the microencapsulation of fish oils with high nutritional value and bioactive properties due to the ability of this drying technique to protect food matrices against lipid autoxidation when using low temperatures. As previously mentioned, PUFAs are very susceptible to autoxidative rancidity caused by numerous external agents such as temperature. In this regard, the freeze-drying technique has been found to be especially respectful with heat-sensitive molecules (Carpena et al. 2021).

Encapsulation of omega-3 PUFAs from fish oil by freeze-drying using liquid nitrogen as a freezing agent yielded greater protection against oxidation than using the traditional cooling system. Melgosa et al. (2019) found a peroxide value (PV) of 4.6 meq O_2/kg oil after the encapsulation process of these fatty acids by freeze-drying using this compound to freeze the sample (value practically identical to that found for the original emulsion). This high protection of the oil against oxidation can be explained by the absence of high temperatures, light, and oxygen, since the process is carried out in an inert atmosphere. This result was contrary to that found when freezing was performed by conventional means, allowing oxygen contact with the sample while the freezing process takes place. Even so, traditional freeze-drying was able to better preserve omega-3 acids against primary oxidation than using spray-drying technology, even after 28 days of dark refrigerated storage. This drying method exposed the fatty acids to elevated and prolonged temperatures likely due to the rapid formation of the particle shell that increased the resistance to evaporation of water. In contrast, the same study reported a superior preservation capacity of omega-3 PUFAs when encapsulated by drying using gas-saturated solutions (PGSS) with the addition of ascorbic acid as an antioxidant.

Despite the advantage of freeze-drying regarding the protection of the encapsulated material against lipid oxidation due to the low temperature used in the process, the structure of the microcapsule formed seems to oxidatively stabilize fish oil worse than spray-drying technology according to the study carried out by Pang et al. (2017). These authors studied the efficiency of these two encapsulation methods and a combination of both to preserve the integrity of fish fatty acids against the autoxidative process. The use of scanning electron microscopy showed that the microstructure obtained by spraying consisted of spherical capsules, with a smooth surface and a low occurrence of dents and cracks compared to the microstructure obtained by freeze-drying. The absence of cracks has been precisely reported to be of paramount importance for the functionality of the wall by limiting the deterioration of fish oil and/or its oxidation during storage, thus prolonging the shelf life of the encapsulated product (Anwar et al. 2010).

4.5 Conclusions

Freeze-drying is an industrial treatment that inevitably alters the structure and characteristics of meat and seafood products. However, it has the capacity to better safeguard the initial sensory quality of the fresh product and mimic its features after rehydration compared to more traditional drying methods, such as sun-drying or hot-air circulation, though some exceptions can be found depending on the sample. Among all positive characteristics reported, the strong and hygienic reduction of a_w is especially interesting as it allows to considerably slow down microbial development without adding an extra load to the food and consequently increasing shelf life. Similarly, low temperatures to which food is exposed delay the onset of oxidation reactions in encapsulated fish oils. However, the excessive production of pores in solid matrices is a negative aspect for conservation of freeze-dried meat and seafood products as it favors the diffusion of oxygen, a strong promoter of autoxidation reactions, inside these foodstuffs. On the other hand, freeze-drying is currently a too expensive process and the high cost of producing dehydrated foods using this technology limits its application. Therefore, both the high costs derived from the use of freeze-drying and its ability to provide quality meat and seafood products should be taken into consideration before carrying out a process that involves its use on an industrial scale.

Acknowledgments

Rubén Agregán and Noemí Echegaray acknowledge to GAIN (AxenciaGalega de Innovación), Xunta de Galicia, for granting with a postdoctoral scholarship (grant number IN606B-2022/005 and IN606B-2022/006, respectively). Jose M. Lorenzo belongs to the competitive reference research group FunMeat, GAIN (GAININ607A2019/01). Jose M. Lorenzo is member of the HealthyMeat network funded by CYTED (ProgramaIberoamericano de Ciencia y Tecnología para el Desarrollo) (Ref. 119RT0568). Paulo E.S. Munekata acknowledges postdoctoral fellowship support from the Ministry of Science and Innovation (MCIN, Spain) "Juan de la Cierva" program (IJC2020-043358-I).

References

Adams, G., Cook, I., and Ward, K.R. (2015). The principles of freeze-drying. In: *Cryopreservation and Freeze-Drying Protocols* (ed. W.F. Wolkers and H. Oldenhof), 121–143. Humana Press https://doi.org/10.1007/978-1-59745-362-2_2.

Agregán, R., Popova, T., López-Pedrouso, M. et al. (2022). Fatty acids. In: *Food Lipids* (ed. J.M. Lorenzo, P.E.S. Munekata, M. Pateiro, et al.), 257–286. Elsevier https://doi.org/10.1016/B978-0-12-823371-9.00015-0.

Akpinar, E.K. (2006). Mathematical modelling of thin layer drying process under open sun of some aromatic plants. *Journal of Food Engineering* 77 (4): 864–870. https://doi.org/10.1016/J.JFOODENG.2005.08.014.

Aksoy, A., Karasu, S., Akcicek, A., and Kayacan, S. (2019). Effects of different drying methods on drying kinetics, microstructure, color, and the rehydration ratio of minced meat. *Foods* 8 (6): 216. https://doi.org/10.3390/foods8060216.

Angeli, J.P.F., Garcia, C.C.M., Sena, F. et al. (2011). Lipid hydroperoxide-induced and hemoglobin-enhanced oxidative damage to colon cancer cells. *Free Radical Biology and Medicine* 51 (2): 503–515. https://doi.org/10.1016/j.freeradbiomed.2011.04.015.

Anwar, S.H., Weissbrodt, J., and Kunz, B. (2010). Microencapsulation of fish oil by spray granulation and fluid bed film coating. *Journal of Food Science* 75 (6): https://doi.org/10.1111/j.1750-3841.2010.01665.x.

Assegehegn, G., Brito-de la Fuente, E., Franco, J.M., and Gallegos, C. (2019). The importance of understanding the freezing step and its impact on freeze-drying process performance. *Journal of Pharmaceutical Sciences* 108 (4): 1378–1395. https://doi.org/10.1016/j.xphs.2018.11.039.

AykınDinçer, E. (2021). Dried meat products obtained by different methods from past to present. *Food Reviews International* 2021: 1–20. https://doi.org/10.1080/87559129.2021.1956944.

Babić, J., Cantalejo, M.J., and Arroqui, C. (2009). The effects of freeze-drying process parameters on Broiler chicken breast meat. *LWT - Food Science and Technology* 42 (8): 1325–1334. https://doi.org/10.1016/j.lwt.2009.03.020.

Berk, Z. (2013). Freeze drying (lyophilization) and freeze concentration. In: *Food Process Engineering and Technology* (ed. Z. Berk), 567–581. Academic Press https://doi.org/10.1016/B978-0-12-415923-5.00023-X.

Cantalejo, M.J., Zouaghi, F., and Pérez-Arnedo, I. (2016). Combined effects of ozone and freeze-drying on the shelf-life of Broiler chicken meat. *LWT - Food Science and Technology* 68: 400–407. https://doi.org/10.1016/j.lwt.2015.12.058.

Carpena, M., Garcia-Oliveira, P., Lourenço-Lopes, C. et al. (2021). Freeze-drying encapsulation as a mechanism of choice in oils: methods and mechanism. In: *Basic Protocols in Encapsulation of Food Ingredients* (ed. A. Gomez-Zavaglia), 91–101. Springer https://doi.org/10.1007/978-1-0716-1649-9_9.

Cheng-Hai, X.U., Shi-Wei, Z., Run-Ling, P., and Zhi-Jun, Z. (2008). Today and tomorrow of vacuum freeze-drying technology (II). *Vacuum* 3: 1–3.

Cittadini, A., Munekata, P.E.S., Pateiro, M. et al. (2022). Encapsulation techniques to increase lipid stability. In: *Food Lipids* (ed. J.M. Lorenzo, P.E.S. Munekata, M. Pateiro, et al.), 413–459. Elsevier https://doi.org/10.1016/B978-0-12-823371-9.00010-1.

Coombs, C.E.O., Holman, B.W.B., Friend, M.A., and Hopkins, D.L. (2017). Long-term red meat preservation using chilled and frozen storage combinations: a review. *Meat Science* 125: 84–94. https://doi.org/10.1016/J.MEATSCI.2016.11.025.

Cozzolino, D. and Murray, I. (2012). A review on the application of infrared technologies to determine and monitor composition and other quality characteristics in raw fish, fish products, and seafood. *Applied Spectroscopy Reviews* 47 (3): 207–218. https://doi.org/10.1080/05704928.2011.639106.

Das, A.K., Nanda, P.K., Chowdhury, N.R. et al. (2021). Application of pomegranate by-products in muscle foods: oxidative indices, colour stability, shelf life and health benefits. *Molecules* 26 (2): 467. https://doi.org/10.3390/MOLECULES26020467.

Domínguez, R., Pateiro, M., Gagaoua, M. et al. (2019). A comprehensive review on lipid oxidation in meat and meat products. *Antioxidants* 8: 429. https://doi.org/10.3390/antiox8100429.

Ekezie, F.G.C., Sun, D.W., and Cheng, J.H. (2019). Altering the IgE binding capacity of king prawn (*Litopenaeus vannamei*) tropomyosin through conformational changes induced by cold argon-plasma jet. *Food Chemistry* 300: 125143. https://doi.org/10.1016/J.FOODCHEM.2019.125143.

Elmas, F., Bodruk, A., Köprüalan, Ö. et al. (2020). Drying kinetics behavior of turkey breast meat in different drying methods. *Journal of Food Process Engineering* 43 (10): e13487. https://doi.org/10.1111/jfpe.13487.

Esua, O.J., Cheng, J.H., and Sun, D.W. (2021). Functionalization of water as a nonthermal approach for ensuring safety and quality of meat and seafood products. *Critical Reviews in Food Science and Nutrition* 61 (3): 431–449. https://doi.org/10.1080/10408398.2020.1735297.

Evangelisti, F., Zunin, P., Boggia, R., and Calcagno, C. (2004). Cholesterol oxidation in meat-based baby foods. *Journal of AOAC International* 87 (2): 505–510. https://academic.oup.com/jaoac/article/87/2/505/5657288.

FAO (2021). *FAO yearbook*. In: *Fishery and Aquaculture Statistics 2019*. www.fao.org/fishery/static/Yearbook/YB2019_USBcard/index.htm.

FAOSTAT (2020). *Food & Agriculture Organization of the United Nations Statistics Division*. FAO Stat: Agriculture Data *Accessed December 13, 2022* https://www.fao.org/faostat/es/#data/QCL.

Gómez, B., Munekata, P.E.S., Gavahian, M. et al. (2019a). Application of pulsed electric fields in meat and fish processing industries: an overview. *Food Research International* 123: 95–105. https://doi.org/10.1016/j.foodres.2019.04.047.

Gómez, B., Munekata, P.E.S., Zhu, Z. et al. (2019b). Challenges and opportunities regarding the use of alternative protein sources: aquaculture and insects. In: *Advances in Food and Nutrition Research*, vol. 89 (ed. F. Toldrá), 259–295. Academic Press Inc. https://doi.org/10.1016/bs.afnr.2019.03.003.

Gong, K.J., Shi, A.M., Liu, H.Z. et al. (2016). Emulsifying properties and structure changes of spray and freeze-dried peanut protein isolate. *Journal of Food Engineering* 170: 33–40. https://doi.org/10.1016/j.jfoodeng.2015.09.011.

Haseley, P. and Oetjen, G.W. (2018). Foundations and process engineering. In: *Freeze-Drying* (ed. P. Haseley and G.W. Oetjen), 1–176. John Wiley & Sons.

Hashim, N., Daniel, O., and Rahaman, E. (2014). A preliminary study: kinetic model of drying process of pumpkins (*Cucurbita moschata*) in a convective hot air dryer. *Agriculture and Agricultural Science Procedia* 2: 345–352. https://doi.org/10.1016/J.AASPRO.2014.11.048.

Hossain, A., Dave, D., and Shahidi, F. (2020). Northern sea cucumber (*Cucumaria frondosa*): a potential candidate for functional food, nutraceutical, and pharmaceutical sector. *Marine Drugs* 18 (5): 274. https://doi.org/10.3390/md18050274.

Immaculate, K., Sinduja, P., Velammal, A., and Patterson, J. (2013). Quality and shelf life status of salted and sun dried fishes of Tuticorin fishing villages in different seasons. *International Food Research Journal* 20 (4): 1855–1859.

Jia, Y., Khalifa, I., Hu, L. et al. (2019). Influence of three different drying techniques on persimmon chips' characteristics: a comparison study among hot-air, combined hot-air-microwave, and vacuum-freeze drying techniques. *Food and Bioproducts Processing* 118: 67–76. https://doi.org/10.1016/j.fbp.2019.08.018.

Karathanos, V.T., Anglea, S.A., and Karel, M. (1996). Structural collapse of plant materials during freeze-drying. *Journal of Thermal Analysis* 47 (0996): 1451–1461.

Karthik, P. and Anandharamakrishnan, C. (2013). Microencapsulation of docosahexaenoic acid by spray-freeze-drying method and comparison of its stability with spray-drying and freeze-drying methods. *Food and Bioprocess Technology* 6 (10): 2780–2790. https://doi.org/10.1007/s11947-012-1024-1.

Karwowska, M., Stadnik, J., Stasiak, D.M. et al. (2021). Strategies to improve the nutritional value of meat products: incorporation of bioactive compounds, reduction or elimination of harmful components and alternative technologies. *International Journal of Food Science and Technology* 56 (12): 6142–6156. https://doi.org/10.1111/IJFS.15060.

Khoshnoudi-Nia, S., Forghani, Z., and Jafari, S.M. (2022). A systematic review and meta-analysis of fish oil encapsulation within different micro/nanocarriers. *Critical Reviews in Food Science and Nutrition* 62: 2061–2082. https://doi.org/10.1080/10408398.2020.1848793.

Kim, B.S., Oh, B.J., Lee, J.H. et al. (2020). Effects of various drying methods on physicochemical characteristics and textural features of yellow croaker (*Larimichthys polyactis*). *Foods* 9 (2): 196. https://doi.org/10.3390/foods9020196.

Liu, Y., Dave, D., Trenholm, S. et al. (2021). Effect of drying on nutritional composition of atlantic sea cucumber (*Cucumaria frondosa*) viscera derived from Newfoundland fisheries. *Processes* 9 (4): https://doi.org/10.3390/pr9040703.

Liu, Y., Zhang, Z., and Hu, L. (2022). High efficient freeze-drying technology in food industry. *Critical Reviews in Food Science and Nutrition* 62 (12): 3370–3388. https://doi.org/10.1080/10408398.2020.1865261.

Ma, Y., Liu, W.W., and Huang, G.H. (2014). Manufacturing research with feasibility of vacuum freeze drying technology for leisure meat products processing. *Advanced Materials Research* 1056: 84–87. https://doi.org/10.4028/www.scientific.net/AMR.1056.84.

Ma, Y., Wu, X., Zhang, Q. et al. (2018). Key composition optimization of meat processed protein source by vacuum freeze-drying technology. *Saudi Journal of Biological Sciences* 25 (4): 724–732. https://doi.org/10.1016/j.sjbs.2017.09.013.

Manuel, H., Thompson, M., and Carroll, K. (2011). Emerging trends in seafood processing technologies. Putting the 'fish' in efficiency. *Journal of Ocean Technology* 6 (4): 15–20.

Mathlouthi, M. (2001). Water content, water activity, water structure and the stability of foodstuffs. *Food Control* 12 (7): 409–417. https://doi.org/10.1016/S0956-7135(01)00032-9.

Meda, L. and Ratti, C. (2005). Rehydration of freeze-dried strawberries at varying temperatures. *Journal of Food Process Engineering* 28 (3): 233–246. https://doi.org/10.1111/J.1745-4530.2005.00404.X.

Melgosa, R., Benito-Román, Ó., Sanz, M.T. et al. (2019). Omega–3 encapsulation by PGSS-drying and conventional drying methods. Particle characterization and oxidative stability. *Food Chemistry* 270: 138–148. https://doi.org/10.1016/j.foodchem.2018.07.082.

Munekata, P.E.S., Rocchetti, G., Pateiro, M. et al. (2020). Addition of plant extracts to meat and meat products to extend shelf-life and health-promoting attributes: an overview. *Current Opinion in Food Science* 31: 81–87. https://doi.org/10.1016/j.cofs.2020.03.003.

Nagwekar, N., Tidke, V., and Thorat, B.N. (2017). Microbial and biochemical analysis of dried fish and comparative study using different drying methods. *Drying Technology* 35 (12): 1481–1491. https://doi.org/10.1080/07373937.2016.1256889.

Ngamwonglumlert, L. and Devahastin, S. (2017). Microstructure and its relationship with quality and storage stability of dried foods. In: *Food Microstructure and Its Relationship With Quality and Stability* (ed. S. Devahastin), 139–159. Woodhead Publishing https://doi.org/10.1016/B978-0-08-100764-8.00008-3.

Niu, J., Zhao, B., Guo, X., and Yin, T. (2019). Effects of vacuum freeze-drying and vacuum spray-drying on biochemical properties and functionalities of myofibrillar proteins from silver carp. *Journal of Food Quality* 2019: 835. https://doi.org/10.1155/2019/9457835.

Nowak, D. and Jakubczyk, E. (2020). The freeze-drying of foods-The characteristic of the process course and the effect of its parameters on the physical properties of food materials. *Foods* 9 (10): 1488. https://doi.org/10.3390/foods9101488.

Onwude, D.I., Hashim, N., and Chen, G. (2016). Recent advances of novel thermal combined hot air drying of agricultural crops. *Trends in Food Science & Technology* 57: 132–145. https://doi.org/10.1016/J.TIFS.2016.09.012.

Pang, Y., Duan, X., Ren, G., and Liu, W. (2017). Comparative study on different drying methods of fish oil microcapsules. *Journal of Food Quality* 2017: https://doi.org/10.1155/2017/1612708.

Pereira, P.M. and Vicente, A.F. (2013). Meat nutritional composition and nutritive role in the human diet. *Meat Science* 93 (3): 586–592. https://doi.org/10.1016/J.MEATSCI.2012.09.018.

Perera, C.O. and Rahman, M.S. (1997). Heat pump dehumidifier drying of food. *Trends in Food Science & Technology* 8 (3): 75–79. https://doi.org/10.1016/S0924-2244(97)01013-3.

Pikal, M.J. and Shah, S. (1990). The collapse temperature in freeze drying: dependence on measurement methodology and rate of water removal from the glassy phase. *International Journal of Pharmaceutics* 62: 165–186.

Rahman, M.S., Al-Amri, O.S., and Al-Bulushi, I.M. (2002). Pores and physico-chemical characteristics of dried tuna produced by different methods of drying. *Journal of Food Engineering* 53 (4): 301–313. https://doi.org/10.1016/S0260-8774(01)00169-8.

Rahman, M.S., Salman, Z., Kadim, I.T. et al. (2005). Microbial and physico-chemical characteristics of dried meat processed by different methods. *International Journal of Food Engineering* 1 (2): 106. https://doi.org/10.2202/1556-3758.1016.

Ratti, C. (2001). Hot air and freeze-drying of high-value foods: a review. *Journal of Food Engineering* 49 (4): 311–319. https://doi.org/10.1016/S0260-8774(00)00228-4.

Sablani, S.S. (2006). Drying of fruits and vegetables: retention of nutritional/functional quality. *Drying Technology* 24 (2): 123–135. https://doi.org/10.1080/07373930600558904.

Santchurn, S.J., Arnaud, E., Zakhia-Rozis, N., and Collignan, A. (2012). Drying: principles and applications. In: *Handbook of Meat and Meat Processing* (ed. Y.H. Hui), 505–521. CRC Press https://doi.org/10.1201/b11479.

Sarkardei, S. and Howell, N.K. (2007). The effects of freeze-drying and storage on the FT-Raman spectra of Atlantic mackerel (*Scomber scombrus*) and horse mackerel (*Trachurus trachurus*). *Food Chemistry* 103 (1): 62–70. https://doi.org/10.1016/j.foodchem.2006.06.065.

Shaviklo, A.R. (2015). Development of fish protein powder as an ingredient for food applications: a review. *Journal of Food Science and Technology* 52 (2): 648–661. https://doi.org/10.1007/s13197-013-1042-7.

Šimat, V., Čagalj, M., Skroza, D. et al. (2021). Sustainable sources for antioxidant and antimicrobial compounds used in meat and seafood products. In: *Advances in Food and Nutrition Research*, vol. 97 (ed. F. Toldrá), 55–118. Academic Press https://doi.org/10.1016/bs.afnr.2021.03.001.

Söbeli, C. and Kayaardı, S. (2020). Optimization of primary freeze drying conditions for powdered chicken meat hydrolysate from mechanically deboned chicken residues. *Drying Technology* 38 (10): 1356–1366. https://doi.org/10.1080/07373937.2019.1640723.

Stapley, A. (2008). Freeze drying. In: *Frozen Food Science and Technology* (ed. J.A. Evans), 248–275. Blackwell Publishing Ltd. https://doi.org/10.1002/9781444302325.ch12.

Tomasevic, I., Djekic, I., Font-i-Furnols, M. et al. (2021). Recent advances in meat color research. *Current Opinion in Food Science* 41: 81–87. https://doi.org/10.1016/J.COFS.2021.02.012.

Umaraw, P., Munekata, P.E.S., Verma, A.K. et al. (2020). Edible films/coating with tailored properties for active packaging of meat, fish and derived products. *Trends in Food Science and Technology* 98: 10–24. https://doi.org/10.1016/j.tifs.2020.01.032.

Uscanga, M.A., Salvador, A., del Camacho, M., M., and Martínez-Navarrete, N. (2021). Impact of freeze-drying shelf temperature on the bioactive compounds, physical properties and sensory evaluation of a product based on orange juice. *International Journal of Food Science and Technology* 56 (10): 5409–5416. https://doi.org/10.1111/ijfs.15086.

Wang, Y., Zheng, Q., Li, W. et al. (2018). Measurement of free water in foods by secondary derivative thermogravimetry. *CYTA—Journal of Food* 16 (1): 438–443. https://doi.org/10.1080/19476337.2017.1416675.

Zhang, J., Cao, J., Pei, Z. et al. (2019). Volatile flavour components and the mechanisms underlying their production in golden pompano (*Trachinotus blochii*) fillets subjected to different drying methods: a comparative study using an electronic nose, an electronic tongue and SDE-GC-MS. *Food Research International* 123: 217–225. https://doi.org/10.1016/j.foodres.2019.04.069.

5

Freeze-Drying of Dairy Products

Suchismita Roy[1] *and Gunvantsinh Rathod*[1,2]

[1] *Food Science Institute, Kansas State University, Manhattan, KS, USA*
[2] *Research and Development Department, Idaho Milk Products, Jerome, ID, USA*

5.1 Introduction

Water is the major component in milk along with other nutrients. It makes a preferred choice for microbes to grow and proliferate. Being one of the best media for microbes to grow makes milk a perishable commodity. Considering water as an essential component to support microbial activity, the removal of water prevents microbial growth by lowering water activity and water availability. Along with microbial activity, chemical and enzymatic deterioration also take place due to higher water activity of milk and milk products. Therefore, milk is subjected to evaporation and spray drying to remove water from ~84–87% to ~5% in milk powder (Walstra et al. 2005). Hence, evaporation followed by spray drying is a very cost-effective process adopted by the dairy industry to convert milk into powder. Since evaporation is done under vacuum, milk is evaporated below its boiling point, therefore heat damage is considerably less with evaporation, but still, it will affect heat-labile protein, and also vacuum evaporation removes characteristic aroma (Schuck et al. 2013). Furthermore, spray drying is done with very high-temperature air (180–200 °C), which goes out at 80–90 °C. However, due to the evaporative cooling of product droplets during drying and low residence time in the dryer, product temperature is not going beyond 60–65 °C (Anandharamakrishnan et al. 2008). Despite these facts, heat-sensitive proteins and components are damaged owing to the heat. Furthermore, during spray drying, moisture in the product droplet travels from the center to the periphery of the droplet, and the periphery is exposed to hot air, which modifies the powder surface. This gives some insoluble particles in

Freeze Drying of Food Products: Fundamentals, Processes and Applications, First Edition.
Edited by Roji Balaji Waghmare, Manoj Kumar, and Parmjit Singh Panesar.

reconstituted milk. Considering this fact, freeze-drying is applied in dairy products, such as dairy cultures, and heat-labile dairy proteins, where heat-induced denaturation is not good for product properties and the product has a very high value in terms of nutrition and cost. In freeze-drying, the product is frozen and water is removed by sublimation, therefore the product is not exposed to any high temperature. Hence, the dairy product's nutritive value, color, and aroma are retained after drying. Furthermore, the product will be easy to reconstitute using water. The basic mechanism and process of freeze-drying are well discussed in earlier sections. Therefore, this section will focus on the selected application of freeze-drying in the dairy industry in this section.

5.2 Application of Freeze-Drying in Different Dairy Products/Ingredients

Freeze-dried food market is growing at 7.7% CAGR during 2022–2030 and will reach 71.7 billion (Brainy Insight 2023). Among freeze-dried dairy products, freeze-dried cheese such as cream cheese, mozzarella cheese for pizza application, freeze-dried yogurt powder, freeze-dried butter powder, and freeze-dried cultures are on the top (Dairy Reporter 2016). The main customer for these freeze-dried dairy products are restaurants (e.g. freeze-dried grated mozzarella for pizzas), bakeries (e.g. prebiotic dairy cultures for bread making), biotech industry (fat-free milk powder used as a saturating agent), camping, mountaineering, hiking activities, military, and emergency food rations where long shelf-life products are required. This section will focus on some selected interventions for freeze-drying selected dairy products.

5.2.1 Milk

Milk is converted to milk powder commonly using vacuum evaporation followed by spray drying using hot air, as spray drying is the most economical process (Hazlett et al. 2021). However, the heating of milk results in some protein denaturation and loss of volatile compounds. Therefore, studies were conducted to dry milk using freeze-drying especially milk from minor species such as camel (Zou et al. 2022) and donkey (Vincenzetti et al. 2018) mainly to avoid heat-induced effects on the bioactive compounds. Heating causes the reaction of the free —SH group with other —SH group on other proteins or other S—S (disulfide) bonds on protein making irreversible protein complexes (Wijayanti et al. 2014). Recently, Rogers et al. (2008) carried out freeze-drying of skim milk using the spray freeze-drying technique and compared it with powder produced by the conventional spray drying technique. They found that spray freeze-dried powder wetted three times faster than the conventional powder. However, for whole milk powder, they saw poor wettability than industrial whole milk powder. Deshwal et al. (2020) have also compared camel milk powder produced using spray drying and freeze-drying. They found better reconstitution for freeze-dried powder than spray dried one. Freeze-dried camel milk powder showed 88.77% solubility, which was higher than spray-dried whole camel milk powder (65.47%) and spray-dried skim milk powder (72.96%).

5.2.2 Cheese

Cheese is converted to powder and used in the preparation of dips, dressing, biscuits, and chips as a flavoring and used in some hot dishes such as soups and spaghetti. During production with conventional drying, some cheese aroma is lost which is balanced by a proportion of aged cheese (Písecký 2005). For cheese powder preparation, cheese is grounded, converted to slurry with the addition of water emulsifying salts and using heat and shear, then subjected to spray drying to get powder (Erbay et al. 2015). Since cheese slurry is thick, foam mate drying was in practice (Izadi et al. 2020). These processes are economical, however, heating employed in the preparation of cheese powder reduced the original cheese aroma intensity. Therefore, freeze-drying was tried to preserve the original aroma of the cheese. Different cheese varieties such as cheddar, mozzarella, and parmesan were freeze-dried and they are available in the market. This freeze-dried cheese can be stored for a longer time. Bedir and Kuleaşan (2019) freeze-dried traditional Turkish cheese and they found the best retention of the original shape and less loss of flavor with freeze-drying. However, they also reported a higher cost of freeze-drying. Other than preserving cheese through freeze-drying, the freeze-drying process is mainly used to preserve cheese starter culture which will be discussed in coming section.

5.2.3 Yogurt

Yogurt is a fermented dairy product prepared from milk by fermentation with *Streptococcus thermophilus* and *Lactobacillus bulgaricus*. The nutritional value, therapeutic and antagonistic benefits, and sensorial attributes make yogurt a highly acceptable product (Kumar and Mishra 2004). It can be stored for a maximum of 20 days in refrigerated conditions (Al-Kadamany et al. 2003; Zhi et al. 2018). This short shelf life of yogurt makes the dehydrated form a preferable option for storage, incorporating it into other products like yogurt-based drinks, ready-to-drink juice, or for direct consumption of yogurt (Kumar and Mishra 2004; Tamime and Robinson 2007; Santos et al. 2018). Freeze-drying is an attractive method for dehydrating the yogurt as the process takes place at a low temperature (−30 °C). The low-temperature processing helps in maintaining viable cell counts after dehydration and preserves the flavor and aroma. Freeze-dried products have a highly porous structure, which helps in easy rehydration (Kumar and Mishra 2004; Santos et al. 2017; Santos et al. 2018).

For manufacturing freeze-dried yogurt, the yogurt samples are kept in the freezer (−25 °C) for 24 hours followed by freeze-drying at −37 °C under vacuum (0.18 mbar) for 24 hours. The dried products are ground, sieved, and packed for storage (Santos et al. 2018). The thickness of the layer of yogurt during freezing and the vacuum of the chamber affects the drying rate (Sharma and Arora 1995). The presence of additives (sugar, fat, nuts, fruits, etc.) can positively influence the survival of the bacteria in the end product which is important for therapeutic benefits (Venir et al. 2007; Sakin-Yilmazer et al. 2014). They can work as cryoprotectants and help in maintaining viable cell counts. Venir et al. (2007) have reported a 1.52 ± 0.39 log reduction for *S. thermophilus* and a 2.33 ± 0.83 log reduction for *Lactobacillus delbrueckii* subsp. *bulgaricus* in a freeze-dried yogurt having 10% sucrose and 10% blueberry pulp. The presence of sugar helps in controlling the viscosity and lowers mobility; lowering the glass transition temperatures of the medium leads to a higher survival rate of bacteria (Carvalho et al. 2004). Wang et al. (2004) reported a higher survival rate of *S. thermophilus* than *L. bulgaricus* due to the difference in their cell structure and size.

The duration required for freeze-drying decreases with increasing surface temperatures and a maximum of 45–50 °C can be applied to maintain the quality of the yogurt (Kumar and Mishra 2004). The total solids content and feed rate positively affect the freeze-drying rate, whereas the product's thickness, chamber pressure, and heating condition can affect it negatively (Sharma et al. 1992).

5.2.3.1 Rehydration of Freeze-Dried Yogurt

It is rehydrated for the consumption of frozen dried yogurt as a traditional one. The consistency, texture, and aroma of yogurt depend on the total solids of the rehydrated product. Although the addition of water depends on the consumer end, 30% total solid gives similar rheological properties as traditional yogurt with 14% total solid (Sakin-Yilmazer et al. 2014).

5.3 Application of Freeze-Drying of Other Dairy Products

Freeze-drying is widely used in the manufacturing of heat-sensitive products such as lactic acid and other cultures, colostrum, and immunoglobulin. They are discussed here.

5.3.1 Lactic Acid and Other Cultures

Lactic acid bacteria (LAB) are widely used in the food industry for their starter or nonstarter applications. LAB are Gram-positive, catalase-negative, nonsporulating, and homo- or heterofermentative microorganisms. The major genera of

LAB are *Lactobacillus*, *Leuconostoc*, *Streptococcus*, *Pediococcus*, and *Enterococcus* (Motta and Gomes 2015). During starter culture application, homofermentative properties are used for rapid acidification, whereas for the nonstarter application, the heterofermentative properties are used to improve the sensorial attributes (Gatti et al. 2014). The method of production and preservation of these starter cultures is chosen based on higher cell viability and activity (Broadbent and Lin 1999) for their industrial end-use. Effects of freeze-drying for different cultures are mentioned in Table 5.1. As per normal industry processes, cultures are propagated for culture viability, their activity, and an increased number of live bacteria. However, routine propagation and maintaining culture is a time-consuming job and need a skilled professional (Hansen 2002). In addition, it also hinders fast-moving food production and does not rule out the chance of culture contamination. Therefore, culture in dried form provides ease of operation as well as eliminates the need for regular propagation to maintain culture activity. The conventional thermal drying process cannot be employed for drying of starter culture as the thermal drying process damages the starter culture. Therefore, freeze-drying is the most common method of preserving the industrial starter culture (Mohammadi-Gouraji et al. 2017). The first step of the lyophilization technique is freezing which is governed by ice crystal size, structure, and distribution. Then, the sublimation of ice crystals to water vapor takes place under a vacuum at a very low temperature. After sublimation, final drying for the unfrozen water takes place at around 20–30 °C (Jennings 1999; Searles et al. 2001; Passot et al. 2012). This step reduces water activity and increases the shelf life of the culture. For the manufacturing of fermented foods, commercial starter cultures are directly added for inoculation to achieve the highest level of safety and flexibility. The viability and activity of the dried culture are extremely important to carry out the fermentation process (Santivarangkna et al. 2008). The freezing process and storage of frozen culture induce the loss of viability of the bacterial cell (Santivarangkna et al. 2008; Fonseca et al. 2015). During the freezing process, the starter cultures are exposed to subzero temperatures. As the temperature decreases, the solute concentration increases leading to the migration of cellular water toward the solution due to osmotic pressure difference. In addition to that, the bacterial cells are compressed by the ice crystals as the temperature decreases leading to cell injury (Mazur et al. 1972; Mazur and Rigopoulos 1983; Takamatsu and Kumagae 2002). If the cooling rate is too fast, intracellular ice formation occurs as the cell fails to lose water rapidly. This leads to the inactivation of cells (Muldrew and McGann 1990, 1994; Gao and Critser 2000). Therefore, the rate of freezing plays an important role in maintaining the stability of the cells (Figure 5.1).

The dehydration of cells changes the stability of the proteins, DNA, and RNA as well as the fluidity of the cellular membrane. To protect the cells from inactivation during freezing and storage, the culture medium's formulation and operating conditions must be optimized.

Table 5.1 Effects of freeze-drying on different cultures.

Culture	Parameter	Effects on properties	Reference
Lactobacillus reuteri ATCC 55730 cells suspended in skim milk	The feed solution with bacteria was frozen at −80 °C and freeze-dried for 17 h at 0.3 bar	The highest survival rate was for cells grown at pH 5 after freeze-drying.	Palmfeldt and Hahn-Hägerdal (2000)
Kefir coculture	The culture was suspended in fermented whey and frozen to −45 °C followed by freeze-drying overnight at 5×10^2 Pa at −45 °C	Feta cheese prepared using this culture was having a better structure and extended shelf life up to 70 d. The aroma was also improved due to the production of different aldehydes, ketones, esters, and organic acids by the kefir culture.	Kourkoutas et al. (2006)
Kefir culture with fermented and unfermented whey as cryoprotectant	Samples were freeze-dried overnight at −45 °C and 5×10^{-3} bar pressure	The survival rate of bacteria was highest (86%) due to the presence of cryoprotective properties and trace nutrients.	Papavasiliou et al. (2008)
Lactobacillus casei ATCC 393 as a starter culture immobilized on casein and apple pieces was freeze-dried. Different cytoprotectants like glucose, lactose, fructose, monosodium glutamate, glycerol, skim milk, and whey were also used in the medium	*L. casei* cells immobilized on apple pieces and casein curd in a MRS broth were collected at growth phase. They were cooled to −45 °C with 1, 3, and 5 °C/min cooling rates followed by freeze-drying for 24 h under 4×10^{-2} mbar pressure at −45 °C	A faster cooling rate increased the cell viability and 5 °C/min cooling rate provided the highest viability (13.8%). Casein was having better immobilized support due to the entrapment of cells in the protein matrix. The highest level of viable cells was achieved when lactose and monosodium glutamate were used (4.8×10^8 and 2.0×10^9 cfu/g respectively).	Dimitrellou et al. (2016)

Mexican kefir grains	Freezing times 3, 5, 7, or 9 h; freezing temperature set to −20, −40, −60, or −80 °C; pressure of 0.2, 0.4, 0.6, or 0.8 mbar; and 5, 10, 15, and 20 h were selected as lyophilization time	The optimal parameters for freeze-drying were a freezing time of 3 h, a freezing temperature of −20 °C, a pressure of 0.6 mbar, and lyophilization of 15 h. The pressure of the drying chamber and lyophilization time affect water activity and the survival rate of the bacteria.	Conde-Islas et al. (2019)
Lactobacillus plantarum was freeze-dried in a medium containing skim milk (10%) along with acid casein, rennet casein, whey powder, and demineralized whey powder (5%). The cryoprotective activity of these dairy byproducts was measured	The culture was frozen at −80 °C for 24 h and then freeze-dried in a lyophilizer for 48 h under a vacuum	Samples with acid casein showed maximum protection for up to 120 d and maximum stability was achieved by whey powder and demineralized whey powder up to 150 d. Rennet casein did not have cryoprotective properties because of the larger pore size leading to syneresis.	Üçok and Sert (2022)
Probiotic starter with *Lactobacillus delbrueckii* ssp. *bulgaricus* (1.0%), *Bifidobacterium bifidum* (6%), *Streptococcus salivarius* ssp. *thermophilus* (80%), *Lactobacillus acidophilus* (13%) was microencapsulated using whey, whey protein concentrate, and sodium alginate as the encapsulating agent.	The solution was frozen at −80 °C for 1 h. The primary drying was carried out at −60 °C under pressure 0.011 mbar for 24 h followed by secondary drying at −75 °C under 0.012 mbar for 1 h	The yield was more with freeze-dried whey solution and alginate carrier. The solubility was more in freeze-dried powder due to the porosity. The survival of the cells and encapsulation efficiency were highest when freeze-drying was used with whey–WPC–alginate as carrier agents. But the spray-dried encapsulation showed better viability than freeze-dried probiotic starter with the same carrier due to the protective effects of denatured proteins.	Obradović et al. (2022)

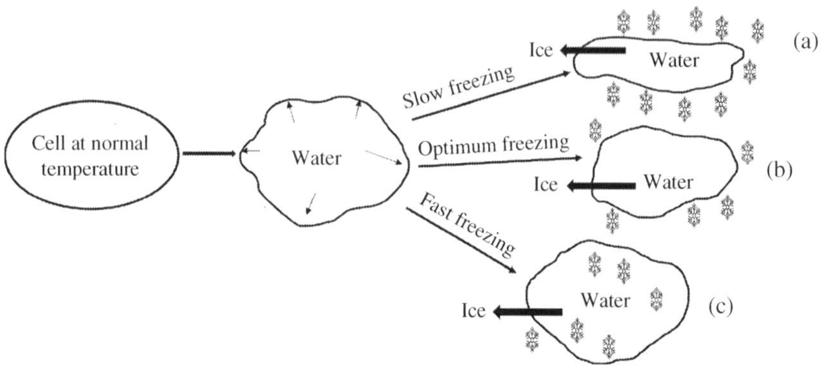

Figure 5.1 Changes in the cells at different freezing rates. (a) Slow freezing rate, (b) optimal freezing rate, and (c) fast freezing rate. Adapted from Santivarangkna et al. (2008).

5.3.1.1 Different Cryoprotectants Used During Freeze-Drying

The optimum formulation and protocol for preparing freeze-dried LAB culture depends on the species and strains (Carvalho et al. 2004). The composition of the drying medium can control the temperatures of the bacterial cells and solutes, and optimizes the viability of the cells during and after freeze-drying as well as in storage (Fonseca et al. 2021). To withstand the effects of freezing, several cryoprotectants are used in the culture medium. Sugars (glucose, fructose, sucrose, trehalose, etc.), polyols (glycerol, mannitol, etc.), and polysaccharides (maltodextrin, fructooligosaccharides, etc.) are used as cryoprotectants. These compounds can help in controlling the freezing point depression, rate of ice crystal formation, nucleation of ice crystals, and mobility of water (Gardiner et al. 2000). They also help in reducing the stress of ice crystals or bacterial cells (Coulibaly et al. 2018). Cryoprotectants can also work as water substitutes in the drying stage of unfrozen water. They can stabilize the intracellular proteins by binding with membrane phospholipids (Leslie et al. 1995). For additional effects, antioxidants (vitamin C, sodium ascorbate, etc.) and surfactants are also used as cryoprotectants (Fonseca et al. 2021).

5.3.2 Application of Freeze-Drying on Colostrum

Colostrum is the mammary gland secretion of cows immediately within 24–72 hours after parturition. It is rich in bioactive components like lactoferrin, immunoglobulins (Ig), lactoperoxidase, oligosaccharides, and growth factors which help in the vitality, growth, and passive immunity of the calf (Playford et al. 2000; Santos et al. 2017). The excess colostrum after feeding the calves can be processed and utilized. The bioactive components in colostrum make it attractive for functional foods and dietary supplement applications (Gapper et al. 2007; Borad and Singh 2018). Freeze-drying is a preferable technique to preserve colostrum and for adding it to other products as an ingredient. The low temperature and rapid sublimation of freeze-drying help prevent the denaturation of protein and preserve sensitive biological components. Abd El-Fattah et al. (2014) freeze-dried bovine colostrum at −40 °C for 10 hours. They found freeze-drying has no significant effect on the concentration of IgG, IgM, IGF-1 (insulin-like growth factor-1), and lactoferrin after production, but there was a significant reduction in IgG after storage at 7 °C for three months. Elfstrand et al. (2002) reported a reduction in immunoglobulin, TGF-β2, and growth factors amount after the freeze-drying of colostrum whey. The high cost of freeze-drying is a critical factor for using it at the farm level to preserve colostrum (Denholm 2022).

5.3.3 Application of Freeze-Drying on Lactoferrin

Lactoferrin is a non-heme iron-binding glycoprotein commonly found in milk, tear, saliva, and other peripheral fluid. It is present in higher concentrations in human milk (2.00–7.00 g/l) and bovine milk (0.02–0.2 g/l) (Masson and Heremans 1971). The lactoferrin obtained from bovine and human milk has similar functionality and is known to have antibacterial, antioxidant, anti-inflammatory, and anticarcinogenic effects and stimulates the growth of beneficial microbiota in the gut (Schanbacher et al. 1993; Wakabayashi et al. 2006). Due to these properties, lactoferrin is widely used in functional foods, infant formula, fermented milk, nutraceuticals, and even cosmetics (Tomita et al. 2009; Brock 2002).

Lactoferrin has three forms depending on the iron saturation level named apo- (0.9% iron content), native- (12.9%), and holo- (99.7%). The physicochemical, rheological, and biological properties of lactoferrin are dependent on their form (Bokkhim et al. 2013). Wang et al. (2017) produced powdered lactoferrin using freeze-drying. They used commercially available liquid bovine lactoferrin solution as the feed material and prefreezing was done at −30 °C for 24 hours followed by drying at 20 °C for 6 hours. They reported negligible denaturation of lactoferrin after drying due to low temperature, but there was a loss in antioxidant activity. Morel et al. (2022) freeze-dried lactoferrin in an industry-scale dryer at −40 °C and found no significant effect on denaturation or iron binding capacity. Hence, we can say that the lactoferrin produced by freeze-drying has similar characteristics as the native lactoferrin. Several researchers have applied freeze-drying for drying various bioactive compounds which are discussed in Table 5.2.

Freeze-drying is applied in combination with other existing techniques for wider application. One of the most important combinations is spray freeze-drying.

Table 5.2 Application of freeze-drying in different bioactive products.

Product	Parameters	Properties	Reference
Milk fat globule membrane (MFGM) isolated from cheese whey	MFGM was suspended in water and freeze-dried	The freeze-dried sample showed higher oxidative stability than spray-dried sample. The phase-transition temperature for freeze-dried MFGM was 37–38 °C, but it was 48 °C in the spray-dried samples.	Zhu and Damodaran (2011)
Human milk	Human milk was frozen at −40 °C for 9 h followed by freeze-drying under vacuum 30 mTorr, at −85 °C and secondary heating at 20, 30, and 40 °C	Freeze-drying at 30 °C can retain the highest content of IgA after freeze-drying. The freeze-dried human milk powder showed a glass transition temperature above 60 °C, so can be stored at ambient temperature with moisture barrier packaging material.	Castro-Albarrán et al. (2016)
Human milk	Human milk was pasteurized at 62.5 °C for 30 min and then freeze-dried at ≤−70 °C and ≤1.33 Pa for 24 h	Human milk oligosaccharide content was not affected by pasteurization or freeze-drying. But the presence of large differences in oligosaccharides content might be a reason for that.	Hahn et al. (2017)
Vitamin D2-encapsulated casein micelle suspension	The solution was frozen at −20 °C	The freeze-dried casein micelles showed higher recovery of vitamin D2 (80%) than spray drying, and they were stable during 4 mo of storage.	Moeller et al. (2018)
Saccharomyces cerevisiae encapsulated in skim milk	The emulsion was sprayed followed by freezing at −30 °C	*S. cerevisiae* encapsulated using 11% skim milk powder showed 76% survival rate.	Cao et al. (2020)
Lactobacillus plantarum encapsulated with denatured whey protein	Feed solution containing bacteria was spray freeze-dried at −10 °C under 107 Pa pressure	Denatured whey protein isolates and fructooligosaccharides as wall material can show 94% encapsulation efficiency in spray freeze-drying process.	Rajam and Anandharamakrishnan (2015)

5.4 Spray Freeze-Drying

Spray drying is the most extensively used method for drying dairy ingredients. However, spray drying may affect the physicochemical properties negatively. The higher temperature (>100 °C) can cause denaturation of the proteins and develops insolubility (Baldwin 2010; Haque and Adhikari 2015). For heat-sensitive products, spray freeze-drying can be used as an alternative to spray drying. Spray freeze-drying is three-step process that involves the atomization of feed into droplets, freezing those droplets in contact with a cryogen, and sublime them at low temperature and pressure (Leuenberger 2002).

5.4.1 Parameter of Spray Freeze-Drying

In the first step of spray freeze-drying, the bulk feed material is converted to droplets through atomization. Different kinds of automizers are used for spraying such as hydraulic, pneumatic, or ultrasonic. The freeze-drying after atomization does not affect the size distribution of the droplets. So, particle formation is very important in the atomization stage as it is reflected in the particle size of the end product (Ishwarya et al. 2015). After atomization, the droplets are frozen after encountering cryogens. The freezing of the particles goes through different stages: (a) initially the droplets are cooled to a super low temperature below the normal freezing point, (b) then ice crystals start growing followed by nucleation, (c) next, the crystal growth slows down due to reduced heat transfer from the cryogenic fluid, and some freezing point depression might occur at this stage, (5) finally, once freezing is complete, the frozen particles cool down to the cryogen temperature (Ishwarya et al. 2015). A schematic diagram of the spray freeze-dryer is provided in Figure 5.2.

5.4.1.1 Several Stages Are Involved in Atomizing the Product During Spray Freeze-Drying

Spray Freezing into Vapor (SPV) In this method, droplets are frozen using the spray of cold desiccated gas. The freezing and nucleation rates influence the particles' morphology (Khwanpruk et al. 2008).

Spray Freezing into Vapor into Liquid (SPV/L) In SPV/L, the atomized droplets are sprayed on a boiling cryogenic liquid and they start solidifying in contact with the vapors and freeze in contact with the liquid (Buxton and Peach 1984).

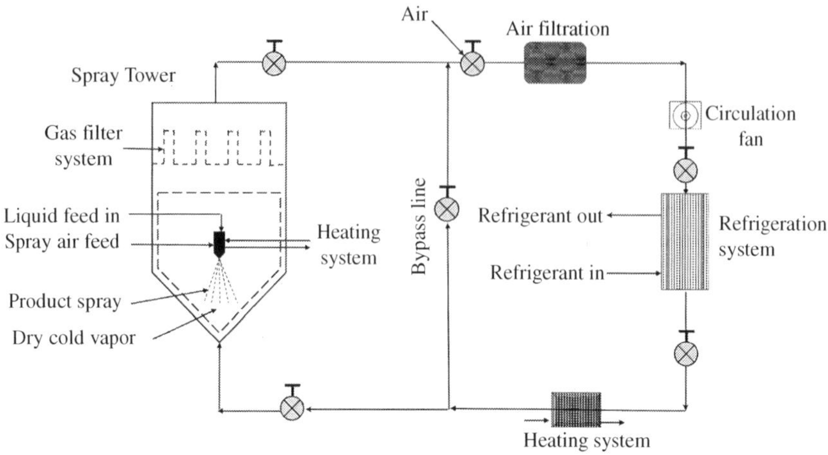

Figure 5.2 Schematic diagram of atmospheric spray freeze-drying apparatus. Adapted from Leuenberger (2002).

Spray Freezing into Liquid (SFL) In this method, the spray nozzle is inserted below the surface of the cryogenic liquid. The droplets freeze instantly after formation. As the atomization of the feed takes place in a liquid environment, small size particles are formed as much higher density and viscosity of a liquid is encountered instead of a gaseous environment (Hu et al. 2003). The increased surface area and high heat transfer coefficient increase the freezing rate. The rapid freezing prevents phase separation in the feed and reduces protein denaturation (Engstrom et al. 2007).

5.4.1.2 Freeze-Drying

The last step is the drying of the frozen particles. There are three methods of drying.

Conventional Freeze-Drying The suspended frozen droplets are dried by a lyophilizer using conventional freeze-drying. The primary drying stage takes place at low pressure and at a temperature lower than the product collapse temperature (T_c) to maintain the structure. The secondary stage drying is performed to remove bound moisture in the evaporation stage.

Atmospheric Freeze-Drying Conventional freeze-drying is performed under a vacuum which increases the operating cost. Hence, the idea of freeze-drying at atmospheric pressure was introduced. The drying rate in freeze-drying is dependent on the temperature of the ice and the vapor pressure gradient between the drying sample and the site of vapor formation (Engstrom et al. 2007). By using this theory, atmospheric freeze-drying was introduced. This method uses a cold gas flow (air, nitrogen, or helium) for the sublimation of moisture at atmospheric pressure (Mumenthaler and Leuenberger 1991). The same group also reported fine, free-flowing powder with high solubility using atmospheric freeze-drying. The main setback of this method is the requirement of a large amount of cold gas and its circulation. Hence, subatmospheric spray freeze-drying is introduced. In subatmospheric freeze-drying, low pressure is used to decrease the density of the gas which leads to fewer requirements of freezing gas and reduces the separation of heavy and light particles. Researchers have noticed better retention of volatiles and shorter drying time when the subatmospheric condition was applied (Leuenberger et al. 2006; Khwanpruk et al. 2008). A schematic diagram of atmospheric freeze-drying process is illustrated here (Figure 5.2).

Spray freeze-drying has been applied to the dairy industry for drying several components which are discussed in Table 5.3.

5.5 Advanced Freeze-Drying Methods

Freeze-drying is used widely in different foods and pharmaceutical applications, but longer drying time and energy-consuming (Cao et al. 2018). Several advanced methods are now combined with freeze-drying to reduce drying time and energy consumption. Infrared drying, high-pressure processing, microwave drying, etc., are some of the methods which have gained interest in recent times.

5.5.1 Infrared Freeze-Drying

During freeze-drying, the sublimation step involves electric heating which takes a long time and higher energy due to low heat conductivity (Wu et al. 2019). Infrared heating is used instead of electric heating to improve efficiency. Infrared radiation is classified into regions as near-infrared (NIR), mid-infrared (MID), and far-infrared (FIR), according to their wavelengths of 0.75–1.4, 1.4–3, and 3–1000 μm respectively (Sakai and Hanzawa 1994). The wavelength varies according to the application. These electromagnetic waves do not need a medium to transport which enhances their thermal efficiency (Sui et al. 2014). Ice and organic compounds absorb the energy and increase the vibration of the molecules resulting in the sublimation of ice. During sublimation, the product temperature should not be more than the product melting temperature to prevent the collapse of the structure and shrinkage (Duan et al. 2007). Antal et al. (2017) reported that mid-infrared (2.4–3 μm) increased the sublimation rate and temperature rose faster

Table 5.3 Effect of spray freeze-drying on various dairy products.

Product	Parameter	Effects on properties	Reference
Homogenized whole milk and reconstituted skim milk	The feed was sprayed by 100 and 150 μm orifices at 21 °C at a rate of 2.5 ml/min and a liquid nitrogen bed was placed beneath the stream of feed.	All the spray-dried powders were highly porous with fine textures. The skim milk powder from spray freeze-drying was highly wettable than spray-dried powder. The whole milk powder from the same method were showing less sinkability and higher wetting time due to its very porous nature and the formation of fine fragments.	Rogers et al. (2008)
Whey protein isolate solution (20% w/v)	Feed was sprayed in a chamber through a nozzle and the chamber was maintained at a partial vacuum and dried using a fluidized bed dryer. The inlet gas temperature was maintained at −10, −15, and −30 °C.	In all the temperatures, large particles with fragments of small particles adhered on the surface. They reported a nondetectable loss of solubility for α-lactalbumin and a 2% loss of β-lactoglobulin after spray freeze-drying.	Anandharamakrishnan et al. (2010)
Lactobacillus plantarum was encapsulated using whey protein isolate, sodium alginate, and fructooligosaccharides	The wall material containing probiotics was sprayed on a liquid nitrogen bed. The sublimation of ice was done at −40 °C for 6 h with a gradual increase of temperature to 10 °C at 107 Pa. The secondary drying was done for 2 h at 40 Pa with an increase in temperature from −10 to 10 °C.	The spray freeze-dried microcapsules were having a fine porous structure with spherical shapes. Good encapsulation efficiency and better stability were achieved with lower processing time than conventional freeze-drying.	Rajam and Anandharamakrishnan (2015)

Cow milk with 2% and 1% lactose; 0%, 1%, 2%, 3.25%, 6%, 10%, 18%, and 33% fat; and 5%, 10%, and 20% maltodextrin	Atomization with an airflow of 5 L/min and liquid flow of 3 ml/min Freeze-drying pressure 150 mTorr $-5 \pm 5\,°C$ for 48 h.	The powder produced by spray freeze-drying had higher porosity and dispersibility than spray-dried powder at all levels of fat contents. Maltodextrin increased lactose crystallization rate during freezing and produced larger particles and decreased dispersibility.	Baldelli et al. (2022)
Micellar casein (MC)	MC (12%) solution was atomized using two nozzles at a feed rate of 6 ml/min and sprayed on a liquid nitrogen container at a height of 10 cm. Liquid nitrogen was volatilized at $-80\,°C$ for 1 h followed by vacuum freeze-drying.	The particle size of spray freeze-dried MC was 86.4, 65.0, and 20.6 μm with decreasing flow rates. The smaller particles were uniformly spherical structures, whereas the larger particles showed irregularity. The smaller particles showed less percentage of colloidal Ca^{+2} and $PO^{-3}{}_{4}$ (<50%) due to weakened hydrophobic interactions and increased dissociation of CCP. This led to improved rehydration than spray-dried micellar casein powders.	Ren et al. (2022)

than traditional freeze-drying in cubes of pears. Wu et al. (2019) dried *Cordyceps militaris* using infrared drying. The samples were frozen at −40 °C followed by drying in a chamber using a power lamp 5 mm below the sample bed. They reported that at 40, 50, 60, and 70 °C temperatures, infrared drying reduced 7.21%, 13.48%, 15.38%, and 17.78% of the drying time as compared to traditional freeze-drying methods respectively. The energy consumption was reduced by 12.09%, 11.88%, 12.12%, and 18.37% at 40, 50, 60, and 70 °C, respectively, from total energy consumption.

5.5.2 High-Pressure Freeze-Drying

High-pressure processing is a widely used nonthermal technique that involves the application of a 100–800 mPa range of pressure with or without heating. When pressure is applied to a system, it is transmitted homogenously in the system and adapts to minimize its effect. The high-pressure processing can inactivate microorganisms by affecting their cell permeability (Balasubramaniam et al. 2015). Many researchers have reported the effects of high-pressure processing on the softening of cell structures in foods and the changes in protein, enzymes, etc. (de Oliveira et al. 2015). These changes can influence the drying rate which can improve freeze-drying. The increase in the cell membrane permeability can enhance the diffusion, heat, and mass transfer during drying (Witrowa-Rajchert et al. 2014). Zhang et al. (2020) reported high-pressure treatment increased the porosity of strawberries leading to a decrease in freeze-drying time from 37 to 28 hours. Ling et al. (2020) pretreated shrimp at 550 Pa for 10 minutes followed by freeze-drying. They reported higher efficiency of HPP-assisted freeze-drying method during the reduction of immobile water due to the effect of pressure on the muscle protein.

5.5.3 Microwave Freeze-Drying (MWFD)

Microwaves are electromagnetic waves having a frequency of 200 MHz–300 GHz (Onwude et al. 2016). Microwaves are used in this method to heat the sample during the drying stage. The sublimation of ice crystals takes place when microwaves are applied to the product and water molecules directly absorb the energy followed by the formation of pores resulting in easy rehydration (Duan et al. 2010). A schematic diagram of MWFD process is provided in Figure 5.3. Plasma formation is a considerable factor while using microwave drying. When the applied powder is much higher than the absorption ability of the product, the excess energy can be ionized (Barbosa-Cánovas and Vega-Mercado 1996). Another important aspect is the dielectric loss factor of ice and water. If excess power is applied, then the ice melts instead of forming vapor, resulting in the absorption

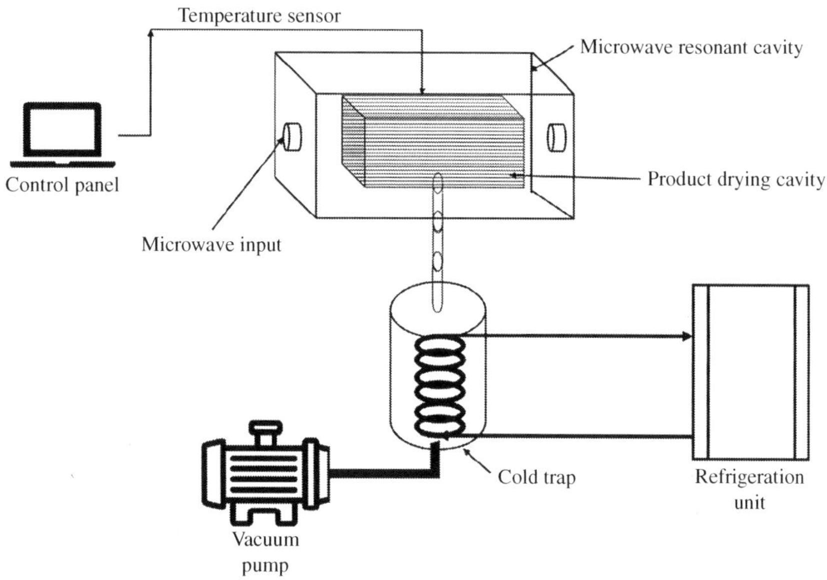

Figure 5.3 Schematic diagram of microwave freeze-drying. Adapted from Duan et al. (2016).

of more energy in the liquid stage due to the higher dielectric loss factor. This can lead to the overheating of the product (Chandrasekaran et al. 2013). Duan et al. (2007) reported that at the desorption stage of drying the temperature of the product rises rapidly when the moisture content is low. Thus, they suggested high microwave power during sublimation and lower power during the desorption stage. They also reported that MWFD can reduce the drying time by half from conventional freeze-drying without affecting product quality. Ambros et al. (2018) used MWFD for drying *Lactobacillus paracasei* ssp. *paracasei* F19 and *Bifidobacterium animalis* ssp. *lactis*. They reported MW-assisted freeze-drying can reduce the drying time to 3–5 hours instead of long hours or days from conventional freeze-drying. Lower chamber pressure is more important to maintain cell membrane integrity than the power input.

5.6 Conclusion

In conclusion, freeze-drying is a highly efficient drying technique for the dairy industry, especially for high-valued, heat-sensitive, and aroma-rich products. It is a superior alternative to conventional drying techniques like vacuum evaporation and spray drying, as it retains the nutritive value, shape, and aroma of dairy products. This makes it an ideal technology for the production of dried cheese and dairy cultures. However, to further enhance the efficiency of freeze-drying in the dairy industry, more work is required in the area of combining it with other advanced techniques such as infrared, microwave, and high-pressure processing. These techniques can help improve the efficiency and reduce the cost of freeze-drying, making it more attractive to the industry. One of the main constraints of freeze-drying is its capital and processing cost, which is why it is currently used mostly for high-valued product applications. However, if the cost and efficiency of freeze-drying can be improved, it has the potential to become a prominent technology in the processing of high-valued bioactive compounds used in different food and pharma formulations. Overall, the dairy industry can benefit greatly from the use of freeze-drying, especially for high-valued products that require preservation of their nutritional and sensory properties. Therefore, further research and investment in freeze-drying technology is crucial for the dairy industry to keep up with the demand for high-quality dairy products in the market.

References

Abd El-Fattah, A.M., Abd Rabo, F.H.R., El-Dieb, S.M., and El-Kashef, H.A.S. (2014). Preservation methods of buffalo and bovine colostrum as a source of bioactive components. *International Dairy Journal* 39 (1): 24–27.

Al-Kadamany, E., Khattar, M., Haddad, T., and Toufeili, I. (2003). Estimation of shelf-life of concentrated yogurt by monitoring selected microbiological and physicochemical changes during storage. *LWT – Food Science and Technology* 36 (4): 407–414.

Ambros, S., Mayer, R., Schumann, B., and Kulozik, U. (2018). Microwave-freeze drying of lactic acid bacteria: influence of process parameters on drying behavior and viability. *Innovative Food Science & Emerging Technologies* 48: 90–98.

Anandharamakrishnan, C., Rielly, C.D., and Stapley, A.G. (2008). Loss of solubility of α-lactalbumin and β-lactoglobulin during the spray drying of whey proteins. *LWT – Food Science and Technology* 41 (2): 270–277.

Anandharamakrishnan, C., Rielly, C.D., and Stapley, A.G. (2010). Spray-freeze-drying of whey proteins at sub-atmospheric pressures. *Dairy Science & Technology* 90 (2): 321–334.

Antal, T., Tarek-Tilistyák, J., Cziáky, Z., and Sinka, L. (2017). Comparison of drying and quality characteristics of pear (*Pyrus communis* L.) using mid-infrared-freeze drying and single stage of freeze drying. *International Journal of Food Engineering* 13 (4).

Balasubramaniam, V.M., Martinez-Monteagudo, S.I., and Gupta, R. (2015). Principles and application of high pressure-based technologies in the food industry. *Annual Review of Food Science and Technology* 6: 435–462.

Baldelli, A., Oguzlu, H., Liang, D.Y. et al. (2022). Spray freeze drying of dairy products: effect of formulation on dispersibility. *Journal of Food Engineering* 335: 111191.

Baldwin, A.J. (2010). Insolubility of milk powder products – a mini review. *Dairy Science & Technology* 90 (2): 169–179.

Barbosa-Cánovas, G.V. and Vega-Mercado, H. (1996). *Dehydration of Foods*. Springer Science & Business Media.

Bedir, T.B. and Kuleaşan, H. (2019). Determination of microbial properties of freeze dried traditional cheese. *Turkish Journal of Agriculture-Food Science and Technology* 7 (4): 688–692.

Bokkhim, H., Bansal, N., GrØndahl, L., and Bhandari, B. (2013). Physico-chemical properties of different forms of bovine lactoferrin. *Food Chemistry* 141 (3): 3007–3013.

Borad, S.G. and Singh, A.K. (2018). Colostrum immunoglobulins: processing, preservation and application aspects. *International Dairy Journal* 85: 201–210.

Brainy Insight, (2023) https://www.globenewswire.com/news-release/2023/01/09/2584832/0/en/Freeze-Dried-Food-Market-Size-to-Reach-USD-71-7-Billion-by-2030-Says-The-Brainy-Insights.html#:~:text=Newark%2C%20Jan.products%2C%20with%20long%20shelf%20life. Accessed on 04.30.2023.

Broadbent, J.R. and Lin, C. (1999). Effect of heat shock or cold shock treatment on the resistance of *Lactococcus lactis* to freezing and lyophilization. *Cryobiology* 39 (1): 88–102.

Brock, J.H. (2002). The physiology of lactoferrin. *Biochemistry and Cell Biology* 80 (1): 1–6.

Buxton, I. R. and Peach, J. M. (1984). *U.S. Patent No. 4,470,202*. Washington, DC: U.S. Patent and Trademark Office.

Cao, X., Zhang, M., Mujumdar, A.S. et al. (2018). Effects of ultrasonic pretreatments on quality, energy consumption and sterilization of barley grass in freeze drying. *Ultrasonics Sonochemistry* 40: 333–340.

Cao, L., Xu, Q., Xing, Y. et al. (2020). Effect of skimmed milk powder concentrations on the biological characteristics of microencapsulated *Saccharomyces cerevisiae* by vacuum-spray-freeze-drying. *Drying Technology* 38 (4): 476–494.

Carvalho, A.S., Silva, J., Ho, P. et al. (2004). Relevant factors for the preparation of freeze-dried lactic acid bacteria. *International Dairy Journal* 14 (10): 835–847.

Castro-Albarrán, J., Aguilar-Uscanga, B.R., Calon, F. et al. (2016). Spray and freeze drying of human milk on the retention of immunoglobulins (IgA, IgG, IgM). *Drying Technology* 34 (15): 1801–1809.

Chandrasekaran, S., Ramanathan, S., and Basak, T. (2013). Microwave food processing – a review. *Food Research International* 52 (1): 243–261.

Conde-Islas, A.Á., Jiménez-Fernández, M., Cantú-Lozano, D. et al. (2019). Effect of the freeze-drying process on the physicochemical and microbiological properties of Mexican kefir grains. *Processes* 7 (3): 127.

Coulibaly, I., Kouassi, E.K., N'guessan, E. et al. (2018). Lyophilization (drying method) cause serious damages to the cell viability of lactic acid bacteria. *Annual Research & Review in Biology* 1–15.

Dairy Reporter (2016) https://www.dairyreporter.com/Article/2016/10/14/What-are-the-top-freeze-dried-dairy-products Accessed on 04.30.2023.

de Oliveira, M.M., Tribst, A.A.L., Júnior, B.R.D.C.L. et al. (2015). Effects of high pressure processing on cocoyam, Peruvian carrot, and sweet potato: changes in microstructure, physical characteristics, starch, and drying rate. *Innovative Food Science & Emerging Technologies* 31: 45–53.

Denholm, K. (2022). A review of bovine colostrum preservation techniques. *Journal of Dairy Research 89* (4): 345–354. https://doi.org/10.1017/S0022029922000711.

Deshwal, G.K., Singh, A.K., Kumar, D., and Sharma, H. (2020). Effect of spray and freeze drying on physico-chemical, functional, moisture sorption and morphological characteristics of camel milk powder. *LWT* 134: 110117.

Dimitrellou, D., Kandylis, P., and Kourkoutas, Y. (2016). Effect of cooling rate, freeze-drying, and storage on survival of free and immobilized *Lactobacillus casei* ATCC 393. *LWT – Food Science and Technology* 69: 468–473.

Duan, X., Zhang, M., and Mujumdar, A.S. (2007). Studies on the microwave freeze drying technique and sterilization characteristics of cabbage. *Drying Technology* 25 (10): 1725–1731.

Duan, X., Zhang, M., Mujumdar, A.S., and Wang, R. (2010). Trends in microwave-assisted freeze drying of foods. *Drying Technology* 28 (4): 444–453.

Duan, X., Yang, X., Ren, G. et al. (2016). Technical aspects in freeze-drying of foods. *Drying Technology* 34 (11): 1271–1285.

Elfstrand, L., Lindmark-Månsson, H., Paulsson, M. et al. (2002). Immunoglobulins, growth factors and growth hormone in bovine colostrum and the effects of processing. *International Dairy Journal* 12 (11): 879–887.

Engstrom, J.D., Simpson, D.T., Cloonan, C. et al. (2007). Stable high surface area lactate dehydrogenase particles produced by spray freezing into liquid nitrogen. *European Journal of Pharmaceutics and Biopharmaceutics* 65 (2): 163–174.

Erbay, Z., Koca, N., Kaymak-Ertekin, F., and Ucuncu, M. (2015). Optimization of spray drying process in cheese powder production. *Food and Bioproducts Processing* 93: 156–165.

Fonseca, F., Cenard, S., and Passot, S. (2015). Freeze-drying of lactic acid bacteria. In: *Cryopreservation and Freeze-Drying Protocols*, 477–488. New York, NY: Springer.

Fonseca, F., Girardeau, A., and Passot, S. (2021). Freeze-drying of lactic acid bacteria: a stepwise approach for developing a freeze-drying protocol based on physical properties. In: *Cryopreservation and Freeze-Drying Protocols*, 703–719. New York, NY: Humana.

Gao, D. and Critser, J.K. (2000). Mechanisms of cryoinjury in living cells. *ILAR Journal* 41 (4): 187–196.

Gapper, L.W., Copestake, D.E., Otter, D.E., and Indyk, H.E. (2007). Analysis of bovine immunoglobulin G in milk, colostrum and dietary supplements: a review. *Analytical and Bioanalytical Chemistry* 389 (1): 93–109.

Gardiner, G.E., O'sullivan, E., Kelly, J. et al. (2000). Comparative survival rates of human-derived probiotic *Lactobacillus paracasei* and *L. salivarius* strains during heat treatment and spray drying. *Applied and Environmental Microbiology* 66 (6): 2605–2612.

Gatti, M., Bottari, B., Lazzi, C. et al. (2014). Invited review: microbial evolution in raw-milk, long-ripened cheeses produced using undefined natural whey starters. *Journal of Dairy Science* 97 (2): 573–591.

Hahn, W.H., Kim, J., Song, S. et al. (2017). The human milk oligosaccharides are not affected by pasteurization and freeze-drying. *The Journal of Maternal-Fetal & Neonatal Medicine* 32 (6): 985–991.

Hansen, E.B. (2002). Commercial bacterial starter cultures for fermented foods of the future. *International Journal of Food Microbiology* 78 (1–2): 119–131.

Haque, M.A. and Adhikari, B. (2015). Drying and denaturation of proteins in spray drying process. *Handbook of Industrial Drying* 33 (10): 971–985.

Hazlett, R., Schmidmeier, C., and O'Mahony, J.A. (2021). Approaches for improving the flowability of high-protein dairy powders post spray drying – A review. *Powder Technology* 388: 26–40.

Hu, J., Johnston, K.P., and Williams, R.O. III (2003). Spray freezing into liquid (SFL) particle engineering technology to enhance dissolution of poorly water soluble drugs: organic solvent versus organic/aqueous co-solvent systems. *European Journal of Pharmaceutical Sciences* 20 (3): 295–303.

Ishwarya, S.P., Anandharamakrishnan, C., and Stapley, A.G. (2015). Spray-freeze-drying: a novel process for the drying of foods and bioproducts. *Trends in Food Science & Technology* 41 (2): 161–181.

Izadi, Z., Mohebbi, M., Shahidi, F. et al. (2020). Cheese powder production and characterization: a foam-mat drying approach. *Food and Bioproducts Processing* 123: 225–237.

Jennings, T.A. (1999). *Lyophilization: Introduction and Basic Principles*. CRC press.

Khwanpruk, K., Anandharamakrishnan, C., Rielly, C., and Stapley, A. (2008). *16th International Drying Symposium (IDS 2008)*, 1066–1072.

Kourkoutas, Y., Kandylis, P., Panas, P. et al. (2006). Evaluation of freeze-dried kefir coculture as starter in feta-type cheese production. *Applied and Environmental Microbiology* 72 (9): 6124–6135.

Kumar, P. and Mishra, H.N. (2004). Yoghurt powder – a review of process technology, storage and utilization. *Food and Bioproducts Processing* 82 (2): 133–142.

Leslie, S.B., Israeli, E., Lighthart, B. et al. (1995). Trehalose and sucrose protect both membranes and proteins in intact bacteria during drying. *Applied and Environmental Microbiology* 61 (10): 3592–3597.

Leuenberger, H. (2002). Spray freeze-drying – the process of choice for low water soluble drugs? *Journal of Nanoparticle Research* 4 (1): 111–119.

Leuenberger, H., Plitzko, M., and Puchkov, M. (2006). Spray freeze drying in a fluidized bed at normal and low pressure. *Drying Technology* 24 (6): 711–719.

Ling, J.G., Xuan, X.T., Yu, N. et al. (2020). High pressure-assisted vacuum-freeze drying: a novel, efficient way to accelerate moisture migration in shrimp processing. *Journal of Food Science* 85 (4): 1167–1176.

Masson, P.L. and Heremans, J.F. (1971). Lactoferrin in milk from different species. *Comparative Biochemistry and Physiology* 1: 119–129.

Mazur, P., Leibo, S.P., and Chu, E.H.Y. (1972). A two-factor hypothesis of freezing injury: evidence from Chinese hamster tissue-culture cells. *Experimental Cell Research* 71 (2): 345–355.

Mazur, P. and Rigopoulos, N. (1983). Contributions of unfrozen fraction and of salt concentration to the survival of slowly frozen human erythrocytes: influence of warming rate. *Cryobiology* 20 (3): 274–289.

Moeller, H., Martin, D., Schrader, K. et al. (2018). Spray-or freeze-drying of casein micelles loaded with Vitamin D2: studies on storage stability and in vitro digestibility. *LWT – Food Science and Technology* 97: 87–93.

Mohammadi-Gouraji, E., Sheikh-Zeinoddin, M., and Soleimanian-Zad, S. (2017). Effects of Persian gum and gum Arabic on the survival of *Lactobacillus plantarum* PTCC 1896, *Escherichia coli*, *Xanthomonas axonopodis*, and *Saccharomyces cerevisiae* during freeze drying. *British Food Journal* 119 (2): 331–341.

Morel, J., Zain, S.N.M., and Archer, R. (2022). Comparison of drying techniques for bovine lactoferrin: iron binding and antimicrobial properties of dried lactoferrin. *International Dairy Journal* 124: 105142.

Motta, A. and Gomes, M.D.S.M. (2015). *Technological and Functional Properties of Lactic Acid Bacteria: The Importance of These Microorganisms for Food*. Revista do Instituto de Latícinios Cândido Tostes.

Muldrew, K. and McGann, L.E. (1990). Mechanisms of intracellular ice formation. *Biophysical Journal* 57 (3): 525–532.

Muldrew, K. and McGann, L.E. (1994). The osmotic rupture hypothesis of intracellular freezing injury. *Biophysical Journal* 66 (2): 532–541.

Mumenthaler, M. and Leuenberger, H. (1991). Atmospheric spray-freeze drying: a suitable alternative in freeze-drying technology. *International Journal of Pharmaceutics* 72 (2): 97–110.

Obradović, N., Volić, M., Nedović, V. et al. (2022). Microencapsulation of probiotic starter culture in protein–carbohydrate carriers using spray and freeze-drying processes: implementation in whey-based beverages. *Journal of Food Engineering* 321: 110948.

Onwude, D.I., Hashim, N., and Chen, G. (2016). Recent advances of novel thermal combined hot air drying of agricultural crops. *Trends in Food Science & Technology* 57: 132–145.

Palmfeldt, J. and Hahn-Hägerdal, B. (2000). Influence of culture pH on survival of *Lactobacillus reuteri* subjected to freeze-drying. *International Journal of Food Microbiology* 55 (1–3): 235–238.

Papavasiliou, G., Kourkoutas, Y., Rapti, A. et al. (2008). Production of freeze-dried kefir culture using whey. *International Dairy Journal* 18 (3): 247–254.

Passot, S., Cenard, S., Douania, I. et al. (2012). Critical water activity and amorphous state for optimal preservation of lyophilised lactic acid bacteria. *Food Chemistry* 132 (4): 1699–1705.

Písecký, J. (2005). Spray drying in the cheese industry. *International Dairy Journal* 15 (6–9): 531–536.

Playford, R.J., Macdonald, C.E., and Johnson, W.S. (2000). Colostrum and milk-derived peptide growth factors for the treatment of gastrointestinal disorders. *The American Journal of Clinical Nutrition* 72 (1): 5–14.

Rajam, R. and Anandharamakrishnan, C. (2015). Spray freeze drying method for microencapsulation of *Lactobacillus plantarum*. *Journal of Food Engineering* 166: 95–103.

Ren, J., Liao, M., Ma, L. et al. (2022). Effect of spray freeze drying on the structural modification and rehydration characteristics of micellar casein powders. *Innovative Food Science & Emerging Technologies* 80: 103093.

Rogers, S., Wu, W.D., Saunders, J., and Chen, X.D. (2008). Characteristics of milk powders produced by spray freeze drying. *Drying Technology* 26 (4): 404–412.

Sakai, N. and Hanzawa, T. (1994). Applications and advances in far-infrared heating in Japan. *Trends in Food Science & Technology* 5 (11): 357–362.

Sakin-Yilmazer, M., Koç, B., Balkir, P., and Kaymak-Ertekin, F. (2014). Rheological behavior of reconstituted yoghurt powder – An optimization study. *Powder Technology* 266: 433–439.

Santivarangkna, C., Kulozik, U., and Foerst, P. (2008). Inactivation mechanisms of lactic acid starter cultures preserved by drying processes. *Journal of Applied Microbiology* 105 (1): 1–13.

Santos, G.D., Silva, J.T.D., Santos, F.H.D.R., and Bittar, C.M.M. (2017). Nutritional and microbiological quality of bovine colostrum samples in Brazil. *Revista Brasileira de Zootecnia* 46: 72–79.

Santos, G.D., Nunes, T.P., Silva, M.A.A.P. et al. (2018). Development and acceptance of freeze-dried yogurt "powder yogurt". *International Food Research Journal* 25 (3): 1159–1165.

Schanbacher, F.L., Goodman, R.E., and Talhouk, R.S. (1993). Bovine mammary lactoferrin: implications from messenger ribonucleic acid (mRNA) sequence and regulation contrary to other milk proteins. *Journal of Dairy Science* 76 (12): 3812–3831.

Schuck, P., le Floch-Fouere, C., and Jeantet, R. (2013). Changes in functional properties of milk protein powders: effects of vacuum concentration and drying. *Drying Technology* 31 (13–14): 1578–1591.

Searles, J.A., Carpenter, J.F., and Randolph, T.W. (2001). The ice nucleation temperature determines the primary drying rate of lyophilization for samples frozen on a temperature-controlled shelf. *Journal of Pharmaceutical Sciences* 90 (7): 860–871.

Sharma, N.K., Arora, C.P., and Mital, B.K. (1992). Influence of concentration of milk solids on freeze-drying rate of yoghurt and its quality 1. *Journal of Food Process Engineering* 15 (3): 187–198.

Sharma, N.K. and Arora, C.P. (1995). Influence of product thickness, chamber pressure and heating conditions on production rate of freeze-dried yoghurt. *International Journal of Refrigeration* 18 (5): 297–307.

Sui, Y., Yang, J., Ye, Q. et al. (2014). Infrared, convective, and sequential infrared and convective drying of wine grape pomace. *Drying Technology* 32 (6): 686–694.

Takamatsu, H. and Kumagae, N. (2002). Survival of biological cells deformed in a narrow gap. *Journal of Biomechanical Engineering* 124 (6): 780–783.

Tamime, A.Y. and Robinson, R.K. (2007). Tamime and Robinson's yoghurt. In: *Science and Technology*. Elsevier.

Tomita, M., Wakabayashi, H., Shin, K. et al. (2009). Twenty-five years of research on bovine lactoferrin applications. *Biochimie* 91 (1): 52–57.

Üçok, G. and Sert, D. (2022). Trial productions of freeze-dried *Lactobacillus plantarum* culture using dairy by-products as cryoprotectants: viability and characterization of cultures. *Food Bioscience* 46: 101541.

Venir, E., Del Torre, M., Stecchini, M.L. et al. (2007). Preparation of freeze-dried yoghurt as a space food. *Journal of Food Engineering* 80 (2): 402–407.

Vincenzetti, S., Cecchi, T., Perinelli, D.R. et al. (2018). Effects of freeze-drying and spray-drying on donkey milk volatile compounds and whey proteins stability. *LWT* 88: 189–195.

Wakabayashi, H., Yamauchi, K., and Takase, M. (2006). Lactoferrin research, technology and applications. *International Dairy Journal* 16 (11): 1241–1251.

Walstra, P., Wouters, J.T., and Geurts, T.J. (2005). Concentration process. In: *Dairy Science and Technology*, 2nde, 297–336. Boca Raton, FL: CRC Press.

Wang, Y.C., Yu, R.C., and Chou, C.C. (2004). Viability of lactic acid bacteria and bifidobacteria in fermented soymilk after drying, subsequent rehydration and storage. *International Journal of Food Microbiology* 93 (2): 209–217.

Wang, B., Timilsena, Y.P., Blanch, E., and Adhikari, B. (2017). Characteristics of bovine lactoferrin powders produced through spray and freeze drying processes. *International Journal of Biological Macromolecules* 95: 985–994.

Witrowa-Rajchert, D., Wiktor, A., Sledz, M., and Nowacka, M. (2014). Selected emerging technologies to enhance the drying process: a review. *Drying Technology* 32 (11): 1386–1396.

Wijayanti, H.B., Bansal, N., and Deeth, H.C. (2014). Stability of whey proteins during thermal processing: a review. *Comprehensive Reviews in Food Science and Food Safety* 13 (6): 1235–1251.

Wu, X.F., Zhang, M., and Bhandari, B. (2019). A novel infrared freeze drying (IRFD) technology to lower the energy consumption and keep the quality of *Cordyceps militaris*. *Innovative Food Science & Emerging Technologies* 54: 34–42.

Zhang, L., Qiao, Y., Wang, C. et al. (2020). Influence of high hydrostatic pressure pretreatment on properties of vacuum-freeze dried strawberry slices. *Food Chemistry* 331: 127203.

Zhi, N.N., Zong, K., Thakur, K. et al. (2018). Development of a dynamic prediction model for shelf-life evaluation of yogurt by using physicochemical, microbiological and sensory parameters. *CyTA Journal of Food* 16 (1): 42–49.

Zhu, D. and Damodaran, S. (2011). Composition, thermotropic properties, and oxidative stability of freeze-dried and spray-dried milk fat globule membrane isolated from cheese whey. *Journal of Agricultural and Food Chemistry* 59 (16): 8931–8938.

Zou, Z., Duley, J.A., Cowley, D.M. et al. (2022). Physicochemical properties and whey proteomes of camel milk powders produced by different concentration and dehydration processes. *Food* 11 (5): 727.

6

Freeze-Drying of Probiotics for the Incorporation in Functional Foods: Drying Process, Viability, and Powder Properties

*Yasemin Ş. Küçükata, Hatice E. Kırtıl, and Ertan Ermiş**

Department of Food Engineering, Faculty of Engineering and Natural Sciences, Istanbul Sabahattin Zaim University, Istanbul, Turkey

6.1 Introduction

Probiotics are defined as live microorganisms administered in adequate amounts as part of foods or supplements to benefit host health (Arellano et al. 2021). Since probiotics improve the intestinal microbial balance, foods containing probiotics are classified as functional foods. The beneficial impact of probiotics on human health includes alleviating inflammatory bowel disease, immune disorders, type 2 diabetes, and atherosclerosis (Arellano et al. 2021). Even though the daily dosage intake or the minimal viable cell number of probiotics is not well defined, it is suggested around $10^8\,g^{-1}$, or ml^{-1}, to exert health benefits in the gastrointestinal system (Her et al. 2015).

Fermented foods (i.e. fermented dairy products) are natural sources of probiotics. On the other hand, freeze-dried and spray-dried probiotic powders in capsules or sachets are available in the market, which is expanding rapidly. Probiotics in the package should maintain viability during transportation and storage. Different drying techniques have been used to obtain probiotic powders. The freeze-drying (FD) method has been preferred over others, such as spray-drying (SD), for higher viability and extended shelf life (Meireles Mafaldo et al. 2022). However, FD takes a long time and is more costly than spray drying. The drying conditions need to be optimized, and appropriate protective/cryoprotectant materials should be used to maintain the cell viability as high as possible during harsh conditions in drying processing steps such as pumping (shear and oxygen stresses),

*Corresponding Author: ertan.ermis@gmail.com

Freeze Drying of Food Products: Fundamentals, Processes and Applications, First Edition.
Edited by Roji Balaji Waghmare, Manoj Kumar, and Parmjit Singh Panesar.

drying (shrinkage and ice crystals), growth to high cell densities, and grinding (shear and heat stress) after drying (Béal and Fonseca 2015; Her et al. 2015).

Microencapsulation might be employed to protect the sensitive probiotic strains using FD and wall materials such as proteins (i.e. whey protein), alginates, polysaccharides, resistant starch, and milk, during processing and gastrointestinal stress conditions (Arellano et al. 2021). Maillard reaction products can improve the solubility, emulsifying properties, and thermal stability of proteins; hence protein and polysaccharide conjugates are functional wall materials for encapsulating probiotics (Li et al. 2023). The rehydration of dried probiotics is a vital step in producing functional foods. The rehydration conditions significantly influence the viability of the probiotics in the food matrix.

This chapter aims to provide a general overview of the FD (lyophilizing) probiotics process, focusing on the significance of protective material, process optimization, characterization, and factors affecting the powder properties and viability of freeze-dried probiotics. It also reviews recent works incorporating freeze-dried probiotic powders into functional food products.

6.2 Functional Foods

Since ancient times, food has been not only for energy and alimentation but also sort of medicine and health-promoting source of life. Greek physician Hippocrates II, the great phenomenon in the history of medicine, once said, "Let food be thy medicine and let medicine be thy food" (King 2020). Food's role in life and its contribution to health through its numerous bioactive properties are gaining the utmost importance (Chugh and Kamal-Eldin 2020). The more the health consciousness develops in society, the more expectation of food efficacy increment. Therefore, there is an increased demand for functional foods (Dinkçi et al. 2019). The term functional food can be simply explained as food for supplemental health use (Lin 2003). Natural foods, foods with additional components, and foods with omitted components can all be functional foods through biotechnological procedures (Figure 6.1) (Ashaolu 2020).

Consuming bioactive and nutritious components, such as metabolites, fatty acids, probiotics, prebiotics, vitamins, polyphenols, and flavonoids, through functional foods is more effective than taking medicine or supplements (Peng et al. 2020). Increasing demand for healthy and functional foods has opened new paths for developing bioactive components like probiotics and nutraceuticals (Reque and Brandelli 2021). Among all the bioactive compounds in functional foods, food products supplemented with probiotic bacteria are of particular interest (Siciliano et al. 2021). Probiotics are best known for restoring gut microbes and controlling gastrointestinal system activities (Lin 2003). A list of probiotic microorganisms can be seen in Table 6.1.

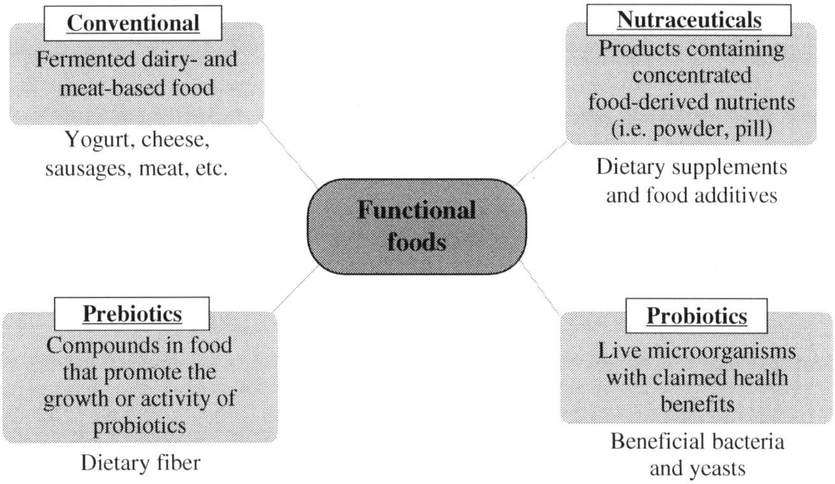

Figure 6.1 Categories of functional foods. *Source:* Adapted from Damián et al. (2022).

Table 6.1 Probiotic microorganisms (Song et al. 2012).

***Lactobacillus* species**	***Bifidobacterium* species**	**Others**
L. acidophilus	*B. adolescentis*	*Bacillus cereus*
L. amylovorus	*B. animalis*	*Clostridium butyricum*
L. brevis	*B. breve*	*Enterococcus faecalis*
L. casei	*B. bifidum*	*Enterococcus faecium*
L. rhamnosus	*B. infantis*	*Escherichia coli*
L. crispatus	*B. lactis*	*Lactococcus lactis* subsp. *cremoris*
L. delbrueckii subsp. *bulgaricus*	*B. longum*	*Lactococcus lactis* subsp. *lactis*
L. fermentum		*Leuconostoc mesenteroides* subsp. *dextranicum*
L. gasseri		*Pediococcus acidilactici*
L. helveticus		*Propionibacterium freudenreichii*
L. johnsonii		*Saccharomyces boulardii*
Lactobacillus lactis		*Streptococcus salivarius* subsp. *thermophilus*
L. paracasei		*Sporolactobacillus inulinus*
L. plantarum		
L. reuteri		
Lactobacillus salivarius		
L. gallinarum		

FAO/WHO's definition of probiotics is "live microorganisms which, when administrated in adequate amounts, confer a health benefit on the host" (FAO/WHO 2001). Foods containing a certain number of live probiotics are considered functional foods that potentially positively affect host health. The commercial interest in probiotic food products and supplements has been growing in recent years. There are numerous probiotics on the market in the form of milk, yogurt, or special freeze-dried pharmaceutical dietary preparations that are also available in tablet form (Zubillaga et al. 2001). The functional foods supplemented with probiotic bacteria directly and indirectly, impact the ecology and health of the gut through the bacteria that live there (Peng et al. 2020). Besides their increasingly growing reputation, probiotic cultures are successfully used in various food matrices to make a variety of functional foods (Table 6.2).

One of the biggest reasons for the expansion of probiotics in the functional food market is people's need for healthy alternatives to their gastrointestinal problems other than medicine and supplements (Reque and Brandelli 2021). In addition to all its favorable properties, it should be kept in mind that there are various parameters for probiotics. The stability and viability of probiotics are two crucial factors; food matrices to deliver probiotics, bacterial ability to adhere intestine, and bacterial strength to gastrointestinal conditions are also essential factors for appropriate and effective probiotic use in functional foods (Siciliano et al. 2021). Besides all, it should be emphasized that

Table 6.2 Some examples of probiotic bacteria used in food products.

Probiotic bacteria	Food products	Reference
Lactobacillus plantarum 299v *Pediococcus acidilactici* HA-6111-2	Orange powder	Barbosa et al. (2015)
Lacticaseibacillus rhamnosus GG	Apple juice, yogurt	Romero-Chapol et al. (2022)
Lactobacillus casei 01	Carrot juice	Petreska-Ivanovska et al. (2014)
Bifidobacterium longum 15708	Cheddar cheese	Amine et al. (2014)
Lactobacillus delbrueckii N102 *Latilactobacillus sakei* H1-5	Dry fermented sausages	Liu et al. (2021)
L. casei ATCC393	Fermented milk	Terpou et al. (2018)
Lactobacillus acidophilus DOWARU	Ice cream	Ferraz et al. (2012)
Bifidobacterium animalis ssp. *lactis* BB12	Kefir	González-Sánchez et al. (2010)
Lactobacillus reuteri E81	Sourdough	Ispirli et al. (2020)

functional foods fortified with probiotics should be available and affordable in the food market (Ashaolu 2020).

Overall, functional food products aim to regulate cellular health and enhance the implementation of cells' endogenous defense mechanisms (Zubillaga et al. 2001). Today, functional foods containing probiotics constitute a developing industry with significant economic interest. The ongoing interest in well-being and healthy lifestyles has led to an expansion of probiotic products in the functional food sector (Siciliano et al. 2021). To meet consumer demands, developing natural, health-supporting, and accessible probiotics and other functional foods as alternative supplements to medicines and drugs should be regarded seriously.

6.3 Cultivation and FD of Probiotics

The trend toward functional foods using probiotic bacteria is increasing as consumers realize the positive effects on health (Fonseca et al. 2020; Haindl et al. 2020; Yao et al. 2020). Lactic acid bacteria (LAB) and related genera (i.e. *Bifidobacterium* species) and some yeasts (i.e. *Saccharomyces cerevisiae*) are generally added to foods as probiotics (Chen et al. 2019; Romero-Chapol et al. 2022; Siaterlis et al. 2009; Vorländer et al. 2020). Besides the advantages of probiotic cultures in these products, some disadvantages limit their use. For this reason, adequate doses of probiotics may not be taken into the body while consuming foods containing these bacteria. During the production and storage period, the cultures may be affected by environmental conditions in the food environment and could not maintain their viability for a long time (Aschenbrenner et al. 2015; Kiepś and Dembczyński 2022). Various applications have been made in different foods to maintain the survival of probiotics from products to reach consumers. These applications include cultivation, drying, and storage period (Haindl et al. 2020; Yao et al. 2020). Cultivation of the probiotics is the first step of the whole process and may ensure to survival of the cells for drying. Specific broths, media, and optimum growth conditions (pH, temperature, etc.) for each probiotic species are used in fermentation (Haindl et al. 2020). Siaterlis et al. (2009) incubated *Lactobacillus rhamnosus* GG (ATCC 53103) and *L. plantarum* (NCIMB 8826) in modified MRS broth (using different ratios of glucose, yeast extract, and peptone) at 37 °C for 15 hours. Then the broth was centrifuged at 3200g and resuspended in PBS. Chen et al. (2019) prepared *Bifidobacterium bifidum* BB01 in MRS medium at 37 °C for 18 hours and harvested by a high-speed centrifuge at 8000 rpm for 15 minutes. In another study, kefir grains were added to UHT (ultra high temperature) milk samples, and this mixture was incubated at 26 °C for 24 hours. Then, kefir grains were washed to separate from the milk residues (Conde-Islas et al. 2019). Haindl et al. (2020) have inoculated *B. longum* ssp. *longum* Reuter

1963 to MRSc medium (MRS medium with 1 g/L cysteine) at 37 °C combined with 0.1 L/min N_2 with a mixing speed of 80 rpm (pH 6.0). Vorländer et al. (2020) activated baker's yeast *S. cerevisiae* onto yeast extract peptone dextrose (YPD) agar plates. Then the cells were suspended in PBS (isotonic phosphate-buffered saline solution, pH 7.4) for the process. *Lactobacillus plantarum* was grown in MRS broth overnight at 37 °C and 140 rpm using a shaker (Oluwatosin et al. 2021). Romero-Chapol et al. (2022) incubated *Lacticaseibacillus rhamnosus* GG cells in MRS broth at 37 °C for 24 hours at 200 rpm.

One of the most used methods recently is FD (lyophilization) as a new concept combined with the microencapsulation technique. FD dramatically increases the survival rates of probiotics in food products and in the gastrointestinal tract (GIT; Fonseca et al. 2020). Although FD is an old method used for the preparation of probiotic preparations, this method has been frequently used in microencapsulation applications in recent years (Aschenbrenner et al. 2015; Kiepś and Dembczyński 2022). FD is a very effective method for preserving sensitive bioactive materials such as probiotic microorganisms by minimizing the damage caused by temperature in living cells due water removal by sublimation (Broeckx et al. 2016). This method is based on freezing microorganisms at a shallow temperature and then drying them under a high vacuum by sublimation far below 0 °C (usually between −50 and −20 °C) (Aschenbrenner et al. 2015; Fonseca et al. 2020).

The FD process consists of three stages. These stages are sublimation-based freezing (crystallization water) and primary (sublimation ice) and secondary drying (desorption water) processes (Aschenbrenner et al. 2015; Broeckx et al. 2016; Rajam and Subramanian 2022). FD's first stage is freezing (to $T \sim -50$ °C), which starts the formation of ice crystals in the liquid solution and causes ice crystals to separate water molecules from the solution. By decreasing the chamber pressure and gradually raising the shelf temperature, which starts the sublimation of ice, the ice crystals are removed from the frozen product, and unfrozen water (15–20%) remains in the product during primary drying (to $T \sim -35/-20$ °C). This unfrozen water is desorbed in secondary drying (H bonds breakage) by adjusting the chamber's pressure and temperature. The product's targeted moisture content (2–10%) is finally achieved, thanks to the higher temperature under vacuum. The working principle of the method is based on the code that the material to be dried is first frozen and then dried by sublimation under a high vacuum. Due to not applying high temperatures during the process, the viability of microorganisms can be preserved at a higher rate than spray drying (Aschenbrenner et al. 2015; Broeckx et al. 2016; Rajam and Subramanian 2022). As the crystals formed due to low temperature and high osmotic pressure during the FD process affect cell viability, cryoprotectants/protective solutions are added to the material before this process (Misra et al. 2022). Different researchers have demonstrated that the results of viability are improved when this temperature is lower. Siaterlis et al. (2009) froze *L. rhamnosus* GG (ATCC 53103) and *L. plantarum* (NCIMB 8826) at −80 °C for 24 hours. After 45 hours of lyophilization, the survival rates of *L. rhamnosus* GG and *L. plantarum* ranged between 40–85% and 60–90%, respectively. *Bifidobacterium bifidum* BB01 cells were lyophilized at −50 °C, 7.23 Pa for 24 hours with a vacuum freeze-dryer, and the viability was detected to 88–90% (Chen et al. 2019). FD of the kefir grains was conducted at −20, −40, −60, or −80 °C; and pressure in 0.2, 0.4, 0.6, or 0.8 mbar (Conde-Islas et al. 2019). In another study, the conditions were at 3700 Pa, 24 hours for *B. longum* ssp. *longum* Reuter 1963 after prefreezing at −80 °C overnight (Haindl et al. 2020). Lyophilization for *S. cerevisiae* was carried out at three different freezing temperatures (−20 and −35 °C in the freezer and −196 °C in liquid nitrogen), stored at −20, −35, or −80 °C for 24 hours and primary drying took 24 hours at 0.220 mbar. For secondary drying, the pressure was 0.002 mbar (Vorländer et al. 2020). *Lactobacillus plantarum* suspensions were frozen at −80 °C for 24 hours, and then they were moved to a freeze-dryer that was ran for the same amount of time at −35 °C and 300 mTorr (Oluwatosin et al. 2022).

In addition to the advantages of the FD method, such as reducing aroma losses, excellent reconstitution properties of the products obtained, and minimizing the losses caused by the movement of solutes in the food, there are also disadvantages, such as high cost and long processing time. While this method is more moderate than spray drying, its use in microencapsulation applications is limited due to economic reasons (Aschenbrenner et al. 2015; Fonseca et al. 2020; Kiepś and Dembczyński 2022).

6.4 Protection of Probiotics During FD

Lyophilization is used to dry microorganisms and preserves cell viability at a high rate. However, this process may cause a decrease in viability because of the denaturation of some sensitive proteins due to the high osmotic pressure during the freezing stage (Broeckx et al. 2016; Fonseca et al. 2020; Yao et al. 2020). The negative effects of FD can be reduced by adding some cryoprotectants. Appropriate cryoprotectants are added to the medium, facilitating the adaptation of probiotics and accumulating in the cell, reducing the osmotic pressure difference with the external environment. The cryoprotectant used should be of food origin, nontoxic, easy to find, cheap, and edible (Aschenbrenner et al. 2015). Polysaccharide, protein, and lipid-based drying matrices are important to utilize for the probiotics during FD, dehydration, storage, and gastrointestinal digestion. For this purpose, polymers such as bacterial cellulose, alginate, skimmed milk powder, whey proteins, betaine, trehalose, adonitol, lactose, dextran, glycerol, and polyethylene glycerol are some of the cryoprotectants used (Bagad et al. 2017; Gwak et al. 2015; Rajam and Subramanian 2022).

Oluwatosin et al. (2021) used 10% of skimmed milk, inulin, maltodextrin, and sucrose as cryoprotectants for the FD of *L. plantarum*. They detected that skimmed milk demonstrated the highest survival (91%), while inulin showed the least (<1% cell survival). Chen et al. (2019) added 5.5% glycine, 0.8%, sodium bicarbonate, 7% xylooligosaccharides, 4.5% arginine, and 25% skim milk to the process as cryoprotectants. The survival rate of *B. bifidum* BB01 was determined as 88–90%. This research showed that optimizing composite cryoprotectant for freeze-dried powder successfully protected *B. bifidum* cells. However, this does not provide all organisms with the same degree of viability. Due to differences in the colony morphology and the cell membrane, *Lactobacillus* is adversely affected during lyophilization compared to *Bifidobacterium* (Chen et al. 2019). This difference in vitality preservation is observed even in the presence of cryoprotectants. Haindl et al. (2020) declared that *B. longum* cells were inactivated during drying; however, storage at +4 °C with the addition of 75% maltodextrin relative to bacterial dry mass totally prevented cell loss. Besides the addition of protective molecules to the medium, adaptation of bacterial cells during the fermentation process (prestressing factors before FD such as acid, temperature, and media composition), process parameter management (the choice of operating conditions such as freezing kinetics), final water content, and storage conditions are the defense mechanisms created to increase bacterial viability during FD (Aschenbrenner et al. 2015; Broeckx et al. 2016). It is also known that prestressing cells before FD boosts cell membrane integrity (Oluwatosin et al. 2022). Cryoprotectants can restrain the negative effects of membrane integrity loss caused by vitrification or glass formation (Broeckx et al. 2016). Haindl et al. (2020) determined that the effects causing viability losses were better understood with measurements of the glass transition temperature and membrane integrity. For *B. longum* Reuter 1963, cultivation at pH 6.0 caused more durable cell membranes, resulting in the preservation of membranes of 43% after drying as opposed to cultivation at free acidification, which preserved only 1.0% of membranes after drying. They found a direct correlation between survival rate and remaining cell membrane integrity under various cultivation conditions. In this study, the best process parameters were cultivation at pH 6.0, a drying process of 3700 Pa, 24 hours, and storage temperatures at −10, +10, and +35 °C. In addition, the survival rate of probiotics during drying depends on the composition of the growth medium and the presence of carbohydrates.

Siaterlis et al. (2009) investigated the effects of some saccharides on the growth of human-derived *L. plantarum* and *L. rhamnosus* GG. Sucrose provided more protection from the studied cryoprotectants than trehalose and sorbitol. The growth of *L. plantarum* tested using 2% of the saccharides in glucose-free MRS and its prebiotic potential was identified. Sucrose and inulin both encouraged cell proliferation, with sucrose doing so better than either (Oluwatosin et al. 2022). The protective effect varies depending on the type, composition, combination, and sugar ratio. The encapsulation technique is also used for freeze-dried probiotics for this purpose. The wall material used in encapsulation provides a physical barrier around the cells to protect the viability of the microorganisms against adverse environmental conditions (thermal, osmotic, mechanical, and oxidative). In this method, a protective film or coating layer is formed around the active microorganism with various substances. With a successful application, the spoilage that microorganisms can cause in the product is prevented, and microorganisms can preserve their vitality at the highest level during storage (Haindl et al. 2020; Yao et al. 2020). The temperature at which the probiotics are frozen is a crucial process parameter that can significantly affect how viable probiotics are during FD.

Previous research demonstrated that freezing probiotic bacteria at lower temperatures could result in higher freezing rates and smaller ice crystals, preventing cellular harm. Additionally, ice crystals during freezing can lead to mechanical and osmotic stresses that compromise the integrity of cell membranes, and the subsequent removal of bound water during desorption can destabilize cellular components like phospholipids and proteins to result in additional viability losses (Aschenbrenner et al. 2015; Broeckx et al. 2016; Rajam and Subramanian 2022). Wang et al. (2020) used C18:1 to increase the survivability of *L. plantarum* during FD. C18:1 effectively maintains the integrity and fluidity of the cell membrane. It may act as a cryoprotectant to preserve the fluidity and integrity of the cell membrane, hence improving *L. plantarum*'s survival rate after FD. This investigation served as a defense mechanism for bacterial survival.

6.5 Stability and Viability Assessment After FD

Probiotics are live microbial supplements that have been shown to have health benefits (Rajam and Subramanian 2022) and can be found in various foods such as non-fermented milk products, fruit juices, and cereals (Saarela et al. 2006). Probiotic bacteria need to meet the satisfaction of safety and functional criteria. On the other hand, from a technical perspective, they should have high cell intensity and maintain viability during the drying processes, storage conditions, and GIT (Ampatzoglou et al. 2010). There are numerous challenges in developing probiotic-containing food products, such as selecting effective strains and ensuring their survival during processing and storage (Celik and O'Sullivan 2013). Encapsulated probiotic powders and probiotic-filled capsules are commercially used for probiotic supplements (Torp et al. 2022). Encapsulation of probiotics increases the survival rate and keeps the viability at recommended level of cfu/g during the storage (Her et al. 2015; Jouki et al. 2021a,b; Obradović et al. 2022).

In the pharmaceutical and food industries, FD (also known as lyophilization) is a preferred preservation technique for the long-term storage of sensitive products like probiotic bacteria (Nguyen et al. 2022). If conditions are optimized, drying cultures can be stored for a long time and transported without refrigeration (Celik and O'Sullivan 2013; Jalali et al. 2012). In the FD process, probiotics are exposed to low temperatures and desiccation (Jalali et al. 2012). While unbound and bound water are removed from the cell, the decrease in water activity during the drying process causes damage to the cellular structures and reduces bacterial viability (Tymczyszyn et al. 2012). During the freezing process, ice crystals can be formed and cause harm to the cell membrane (Meireles Mafaldo et al. 2022). Even though FD is known to be stressful for live bacteria, this technique is nonetheless regarded as a suitable strategy for providing prolonged shelf life for the most of probiotic microorganisms (Arellano-Ayala et al. 2021).

6.5.1 Factors Affecting Survival and Stability

The viability of probiotic bacteria is a crucial parameter during the FD process and storage. Oxygen concentration, water activity, temperature, and storage time affect probiotics' stability and vitality (Celik and O'Sullivan 2013). The variables affecting probiotic survival during drying and storage are discussed by Ermis (2021).

Moisture content is an important parameter affecting the survival rate after drying and the degree of inactivation during storage (Tymczyszyn et al. 2012). The required moisture content of probiotic powders was reported as 4% or below to improve the viability during storage (Heidebach et al. 2010; Ying et al. 2010). The water activity of freeze-dried probiotic powders ranged from 0.24 to 0.27 (Savedboworn et al. 2019). The moisture content, glass transition temperature (T_g), and storage temperature parameters should be taken into consideration together to evaluate the viability of probiotic cells in freeze-dried powder form. The storage temperature should not be close to the T_g to maintain the viable cell number of probiotic bacteria (Savedboworn et al. 2019).

Encapsulated and dried probiotics should be stored in an environment without light and moisture exposure (Rajam and Subramanian 2022). Water activity is one of the significant parameters affecting freeze-dried probiotics. It is essential not only for durability but also for color and solubility. Alteration in color and solubility at higher a_w can be explained by nonenzymatic browning as the Maillard reaction (Celik and O'Sullivan 2013). The survival kinetics revealed that probiotics' viability decreases more quickly over the water activity of 0.33, which is essential for keeping them alive (Rascón et al. 2018). The moisture level of freeze-dried probiotic bacteria is critical for stability and vitality during storage. The moisture content values are proportional to the cryoprotectant molecular weights, with the lowest molecular weight resulting in the highest moisture content. Generally, it is suggested that for probiotic bacteria, moisture content should be under 5% (Oluwatosin et al. 2022). Relative humidity (RH) is another parameter affecting cell viability. The cells were observed to lose their vitality as the RH and temperature increased (Tymczyszyn et al. 2012). *Lactobacillus salivarius* subsp. *salivarius* showed higher viability when stored at 2.8% and 5.6% RH than stored at 8.8% RH (Zayed and Roos 2004). Maintaining probiotics in a dry state with low water content is vital to improve long-term stability (Broeckx et al. 2016). It should be noted that 0% moisture is unnecessary because of the harmful effect of over-drying on the cells. Another reason might be attributed to the free radicals formation from lipid oxidation and membrane damage caused by phospholipid degradation leading to a decrease in the viability of freeze-dried probiotics during storage (Savedboworn et al. 2019).

6.5.2 Protective Agents to Improve Viability

The probiotic strain stability during storage is an essential concern for the functional food industry. Protecting and maintaining the viability of cells by protective agents is a crucial challenge for promoting cellular vitality during storage (Savini et al. 2010). Health-promoting impact of dried probiotic bacteria depends on high viability in foods (Santivarangkna et al. 2011). The most thoroughly researched techniques for protecting probiotic microorganisms are the inclusion of protectants and the alteration of processing parameters (Broeckx et al. 2016).

Maintaining the viability and activity of freeze-dried probiotic bacteria after storage, finding proper protective agents, and controlling the drying process are necessary (Bergenholtz et al. 2012). Cryoprotectants protect the cell in the freezing phase, while cryoprotectants protect it in the drying phase (Broeckx et al. 2016). The use of cryoprotective substances such as skim milk, whey milk, sucrose, lactose, trehalose, sorbitol, proteins, amino acids, dietary fibers, or prebiotics is the subject of extensive research to increase probiotic cell viability in food products (Ampatzoglou et al. 2010). Some studies showed that algae and microalgae could also be potential preservatives for probiotics (Kuo et al. 2022; Meireles Mafaldo et al. 2022). Cryoprotectants are added to the carrier medium to promote cell viability during FD and to stabilize them during storage (Rajam and Subramanian 2022). The choice of wall material for probiotic microencapsulation is critical since the chemicals used to encapsulate probiotics must be food-grade, biodegradable, and capable of forming a protective wall between the cell and its surroundings (Rajam and Subramanian 2022). These protective materials during FD lead to improved viability and high yields (Broeckx et al. 2016). It should be noted that all these cryoprotectants can be strain dependent.

Skim milk is a convenient cryoprotectant that enhances the viability of probiotic bacteria during the FD process. It was found that adding 6% skim milk to medium as a cryoprotective agent increased the vitality of bacteria by up to 20% compared to the absence of any other cryoprotective substance (Jalali et al. 2012). Oluwatosin et al. (2021) showed that probiotics treated with 10% skimmed milk showed 91% viability, while those treated with distilled water had only 22%. Under refrigerated settings, a high degree of product stability was observed, with no decrease in cell viability. It is reported that 8% of skim milk can protect the cell wall and the probiotic *Enterococcus faecalis* during and after FD (Romyasamit et al. 2021). It has been observed that using trehalose and lactose together with skim milk preserves cell viability at room temperature for 39 weeks (Broeckx et al. 2016). Due to the commonness of milk-based probiotic formulations, new probiotic protective agent demands are increased. For customers with lactose intolerance or allergies, non-milk-based probiotic formulations are crucial (Savini et al. 2010). Using dairy products as protective agents is not always adequate owing to their short shelf-life and refrigeration requirements (Kuo et al. 2022). Therefore, it is critical to find and optimize suitable protective agents and convenient parameters for the vitality of probiotic bacteria.

Trehalose, a non-reducing disaccharide, is another suitable cryoprotectant for FD probiotics and is quite effective when combined with other excipients (Celik and O'Sullivan 2013). Several studies show a higher cell number of bacteria when trehalose is added as a cryoprotective agent to the medium. When 24% skim milk, 4% sucrose, 0.3% ascorbic acid, and 5% trehalose were added as cryoprotectants, the initial viable count of *B. longum* was 10^{10} cfu/g, and it decreased only to 10^{9} cfu/g after 56 days of storage at 4 °C (Izquierdo-López et al. 2017). This indicates that cell viability and counts can be maintained by combining different cryoprotectants. In the presence of 10% cellobiose and trehalose, the activity of β-glucosidase, produced by *Bifidobacterium*, was observed to be 98.1% and 97.6%, respectively. In this study, it should be noted that cellobiose is a good cryoprotectant that can protect cells during the drying stages (Basholli-Salihu et al. 2014). Besides being economically reasonable, it was found that during room temperature, maltodextrin, glycerol, and mannitol provide significant stabilization for the viability of the cells (Arellano-Ayala et al. 2021; Savini et al. 2010). Sucrose, on the other hand, is suggested to have a noteworthy lyo- and cryoprotective effect on cell survival (Bergenholtz et al. 2012). Compared to cells without cryoprotectants, adding cryoprotectants dramatically improved the survival of *L. reuteri* at both 4 and 30 °C. The survival rates of cells treated with combined cryoprotectants (8–24% trehalose, 0.13–0.53% Na_2HPO_4, 3–12% lactose, 15–25% skim milk) after six months in storage were 96.4% at 4 °C and 73.8% at 30 °C, respectively (Shu et al. 2018). According to Rascon et al. (2018), hypertonic sucrose solutions had the potential as a cryoprotective agent in infusing fruits loaded with the *L. rhamnosus*. Freeze-dried banana slices infused with probiotic bacteria were successful due to cell count remaining constant at 10^{9} cfu/g for 20–28 days. In another study, apple slices were inoculated with *L. plantarum* and dried using different drying methods. When the bacterial loads of apple slices using sucrose solution as cryoprotectant were examined after drying, it was observed that the best efficiency was obtained from FD (Cui et al. 2018). It is reported that ascorbic acid, skim milk, and sorbitol helped freeze-dried probiotics to remain viable and maintain their probiotic qualities while being stored. In the control group (without protectants), the survivability percentage of strains was 70%, while strains with excipients had a long-term survival rate of 73–93% (Bagad et al. 2017). Hu et al. (2022) showed that isomaltose oligosaccharide (IMO) increases the survival rate of *Pediococcus pentosaceus* by 11.02% during FD and provides protection in different temperatures storage. It was discovered that IMO might boost the survival rate of *P. pentosaceus* during storage and ideal temperatures were between −20 and 4 °C. da Silva Guedes et al. (2019) evaluated the potential use of brewer yeast β-glucans (YβG) as cryoprotective on different *Lactobacillus*, and they found YβG were effective protectives for *L. plantarum*. YβG added probiotics maintained the initial probiotic counts for all strains evaluated for up to 60 days of storage. It should be mentioned that the activity and interaction between the microorganism and cryoprotectants seen *in vitro* may differ when tested *in vivo*. The significance of incorporating functional foods fortified with freeze-dried probiotics in emergency relief strategies is to ensure the provision of essential nutrients and support digestive health among affected populations. It may explore the practical implications of this technology in facilitating the distribution, storage, and consumption of nutritionally fortified products during critical times. The recent focus has also brought attention to the significance of managing various emergency situations and events (Wang et al., 2021). Primarily used cryoprotective agents and viability rates of probiotics are given in Table 6.3.

Although the FD technique is known to be challenging for live bacteria, it is nonetheless regarded as a suitable method for ensuring a longer shelf life for most probiotic products (Arellano-Ayala et al. 2021). In the vast majority of studies, it has been shown that the inclusion of certain cryoprotectants with prebiotic qualities in the medium (such as skim milk) can increase the viability of cells throughout the FD process in addition to their potential function with probiotics (Fatemeh 2011). Despite the availability of a few marketed probiotic medicines, sustaining cell viability for an extended length of time remains challenging. Extending cell viability during storage, processing, and digestion should be investigated for effective probiotic production (Rajam and Subramanian 2022). Therefore, besides *in vitro* studies, more *in vivo* studies are needed on the subject.

6.6 Properties of Freeze-Dried Probiotic Powders

The liquid probiotic preparations require low storage and transport temperature to reduce the effect of environmental stresses such as pH, water activity, and dissolved oxygen acting on cells. Drying probiotics into powder form is a helpful way to minimize the costs and exhibits convenience in handling, storage, and transportation (Ermis 2021). However, it is challenging to produce probiotic powders containing a sizeable bacterial population having high viability during the dehydration process and storage time in complex environments. For this reason, encapsulation techniques are employed to maintain microbial bioactivity by using protective materials

Table 6.3 Different cryoprotectants and their effect on the survival rate of probiotic bacteria in various storage conditions.

Microorganism	Cryoprotective agent (g/ml)	Storage condition	Survival rate (%)	Reference
Lactobacillus paracasei	8% Trehalose	4 °C, 3 months	76	Jalali et al. (2012)
	6% Skim milk			
Lactobacillus delbrueckii	4% Sodium ascorbate		72	
Bifidobacterium infantis	5% Cellobiose	37 °C, 48–72 h	86	Basholli-Salihu et al. (2014)
Enterococcus faecalis	8% Skim milk	4 °C, 30 days	95–99	Romyasamit et al. (2021)
Lactobacillus plantarum	10% Skim milk	4 °C, 12 weeks	91	Oluwatosin et al. (2021)
L. plantarum	32% Sucrose	35 °C, 3 months	34	Strasser et al. (2009)
	32% Trehalose		40	
L. plantarum	8% Brewer yeast β-glucan	4 °C, 120 days	87	da Silva Guedes et al. (2019)
Pediococcus pentosaceus	8% Fructose	4 °C, 120 days	77	Bagad et al. (2017)
	2.5% Ascorbic acid		93	
Lactobacillus acidophilus	13% Trehalose	4 °C, 180 days	93	Shu et al. (2018)
	0.33% Na_2HPO_4			
	7.5% Lactose			
	21% Skim milk			
P. pentosaceus	1% Isomaltose oligosaccharide	25 °C 9 months	63	Hu et al. (2022)
		4 °C	87	
		−20 °C	84	
		−80 °C	86	

(Wang et al. 2022). Various wall materials such as whey protein concentrate, whey, hydrocolloids (i.e. maltodextrin, gums, and agarose-based solutions), and sodium alginate could be used to improve the powder properties (Alehosseini et al. 2019; Jouki et al. 2021a; Obradović et al. 2022).

The amount of total soluble solids in the food mixture is the key factor affecting the powder yield after FD. More soluble solids result in an increased amount of powder. The size of the probiotic cells (1–5 μm) limits the cell loading for beads/capsules at the nanoscale (Frakolaki et al. 2021). The size of the particles containing probiotic cells ranges from 2 to 1000 μm after FD depending on the milling/

grinding method used, such as spiral jet mill (Jiang et al. 2020; Picot and Lacroix 2003). Larger particles confer better protection for the probiotics, while they show poor dispersibility in the food matrix. An optimal microcapsule size should be identified to provide reasonable protection, improved shelf stability, and good food product distribution.

In general, physical, physicochemical, and biological properties of probiotic powders such as flowability, bulk density, size distribution, reconstitution (wettability, dispersibility, stability), chemical properties, particles' microstructural and surface properties, and viability influence the functionality during processing, storage (under varying temperatures and moisture content), and digestion in the host. The formulation of the mixture, the drying method used, and delivery forms (powder beads, tablets, capsules, liquids, etc.) play vital roles in chemical structure, powder quality, and particle properties. The cell viability and performance (i.e. adhesion capacity and growth/colonization) of dried probiotics in the powder state are influenced by biological (e.g. surface membrane adhesion and aggregation), physical (e.g. crystal formation and size and storage temperature), and physicochemical (e.g. oxygen permeability and water activity) factors (Ali et al. 2020). Further characterization of probiotic powders, such as thermal profiling and particle-specific surface area, which is related to the developed surface to react with dispersing liquid media, should be done to evaluate the functional properties better. Increased surface area-to-volume ratio of solid particles of small size accelerates the oxidation and powder permeability affecting powder stability during storage. Particle size, shape, and wettability properties can be correlated to this ratio. The moisture absorption and oxygen transfer rate from the surrounding air are altered by the degree of porosity, permeability, and surface hydrophobicity of the particles (Ali et al. 2020).

Adding encapsulated probiotic cells to yogurt yielded yogurt powder with increased bulk density (around 0.29 g/cm^3). Yogurt powder without probiotics provided porous and rougher particles (Jouki et al. 2021b). The yield and the bulk density of skim milk powder inoculated with probiotics were reported as around 15% and 16.67 g/ml, respectively (Hameed et al. 2021). Plant byproducts such as grape pulp, pomegranate, and beetroot peel extracts can produce synbiotic powder using FD technology (Jouki et al. 2021b).

6.6.1 Comparison of Spray-Dried and Freeze-Dried Probiotic Powders

Probiotic cells can be microencapsulated using FD and SD techniques (Vaessen et al. 2020). Spray and FD techniques were used by Luangthongkam et al. (2021) to produce limestone powder containing *Bacillus amyloliquefaciens*. They report 100% survival after spray drying when adjusting the outlet temperature at 80 °C, while it is reported as 60% after FD. In addition, a lower temperature (4 °C) resulted in more extended stability during storage. Her et al. (2015) employed spray freeze-drying (SFD) to produce *Lactobacillus casei* fine probiotic powder. They reported varying particle shapes and sizes (average 24.8 μm) with porous structures depending on the drying conditions. Previous studies report that irregular shapes and surfaces (porous and spongy flakes) with sharp edges and spikes (Figure 6.2) were produced from the FD process (Jiang et al. 2020; Obradović et al. 2022; Rajam and Anandharamakrishnan 2015; Ying et al. 2010). This irregularity might be attributed to the milling process after the FD of probiotic suspensions. Jouki et al. (2021b) reported the particle size of freeze-dried alginate microcapsules of *L. plantarum* ranged from 350 to 1000 μm.

SFD has been introduced to eliminate the adverse effect of heat during SD and develop a fine powder with a controlled capsule size and specific surface area. Figure 6.3 exhibits the processing steps of SD, FD, and SFD. The SFD method includes spraying liquid media containing probiotic cells through an atomization nozzle into the cold vapor phase of a cryogenic liquid, such as liquid nitrogen.

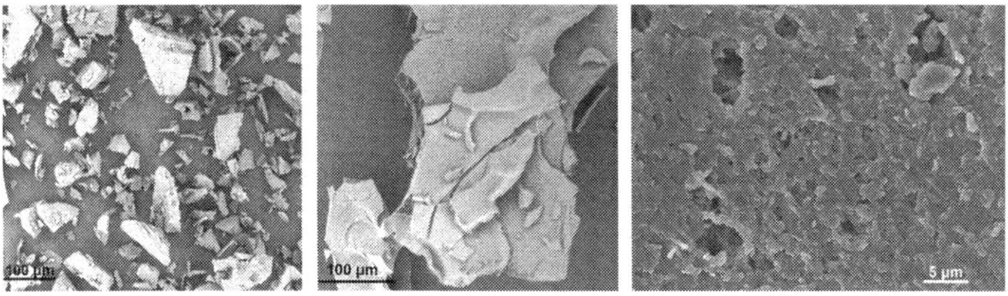

Figure 6.2 SEM images of freeze-dried probiotic powder particles (protective materials: gum Arabic and maltodextrin). *Source:* Shobuz Mahmud et al. 2022/Reproduced with permission from John Wiley & Sons.

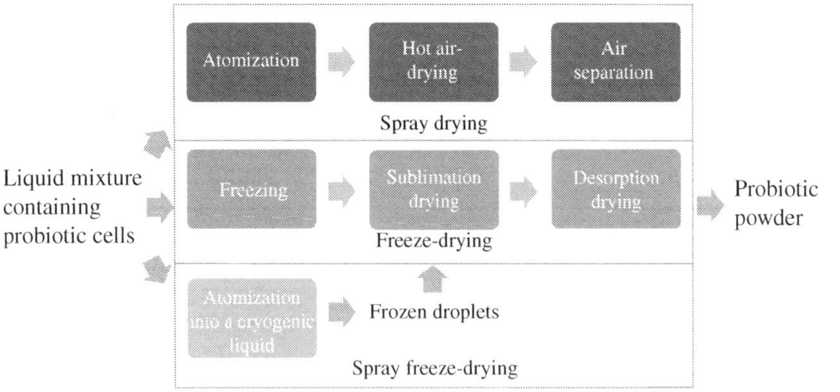

Figure 6.3 The comparison of processing steps of SD, FD, and SFD. *Source:* Adapted from Frakolaki et al. (2021).

Dispersed frozen droplets in cryogenic liquid are dried through the FD technique (Kiepś and Dembczyński 2022). Even though the SFD requires less time than FD, the operating cost is 30–50 times higher than SD (Frakolaki et al. 2021). In addition, the encapsulation efficiency was found to be lower compared to regular FD due to the additional stress factors that occurred during atomization and freezing (Kiepś and Dembczyński 2022).

6.7 Rehydration of Freeze-Dried Probiotics

Rehydration of probiotic powders is required before adding them to food formulations. The wetting time and the solubility analyses are often performed to assess the solubility (Hameed et al. 2021; Jouki et al. 2021b). The hydration of freeze-dried probiotic cells has a significant impact on their viability. Regardless of the drying process, rehydration is a critical stage in the recovery of dehydrated microorganisms. Even when all precautions are taken to ensure survival during freezing, drying, and storing, an inadequate rehydration phase may result in poor cell viability (Chen et al. 2005). Thus, retaining the viability of the cells is essential for effective rehydration (Arellano-Ayala et al. 2021). Careful selection of medium or excipients containing the probiotic culture can be used to identify the least amount of harm caused to microorganisms during rehydration (Nagashima et al. 2013).

The rehydration method and the surrounding media's conditions determine the cell's potential after rehydration (Arellano-Ayala et al. 2021). Slow rehydration of freeze-dried probiotics provides slow water flow through the cell membrane, thus maintaining bacterial viability (Broeckx et al. 2016). However, the viability of probiotic powders by rehydration depends on the strain and the medium in which it is dissolved. When freeze-dried cells were rehydrated directly with water instead of skim milk solution, the death rate increased by 0.65 log cfu/ml (Fatemeh 2011). Skim milk protects cells by stabilizing the components of the cell membrane, forming a porous structure in the freeze-dried product that facilitates rehydration, and having proteins that act as a protective layer for the cells (Jofré et al. 2015). Rehydration of freeze-dried *L. brevis* cells in distilled water resulted in lower recovery than rehydration in a medium containing sugar and minerals (Zhao and Zhang 2005). *Enterococcus faecalis* isolates were freeze-dried with skim milk powder and stored at 4 °C for 30 days. It might imply that freeze-dried *E. faecalis* with skim milk has a high potential for regrowth following rehydration (Romyasamit et al. 2021). On the contrary, no difference in the recovery of the *Lactobacillus helveticus* cells was observed between distilled water and 10% (w/v) skimmed milk medium. This result was attributed to the preservation of probiotics during rehydration by milk components found in kefir (Chen et al. 2005).

Another crucial aspect of probiotic functioning is its capacity to stick to mucosal surfaces and the intestinal epithelium, which is necessary for its establishment in the human GIT (Schillinger et al. 2005). There are specific *in vitro* tests determined by FAO (The Food and Agriculture Organization and WHO (the World Health Organization)). These tests can be listed as resistance to gastric and bile acidity, adherence to mucus and human epithelial cells, antimicrobial activity against potentially pathogenic bacteria, and bile salt hydrolase activity (FAO/WHO 2022). A key aspect in maintaining freeze-dried probiotics' viability and physiological activity is the careful selection of appropriate cryoprotectants when creating new probiotic supplements. Protecting cells from osmotic stress during rehydration and maintaining probiotic viability is essential for such formulations (Arellano-Ayala et al. 2021). In order to provide a favorable environment for probiotics and restore their effectiveness, it is crucial to confirm the strain's adaptation to rehydration in GIT.

6.8 Conclusion and Future Trends

The viability of probiotic bacteria is essential for ensuring the functional activity of probiotic foods. Protecting the stability and viability of probiotics during and after FD is a severe industry challenge. FD is one of the main techniques for keeping probiotics for prolonged storage. The activity of dried probiotics can be improved by selecting the best bacterial strain and modifying the production, processing, storage, and rehydration parameters with precise approaches (i.e. the use of adequate cryoprotection, optimization of freezing rate and freezing temperature, and adaptation procedures applied before freezing to prepare the cells to freezing stress). In addition, investigating the effect of various cryoprotectants on probiotics is necessary for freeze-dried probiotics. Optimizing and maintaining probiotics' stability with newly improved drying and storage techniques is possible. Advanced fermentation, encapsulation, drying, rehydration, and storage technologies can be used to prepare freeze-dried probiotics with prolonged stability. However, further research needs to be done to test new ingredients, improve the cryotolerance of probiotic bacteria, and improve the stability and functionality of probiotic powders.

References

Alehosseini, A., Gomez del Pulgar, E.-M., Fabra, M.J. et al. (2019). Agarose-based freeze-dried capsules prepared by the oil-induced biphasic hydrogel particle formation approach for the protection of sensitive probiotic bacteria. *Food Hydrocolloids* 87: 487–496. https://doi.org/10.1016/j.foodhyd.2018.08.032.

Ali, M.A., Razafindralambo, H.L., Conti, G., and De Coninck, J. (2020). Bulk and surface wettability characteristics of probiotic powders in their compressed disc and packed-bed column forms. *ACS Omega* 5: 22348–22355. https://doi.org/10.1021/acsomega.0c02728.

Amine, K.M., Champagne, C.P., Raymond, Y. et al. (2014). Survival of microencapsulated Bifidobacterium longum in Cheddar cheese during production and storage. *Food Control* 37: 193–199. https://doi.org/10.1016/j.foodcont.2013.09.030.

Ampatzoglou, A., Schurr, B., Deepika, G., Baipong, S., Charalampopoulos, D., 2010. Influence of fermentation on the acid tolerance and freeze drying survival of *Lactobacillus rhamnosus* GG. *Biochemical Engineering Journal* 52, 65–70. doi:https://doi.org/10.1016/j.bej.2010.07.005

Arellano, K., Park, H., Kim, B. et al. (2021). Improving the viability of freeze-dried probiotics using a lysine-based rehydration mixture. *Microbiology and Biotechnology Letters* 49: 1–10. https://doi.org/10.48022/mbl.2012.12007.

Arellano-Ayala, K., Lim, J., Yeo, S. et al. (2021). Rehydration before application improves functional properties of lyophilized *Lactiplantibacillus plantarum* HAC03. *Microorganisms* 9: 1013. https://doi.org/10.3390/microorganisms9051013.

Aschenbrenner, M., Foerst, P., and Ulrich, K. (2015). Freeze-drying of probiotics. In: *Advances in Probiotic Technology* (ed. C. Santivarangkna), 187–220. CRC Press https://doi.org/10.1201/b18807-15.

Ashaolu, T.J. (2020). Immune boosting functional foods and their mechanisms: a critical evaluation of probiotics and prebiotics. *Biomedicine & Pharmacotherapy* 130: 110625. https://doi.org/10.1016/j.biopha.2020.110625.

Bagad, M., Pande, R., Dubey, V., and Ghosh, A.R. (2017). Survivability of freeze-dried probiotic *Pediococcus pentosaceus* strains GS4, GS17 and *Lactobacillus gasseri* (ATCC 19992) during storage with commonly used pharmaceutical excipients within a period of 120 days. *Asian Pacific Journal of Tropical Biomedicine* 7: 921–929. https://doi.org/10.1016/j.apjtb.2017.09.005.

Barbosa, J., Borges, S., Amorim, M. et al. (2015). Comparison of spray drying, freeze drying and convective hot air drying for the production of a probiotic orange powder. *Journal of Functional Foods* 17: 340–351. https://doi.org/10.1016/j.jff.2015.06.001.

Basholli-Salihu, M., Mueller, M., Salar-Behzadi, S. et al. (2014). Effect of lyoprotectants on β-glucosidase activity and viability of *Bifidobacterium infantis* after freeze-drying and storage in milk and low pH juices. *LWT - Food Science and Technology* 57: 276–282. https://doi.org/10.1016/j.lwt.2014.01.011.

Béal, C., Fonseca, F., 2015. Freezing of probiotic bacteria, in: Petra ger Chalat Santivarangkna Advances in Probiotic Technology. CRC Press pp. 179–212. https://doi.org/10.1201/b18807-14

Bergenholtz, Å.S., Wessman, P., Wuttke, A., and Håkansson, S. (2012). A case study on stress preconditioning of a *Lactobacillus* strain prior to freeze-drying. *Cryobiology* 64: 152–159. https://doi.org/10.1016/j.cryobiol.2012.01.002.

Broeckx, G., Vandenheuvel, D., Claes, I.J.J. et al. (2016). Drying techniques of probiotic bacteria as an important step towards the development of novel pharmabiotics. *International Journal of Pharmaceutics* 505: 303–318. https://doi.org/10.1016/j.ijpharm.2016.04.002.

Celik, O.F. and O'Sullivan, D.J. (2013). Factors influencing the stability of freeze-dried stress-resilient and stress-sensitive strains of bifidobacteria. *Journal of Dairy Science* 96: 3506–3516. https://doi.org/10.3168/jds.2012-6327.

Chen, H.-C., Lin, C.-W., and Chen, M.-J. (2005). The effects of freeze drying and rehydration on survival of microorganisms in Kefir. *Asian-Australasian Journal of Animal Sciences* 19: 126–130. https://doi.org/10.5713/ajas.2006.126.

Chen, H., Tian, M., Chen, L. et al. (2019). Optimization of composite cryoprotectant for freeze-drying *Bifidobacterium bifidum* BB01 by response surface methodology. *Artificial Cells, Nanomedicine, and Biotechnology* 47: 1559–1569. https://doi.org/10.1080/21691401.2019.1603157.

Chugh, B. and Kamal-Eldin, A. (2020). Bioactive compounds produced by probiotics in food products. *Current Opinion in Food Science* 32: 76–82. https://doi.org/10.1016/j.cofs.2020.02.003.

Conde-Islas, A., Jiménez-Fernández, M., Cantú-Lozano, D. et al. (2019). Effect of the freeze-drying process on the physicochemical and microbiological properties of Mexican kefir grains. *Processes* 7: 127. https://doi.org/10.3390/pr7030127.

Cui, L., Niu, L., Li, D. et al. (2018). Effects of different drying methods on quality, bacterial viability and storage stability of probiotic enriched apple snacks. *Journal of Integrative Agriculture* 17: 247–255. https://doi.org/10.1016/S2095-3119(17)61742-8.

Damián, M.R., Cortes-Perez, N.G., Quintana, E.T. et al. (2022). Functional foods, nutraceuticals and probiotics: a focus on human health. *Microorganisms* 10: 1065. https://doi.org/10.3390/microorganisms10051065.

Dinkçi, N., Akdeniz, V., and Akalin, A.S. (2019). Survival of probiotics in functional foods during shelf life. In: *Food Quality and Shelf Life* (ed. C.M. Galanakis), 201–233. Elsevier https://doi.org/10.1016/b978-0-12-817190-5.00006-9.

Ermis, E. (2021). A review of drying methods for improving the quality of probiotic powders and characterization. *Drying Technology* 40: 1–18. https://doi.org/10.1080/07373937.2021.1950169.

FAO/WHO 2001. Health and nutritional properties of probiotics in food including powder milk with live lactic acid bacteria. Food and Agriculture Organization of the United Nations and World Health Organization Expert Consultation Report, Cordoba, Argentina.

Fatemeh, S. (2011). Optimization of a cryoprotective medium and survival of freeze-dried *Bifidobacterium infantis* 20088 throughout storage, rehydration and gastrointestinal tract transit for infant formula probiotic applications. *African Journal of Microbiology Research* 5: https://doi.org/10.5897/ajmr11.319.

Ferraz, J.L., Cruz, A.G., Cadena, R.S. et al. (2012). Sensory acceptance and survival of probiotic bacteria in ice cream produced with different overrun levels. *Journal of Food Science* 77: S24–S28. https://doi.org/10.1111/j.1750-3841.2011.02508.x.

Fonseca, F., Girardeau, A., and Passot, S. (2020). Freeze-drying of lactic acid bacteria: a stepwise approach for developing a freeze-drying protocol based on physical properties. *Cryopreservation and Freeze-Drying Protocol* https://doi.org/10.1007/978-1-0716-0783-1_38.

Frakolaki, G., Giannou, V., Kekos, D., and Tzia, C. (2021). A review of the microencapsulation techniques for the incorporation of probiotic bacteria in functional foods. *Critical Reviews in Food Science and Nutrition* 61: 1515–1536. https://doi.org/10.1080/10408398.2020.1761773.

González-Sánchez, F., Azaola, A., Gutiérrez-López, G.F., and Hernández-Sánchez, H. (2010). Viability of microencapsulated Bifidobacterium animalis ssp. lactis {BB}12 in kefir during refrigerated storage. *International Journal of Dairy Technology* 63: 431–436. https://doi.org/10.1111/j.1471-0307.2010.00604.x.

Gwak, H.J., Lee, J.-H., Kim, T.-W. et al. (2015). Protective effect of soy powder and microencapsulation on freeze-dried *Lactobacillus brevis* WK12 and *Lactococcus lactis* WK11 during storage. *Food Science and Biotechnology* 24: 2155–2160. https://doi.org/10.1007/s10068-015-0287-5.

Haindl, R., Neumayr, A., Frey, A., and Kulozik, U. (2020). Impact of cultivation strategy, freeze-drying process, and storage conditions on survival, membrane integrity, and inactivation kinetics of *Bifidobacterium longum*. *Folia Microbiologia (Praha)* 65: 1039–1050. https://doi.org/10.1007/s12223-020-00815-3.

Hameed, A.M., Elkhtab, E., Mostafa, M.S. et al. (2021). Amino acids, solubility, bulk density and water holding capacity of novel freeze-dried cow's skimmed milk fermented with potential probiotic *Lactobacillus plantarum* Bu-Eg5 and *Lactobacillus rhamnosus* Bu-Eg6. *Arabian Journal of Chemistry* 14: 103291. https://doi.org/10.1016/j.arabjc.2021.103291.

Heidebach, T., Först, P., and Kulozik, U. (2010). Influence of casein-based microencapsulation on freeze-drying and storage of probiotic cells. *Journal of Food Engineering* 98: 309–316. https://doi.org/10.1016/j.jfoodeng.2010.01.003.

Her, J.Y., Kim, M.S., and Lee, K.G. (2015). Preparation of probiotic powder by the spray freeze-drying method. *Journal of Food Engineering* 150: 70–74. https://doi.org/10.1016/j.jfoodeng.2014.10.029.

Hu, D., Wu, J., Li, J. et al. (2022). Effect of isomaltose oligosaccharide on the freeze-drying process and shelf-life of *Pediococcus pentosaceus*. *LWT* 163: 113583. https://doi.org/10.1016/j.lwt.2022.113583.

Ispirli, H., Özmen, D., Yilmaz, M.T. et al. (2020). Impact of glucan type exopolysaccharide ({EPS}) production on technological characteristics of sourdough bread. *Food Control* 107: 106812. https://doi.org/10.1016/j.foodcont.2019.106812.

Izquierdo-López, D., Goulet, J., and Ratti, C. (2017). Foam-mat freeze-drying of *Bifidobacterium longum* RO175: viability and refrigerated storage stability. *Journal of Food Science* 82: 90–96. https://doi.org/10.1111/1750-3841.13571.

Jalali, M., Abedi, D., Varshosaz, J. et al. (2012). Stability evaluation of freeze-dried *Lactobacillus paracasei* subsp. *tolerance* and *Lactobacillus delbrueckii* subsp. *bulgaricus* in oral capsules. *Research in Pharmaceutical Sciences* 7: 31–36.

Jiang, N., Dev Kumar, G., Chen, J. et al. (2020). Comparison of concurrent and mixed-flow spray drying on viability, growth kinetics and biofilm formation of *Lactobacillus rhamnosus* GG microencapsulated with fish gelatin and maltodextrin. *LWT* 124: 109200. https://doi.org/10.1016/j.lwt.2020.109200.

Jofré, A., Aymerich, T., and Garriga, M. (2015). Impact of different cryoprotectants on the survival of freeze-dried *Lactobacillus rhamnosus* and *Lactobacillus casei/paracasei* during long-term storage. *Beneficial Microbes* 6: 381–386. https://doi.org/10.3920/bm2014.0038.

Jouki, M., Khazaei, N., Rashidi-Alavijeh, S., and Ahmadi, S. (2021a). Encapsulation of *Lactobacillus casei* in quince seed gum-alginate beads to produce a functional synbiotic drink powder by agro-industrial by-products and freeze-drying. *Food Hydrocolloids* 120: 106895. https://doi.org/10.1016/j.foodhyd.2021.106895.

Jouki, M., Khazaei, N., Rezaei, F., and Taghavian-Saeid, R. (2021b). Production of synbiotic freeze-dried yoghurt powder using microencapsulation and cryopreservation of *L. plantarum* in alginate-skim milk microcapsules. *International Dairy Journal* 122: 105133. https://doi.org/10.1016/j.idairyj.2021.105133.

Kiepś, J. and Dembczyński, R. (2022). Current trends in the production of probiotic formulations. *Foods* 11: 2330. https://doi.org/10.3390/foods11152330.

King, H. (2020). Let food be thy medicine. In: *Hippocrates Now: The 'Father of Medicine' in the Internet Age*, 111–132. Bloomsbury Academic https://doi.org/10.5040/9781350005921.ch-006.

Kuo, C.-C., Clark, S., Qin, H., and Shi, X. (2022). Development of a shelf-stable, gel-based delivery system for probiotics by encapsulation, 3D printing, and freeze-drying. *LWT* 157: 113075. https://doi.org/10.1016/j.lwt.2022.113075.

Li, W., Zhao, Y., Zhao, Y. et al. (2023). Food hydrocolloids improving the viability of *Lactobacillus plantarum* LP90 by carboxymethylated dextran-whey protein conjugates: the relationship with glass transition temperature. *Food Hydrocolloids* 134: 108102. https://doi.org/10.1016/j.foodhyd.2022.108102.

Lin, D.C. (2003). Probiotics as functional foods. *Nutrition in Clinical Practice* 18: 497–506. https://doi.org/10.1177/0115426503018006497.

Liu, Y., Wan, Z., Yohannes, K.W. et al. (2021). Functional characteristics of Lactobacillus and yeast single starter cultures in the ripening process of dry fermented sausage. *Frontiers in Microbiology.* 11: https://doi.org/10.3389/fmicb.2020.611260.

Luangthongkam, P., Blinco, J.-A., Dart, P. et al. (2021). Comparison of spray-drying and freeze-drying for inoculum production of the probiotic *Bacillus amyloliquefaciens* strain H57. *Food and Bioproducts Processing* 130: 121–131. https://doi.org/10.1016/j.fbp.2021.09.010.

Mahmud, S., Khan, S., Khan, M.R. et al. (2022). Viability and stability of microencapsulated probiotic bacteria by freeze-drying under *in vitro* gastrointestinal conditions. *Journal of Food Processing & Preservation* https://doi.org/10.1111/jfpp.17123.

Meireles Mafaldo, Í., de Medeiros, V.P.B., da Costa, W.K.A. et al. (2022). Survival during long-term storage, membrane integrity, and ultrastructural aspects of *Lactobacillus acidophilus* 05 and *Lacticaseibacillus casei* 01 freeze-dried with freshwater microalgae biomasses. *Food Research International* 159: 111620. https://doi.org/10.1016/j.foodres.2022.111620.

Misra, S., Pandey, P., Dalbhagat, C.G., and Mishra, H.N. (2022). Emerging technologies and coating materials for improved probiotication in food products: a review. *Food and Bioprocess Technology* 15: 998–1039. https://doi.org/10.1007/s11947-021-02753-5.

Nagashima, A.I., Pansiera, P.E., Baracat, M.M., and Gómez, R.J.H.C. (2013). Development of effervescent products, in powder and tablet form, supplemented with probiotics *Lactobacillus acidophilus* and *Saccharomyces boulardii*. *Food Science and Technology* 33: 605–611. https://doi.org/10.1590/s0101-20612013000400002.

Nguyen, T.-T., Nguyen, P.-T., Nguyen, T.-T.-V. et al. (2022). Correlation between the amount of extracellular polymeric substances and the survival rate to freeze-drying of probiotics. *Current Microbiology* 79: https://doi.org/10.1007/s00284-022-02862-0.

Obradović, N., Volić, M., Nedović, V. et al. (2022). Microencapsulation of probiotic starter culture in protein–carbohydrate carriers using spray and freeze-drying processes: implementation in whey-based beverages. *Journal of Food Engineering* 321: 110948. https://doi.org/10.1016/j.jfoodeng.2022.110948.

Oluwatosin, S.O., Tai, S.L., and Fagan-Endres, M.A. (2021). Sucrose, maltodextrin and inulin efficacy as cryoprotectant, preservative and prebiotic – towards a freeze dried *Lactobacillus plantarum* topical probiotic. *Biotechnology Reports (Amsterdam, Netherlands)* 33: e00696–e00696. https://doi.org/10.1016/j.btre.2021.e00696.

Oluwatosin, S.O., Tai, S.L., and Fagan-Endres, M.A. (2022). Sucrose, maltodextrin and inulin efficacy as cryoprotectant, preservative and prebiotic – towards a freeze dried *Lactobacillus plantarum* topical probiotic. *Biotechnology Reports* 33: e00696. https://doi.org/10.1016/j.btre.2021.e00696.

Peng, M., Tabashsum, Z., Anderson, M. et al. (2020). Effectiveness of probiotics, prebiotics, and prebiotic-like components in common functional foods. *Comprehensive Reviews in Food Science and Food Safety* 19: 1908–1933. https://doi.org/10.1111/1541-4337.12565.

Petreska-Ivanovska, T., Petrushevska-Tozi, L., Grozdanov, A. et al. (2014). From optimization of synbiotic microparticles prepared by spray-drying to development of new functional carrot juice. *Chemical Industry and Chemical Engineering Quarterly* 20: 549–564. https://doi.org/10.2298/ciceq130218036p.

Picot, A. and Lacroix, C. (2003). Effects of micronization on viability and thermotolerance of probiotic freeze-dried cultures. *International Dairy Journal* 13: 455–462. https://doi.org/10.1016/S0958-6946(03)00050-5.

Rajam, R. and Anandharamakrishnan, C. (2015). Spray freeze drying method for microencapsulation of *Lactobacillus plantarum*. *Journal of Food Engineering* 166: 95–103. https://doi.org/10.1016/j.jfoodeng.2015.05.029.

Rajam, R. and Subramanian, P. (2022). Encapsulation of probiotics: past, present and future. *Beni-Suef University Journal of Basic and Applied Sciences* 11: https://doi.org/10.1186/s43088-022-00228-w.

Rascón, M.P., Huerta-Vera, K., Pascual-Pineda, L.A. et al. (2018). Osmotic dehydration assisted impregnation of *Lactobacillus rhamnosus* in banana and effect of water activity on the storage stability of probiotic in the freeze-dried product. *LWT* 92: 490–496. https://doi.org/10.1016/j.lwt.2018.02.074.

Reque, P.M. and Brandelli, A. (2021). Encapsulation of probiotics and nutraceuticals: applications in functional food industry. *Trends in Food Science and Technology* 114: 1–10. https://doi.org/10.1016/j.tifs.2021.05.022.

Romero-Chapol, O.O., Varela-Pérez, A., Castillo-Olmos, A.G. et al. (2022). Encapsulation of *Lacticaseibacillus rhamnosus* GG: probiotic survival, *in vitro* digestion and viability in apple juice and yogurt. *Applied Sciences* 12: 2141. https://doi.org/10.3390/app12042141.

Romyasamit, C., Saengsuwan, P., Boonserm, P. et al. (2021). Optimization of cryoprotectants for freeze-dried potential probiotic *Enterococcus faecalis* and evaluation of its storage stability. *Drying Technology* 40: 2283–2292. https://doi.org/10.1080/07373937.2021.1931294.

Saarela, M., Virkajärvi, I., Alakomi, H.-L. et al. (2006). Stability and functionality of freeze-dried probiotic *Bifidobacterium* cells during storage in juice and milk. *International Dairy Journal* 16: 1477–1482. https://doi.org/10.1016/j.idairyj.2005.12.007.

Santivarangkna, C., Aschenbrenner, M., Kulozik, U., and Foerst, P. (2011). Role of glassy state on stabilities of freeze-dried probiotics. *Journal of Food Science* 76: R152–R156. https://doi.org/10.1111/j.1750-3841.2011.02347.x.

Savedboworn, W., Teawsomboonkit, K., Surichay, S. et al. (2019). Impact of protectants on the storage stability of freeze-dried probiotic *Lactobacillus plantarum*. *Food Science and Biotechnology* 28: 795–805. https://doi.org/10.1007/s10068-018-0523-x.

Savini, M., Cecchini, C., Verdenelli, M.C. et al. (2010). Pilot-scale production and viability analysis of freeze-dried probiotic bacteria using different protective agents. *Nutrients* 2: 330–339. https://doi.org/10.3390/nu2030330.

Schillinger, U., Guigas, C., and Heinrich Holzapfel, W. (2005). *In vitro* adherence and other properties of lactobacilli used in probiotic yoghurt-like products. *International Dairy Journal* 15: 1289–1297. https://doi.org/10.1016/j.idairyj.2004.12.008.

Shu, G., Wang, Z., Chen, L. et al. (2018). Characterization of freeze-dried *Lactobacillus acidophilus* in goat milk powder and tablet: optimization of the composite cryoprotectants and evaluation of storage stability at different temperature. *LWT* 90: 70–76. https://doi.org/10.1016/j.lwt.2017.12.013.

Siaterlis, A., Deepika, G., and Charalampopoulos, D. (2009). Effect of culture medium and cryoprotectants on the growth and survival of probiotic *Lactobacilli* during freeze drying. *Letters in Applied Microbiology* 48: 295–301. https://doi.org/10.1111/j.1472-765x.2008.02529.x.

Siciliano, R.A., Reale, A., Mazzeo, M.F. et al. (2021). Paraprobiotics: a new perspective for functional foods and nutraceuticals. *Nutrients* 13: 1225. https://doi.org/10.3390/nu13041225.

da Silva Guedes, J., Pimentel, T.C., Diniz-Silva, H.T. et al. (2019). Protective effects of β-glucan extracted from spent brewer yeast during freeze-drying, storage and exposure to simulated gastrointestinal conditions of probiotic lactobacilli. *LWT* 116: 108496. https://doi.org/10.1016/j.lwt.2019.108496.

Song, D., Ibrahim, S., and Hayek, S. (2012). Probiotics. In: *Recent Application of Probiotics in Food and Agricultural Science* (ed. E.C. Rigobelo). INTECH https://doi.org/10.5772/50121.

Strasser, S., Neureiter, M., Geppl, M. et al. (2009). Influence of lyophilization, fluidized bed drying, addition of protectants, and storage on the viability of lactic acid bacteria. *Journal of Applied Microbiology* 107: 167–177. https://doi.org/10.1111/j.1365-2672.2009.04192.x.

Terpou, A., Nigam, P.S., Bosnea, L., and Kanellaki, M. (2018). Evaluation of Chios mastic gum as antimicrobial agent and matrix forming material targeting probiotic cell encapsulation for functional fermented milk production. *LWT* 97: 109–116. https://doi.org/10.1016/j.lwt.2018.06.045.

Torp, A.M., Bahl, M.I., Boisen, A., and Licht, T.R. (2022). Optimizing oral delivery of next generation probiotics. *Trends in Food Science and Technology* 119: 101–109. https://doi.org/10.1016/j.tifs.2021.11.034.

Tymczyszyn, E.E., Sosa, N., Gerbino, E. et al. (2012). Effect of physical properties on the stability of *Lactobacillus bulgaricus* in a freeze-dried galacto-oligosaccharides matrix. *International Journal of Food Microbiology* 155: 217–221. https://doi.org/10.1016/j.ijfoodmicro.2012.02.008.

Vaessen, E.M.J., den Besten, H.M.W., Leito, K.M.N., and Schutyser, M.A.I. (2020). Pulsed electric field pre-treatment for enhanced bacterial survival after drying: effect of carrier matrix and strain variability. *Innovative Food Science and Emerging Technologies* 66: 102515. https://doi.org/10.1016/j.ifset.2020.102515.

Vorländer, K., Kampen, I., Finke, J.H., and Kwade, A. (2020). Along the process chain to probiotic tablets: evaluation of mechanical impacts on microbial viability. *Pharmaceutics* 12: https://doi.org/10.3390/pharmaceutics12010066.

Wang, G., Chen, P., Yu, X. et al. (2020). C18:1 improves the freeze-drying resistance of *Lactobacillus plantarum* by maintaining the cell membrane. *ACS Applied Bio Materials* 3: 4933–4940. https://doi.org/10.1021/acsabm.0c00444.

Wang, G., Chen, Y., Xia, Y. et al. (2022). Characteristics of probiotic preparations and their applications. *Foods* 11: 2472. https://doi.org/10.3390/foods11162472.

Wang, Z., Gong, J., Wang, Q., & Qiao, X. (2021). Emergency management science and technology: An international transdisciplinary platform. *Emergency Management Science and Technology*, 1(1), 1–3.

Yao, M., Xie, J., Du, H. et al. (2020). Progress in microencapsulation of probiotics: a review. *Comprehensive Reviews in Food Science and Food Safety* 19: 857–874. https://doi.org/10.1111/1541-4337.12532.

Ying, D.Y., Phoon, M.C., Sanguansri, L. et al. (2010). Microencapsulated *Lactobacillus rhamnosus* GG powders: relationship of powder physical properties to probiotic survival during storage. *Journal of Food Science* 75: E588–E595. https://doi.org/10.1111/j.1750-3841.2010.01838.x.

Zayed, G. and Roos, Y.H. (2004). Influence of trehalose and moisture content on survival of *Lactobacillus salivarius* subjected to freeze-drying and storage. *Process Biochemistry* 39: 1081–1086. https://doi.org/10.1016/s0032-9592(03)00222-x.

Zhao, G. and Zhang, G. (2005). Effect of protective agents, freezing temperature, rehydration media on viability of malolactic bacteria subjected to freeze-drying. *Journal of Applied Microbiology* 99: 333–338. https://doi.org/10.1111/j.1365-2672.2005.02587.x.

Zubillaga, M., Weill, R., Postaire, E. et al. (2001). Effect of probiotics and functional foods and their use in different diseases. *Nutrition Research* 21: 569–579. https://doi.org/10.1016/s0271-5317(01)00281-0.

7

Freeze-Drying Effect on Nutrients and Their Stability

Yogesh Gat and Papiha Gawande

Department of Food Engineering and Technology, Institute of Chemical Technology, Mumbai, Maharashtra, India

7.1 Introduction

The low operating temperature of freeze drying, also called as lyophilization, is its most important and one of the defining features. The freeze-drying process relies on solvent sublimation to remove water, but it typically differs from the dehydration process. The difference in vapor pressure between the ice front and the surroundings is what drives water removal in freeze-drying processes. In terms of the physical states of ice and water vapor, they coexist at 0 °C and 6.1 mbar. Below this temperature, the system's water can either be in a solid or vapor state; hence, the only way to remove water is through sublimation. The freeze-drying technique was first used only in medicinal applications because it was regarded as expensive. High-value foods are those that have an appraisal that is above average. Nutraceuticals, baby food, highly perishable goods, spices, and special-use foods like military rations and camping meals are a few examples of such high-value foods. Since freeze drying is done at low temperatures, items dried using this method exhibit greater color, aroma, and nutritional content retention. Recently, freeze drying has gained popularity for the preservation of valuable materials, producing superior results overall to conventional drying techniques.

The freeze-drying process can be performed either in a vacuum (VFD) or at air pressure (AFD). However, the freezing stage is an initial step that both procedures have in common. It is vital to look more closely at the effect of freezing on the drying step and the final product quality. Each step during the freezing process heavily influences the inherent quality parameters of the commodity. Different freezing techniques, such as liquid nitrogen and blast chilling, could be employed.

Freeze Drying of Food Products: Fundamentals, Processes and Applications, First Edition.
Edited by Roji Balaji Waghmare, Manoj Kumar, and Parmjit Singh Panesar.

For example, dipping a product into the liquid N_2 causes a quick cooling process, whereas using a household freezer would provide a slow one. The biggest limitation of freeze drying, which yields a dehydrated product despite its many advantages, is the high cost associated with the lengthy procedure, which has prevented its use in the food business. However, in recent years, consumer interest in nutritious food has grown, and hence interest in process optimization and identifying alternatives to reducing time and, consequently, the energy cost without compromising the product's quality. Although it can also be the one that most jeopardizes the product's characteristics, shelf temperature appears to have the biggest effect on process time of all the process factors.

7.1.1 Food Quality Changes During Drying of Food

There are three categories of food quality: physical, chemical, and nutritional. Foods' color, flavor, rehydration ability, texture, water activity, bulk properties, flowability, and retention of volatile chemicals and nutrients are some of the major attributes that can be impacted by drying methods. In terms of nutritional characteristics, O_2, high temperatures, and cell deterioration are frequently adversaries of bioactive component retention throughout processing. Therefore, during dehydration, the stability of the beneficial chemicals in plant-based diets may be impacted. The polyphenol oxidase activity raises the possibility that phenolic substances are susceptible to enzymatic breakdown. Additionally, oxidation and heat reactions can degrade the carotenoid, as it is highly unsaturated. Carotenoids are broken down mostly by a process called oxidation. This process is self-perpetuating and only begins after a certain amount of time. Additionally, during the drying process, temperature and moisture content have a particularly large impact on vitamin C loss and carotenoids. Due to its labile nature relative to other food nutrients, vitamin C is typically chosen as a reference index of nutrient quality. Therefore, if ascorbic acid (AA) content is retained in food, other nutrients like bioactive compounds and minerals will be retained as well.

High-quality characteristics like nutrient retention, rehydration, texture, and color are used to identify freeze-dried foods. Therefore, the most suitable drying technique that has been successfully applied to produce high-quality vegetables, dehydrated fruits, and meats is likely freeze drying. Freeze drying is an extremely energy-intensive process. As a result, the use of freeze-drying in the food business is limited to high-value items only. To increase the scope of application of freeze drying, numerous efforts have been made to alter and enhance the overall drying process. A variety of drying method combinations have been utilized in recent years to dehydrate vegetables and fruits in order to minimize the drawbacks of a single drying technique, such as excessive energy consumption, high cost, and poor drying efficiency (Pei et al. 2014).

The structural and physical changes (shrinkage) that take place during drying might have an impact on the final quality of dried fruits. The rehydration capacity of a dehydrated product is determined by true and apparent densities, porosity, and other factors. These characteristics may change during the drying process, giving products different textures, optical, thermal, and nutritional qualities. The solids in a variety of food products are in an unstable amorphous condition that is highly susceptible to variations in moisture levels and temperature. The matrix shifts from a glassy to a rubbery state at the glass transition temperature (T_g), which varies based on the food material and influences physical and chemical changes during food preparation and storage. The amount of water present in a solution or in a solid material is expressed by the coefficient of activity. The T_g and water activity are the characteristics that are traditionally assessed in the evolution of the quality of dehydrated foods. The degree of shrinkage and the T_g are related in such a way that the material will change size only when its temperature is higher than the T_g and also when it has a certain amount of water in it. The viscosity is significantly lowered at temperatures over T_g, which helps the product to deform.

7.1.2 Freezing Process

Ice nucleation is the combination of water molecules into a well-ordered particle that will act as a location for crystal development. It is the initial phase of the freezing stage. Often, suspended particles may serve as the building blocks of ice crystals (heterogeneous nucleation). The expansion of an ice nucleus caused by the addition of water molecules as intended is known as crystal growth. The freezing process is a random process, hence difficult to predict, and it is dependent on the properties of the food matrix. Before ice nucleation begins, water is first chilled to its supercooling temperature (a temperature only a little below its real freezing point). From this point, the freezing temperature stays constant at 0 °C until all the water gets frozen, and the heat transmission comprises latent heat removal for ice crystal formation (Figure 7.1).

The effect of supercooling can be seen even if the product is not extremely cooled. Then, ice nucleation begins as ice crystals form, but the product temperature continues to drop. This occurs because the concentration of solutes in the remaining liquid water rises as the free water crystallizes, lowering the freezing point. For most food products, the increase in ice content below 3 °C is negligible. The freezing concentration of solutes causes a major shift in the food matrix because the solutes could interact, causing changes, and the pH could drastically vary, impacting the quality of the result.

The quality of food products is mostly impacted by the crystallization process. Crystallization begins outside of the cell, which typically results in a higher freezing point. The crystallization rate is determined by the rate of heat removal and the speed of water diffusion into the intracellular space from the cell. The difference in temperature between the product and the surface as well as heat transfer resistance affect the freezing rate since it is dependent on the heat flux transferred

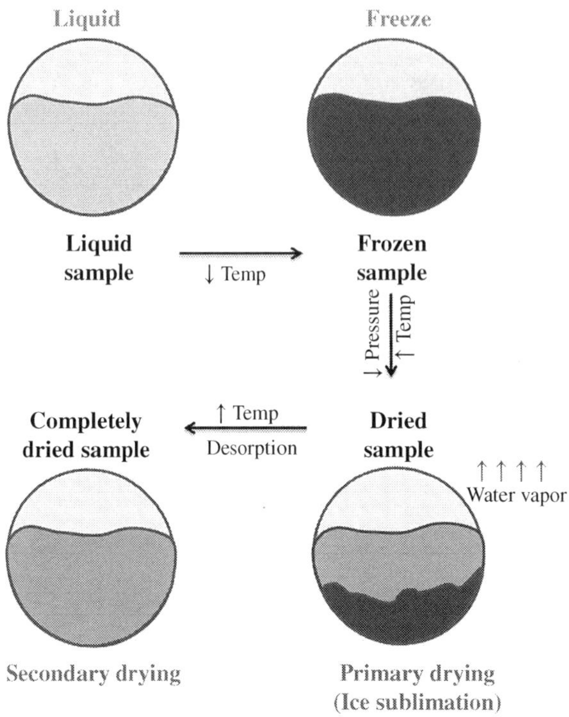

Figure 7.1 Schematic diagram of freeze drying.

between the cooling surface and the product. When the rate of freezing is slower as compared to the rate of crystal growth, fewer ice nuclei form, which results in a small number of comparably big crystals. On the other side, a greater quantity of tiny crystals tends to develop once the freezing rate is faster.

Nutritional compound degradation rates often increase in matrixes with higher water activity due to the improved compound mobility and favored chemical reactions resulting from reduced viscosity. It is important to remember that each substance has unique stability properties with regard to pH and sensitivity to heat, light, and O_2. Hence, it is necessary to determine the compounds that are important before selecting and designing processing based on their properties (Tumbas-Saponjac et al. 2015).

7.1.3 Combined Application of Different Freeze-Drying Methods

Despite being commonly utilized in the food business, freeze drying requires more energy, which drives up the cost of the final product. Therefore, it is crucial to create innovative freeze-drying technologies while lowering energy usage and manufacturing costs. In recent years, combining freeze drying with other drying techniques, such as microwave or ultrasonic combined freeze drying, has proven to be an efficient way to lower the price of freeze-dried food. Following are some of the combined applications of the freeze-drying method.

7.1.3.1 Microwave-assisted Freeze Drying (MD-FD)

The drying technique has the advantages of rapid drying time and high-quality final products. MD-FD has the ability to speed up drying, create goods of high quality, reduce the loss of antioxidant components, enhance tissue quality, and use less energy. In comparison to the freeze-drying process, MD-FD freeze-dried products showed greater sensory quality, medium hardness and brittleness, higher antioxidant component retention, fewer color differences, and a reduction in overall energy consumption as well as drying time (Jiang et al. 2017a). Compared to constant-speed drying, MD-FD has a greater impact on drying rate. MD-FD drying is accelerated by a reduction in material cavity pressure and thickness. The combination of biological and thermal effects of MD-FD also results in a considerable bactericidal effect on food-related materials.

7.1.3.2 Foam-mat Freeze Drying (FM-FD)

Before freeze drying, a foaming agent can be added to the sample (liquid) to turn it into foam and increase the overall drying surface area, thereby speeding up the drying process. Additionally, as this method's drying temperature decreases relative to standard drying temperatures, heat-sensitive ingredients can avoid denaturation or volatilization, preserving the flavor and/or color of food. The results presented by Darniadi et al. (2018) comparing the spray drying (SD) and FM-FD methods for preparing blueberry powder indicated that FM-FD had a yield of over 72%, which was much higher than the spray drying.

7.1.3.3 Far-infrared Radiation-assisted Freeze Drying (FIR-FD)

Fast heating speed, cheap production costs, and great heating efficiency are the crucial benefits of far-infrared heating technology. Under temperatures of 32 and 37 °C, the heating time of the FIR-FD heater was 25% less than that of the freeze-dried samples. Also, the temperature of the FIR-FD heater had no noticeable impact on the product quality. Using far-infrared radiation heating technology, the drying process can be greatly accelerated, along with increased product quality (Liu et al. 2022).

7.2 Vitamin

Vitamins are important nutritional components that an organism needs in very small amounts, and their absence causes deficiencies in certain disorders. Vitamins are essential for stability and preventive qualities. They are only needed at extremely low levels, unlike the majority of other nutrients. However, many of them are not synthesized by our body, so they must be additionally provided through our diet. Vitamins are typically categorized according to their soluble nature. Vitamins A, D, E, and K (fat-soluble vitamins) are transported and used by the body when there are dietary fats present. Extra fat-soluble vitamins are kept in reserve by our body for use when needed, whereas water-soluble vitamins (B and C) cannot be stored by the body. Table 7.1 describes the sources, stability, and functions of fat- and water-soluble vitamins.

The stability of the vitamin may or may not be impacted by the processing circumstances. For instance, niacin is resilient to exposure to heat, light, and oxygen. It is stable under alkaline, acidic, and neutral pH conditions. AA, on the other hand, is only stable at acidic pH levels and becomes unstable in the presence of heat, light, and oxygen. Chemical oxidation, which is encouraged by metal ions in solution, has been discovered to be the main cause of AA breakdown in the dark. Ascorbate oxidase is an enzyme that can break down vitamin C through enzymatic activity in the presence of oxygen. Raising the hot-air drying (HAD) temperature reduces the vitamin C content of tomatoes that have been sliced into pieces, especially in the early drying process. But distinct degradation ranges were seen in a study when operated at various temperatures, indicating that various degradation pathways might exist depending on the operating conditions. The use of inert carrier gases during the process to lessen degradation reactions is suggested by the relationship between increased air exposure and greater vitamin C degradation. Both aerobic and anaerobic processes can lead to the breakdown of vitamin C, while the latter normally happens at a slower rate.

Semi-dehydroascorbic acid is produced on the aerobic pathway by the transfer of one electron, and dehydroascorbic acid (DHAA), which is very unstable, is produced by further loss of an additional electron. The synthesis of DHAA without the development of the intermediate could occur directly through the loss of two electrons. The 2,3-diketogluonic acid that results from the hydrolysis of the lactone bridge in the DHAA, which is very susceptible to this reaction, is what causes vitamin C loss. The breakdown mechanism for anaerobic organisms has not yet been fully established. However, it has been suggested that the 1,4-lactone bridge may be directly cleaved without first being oxidized to DHAA. The anaerobic pathway rate is optimum in environments with a pH of 3–4, maybe because the lactone ring opens up in these conditions (Herbig and Renard 2017).

The pH of the matrix changes during freezing, which may benefit this chemical pathway in low oxygen partial pressure operations. Other intrinsic factors could impact the pathways through which AA degrades since food matrices are complex. Marques et al. (2009) observed that acerola berry maturity affected the AA content following vacuum freeze drying. Green (unripe) acerola berries lost 69.3% of their vitamin C, compared to 13.0% in yellow-reddish acerola and 51.6% in red (ripe) acerola. AA's sensitivity to pH, metallic ions, and moisture content fluctuates depending on the stages of fruit development, which may be the main cause of this variance in

Table 7.1 Sources, stability, and functions of fat-soluble and water-soluble vitamins.

Vitamins	Source	Stability	Functions
Fat-soluble vitamins			
Vitamin A	Fish liver oil, green leafy vegetables Yellow fruits like papaya, mango, carrot, spinach, fortified margarine, and vegetable oil	• Stable to heat in the absence of oxygen • Susceptible to oxidation in the presence of light and oxygen • Unstable under the influence of mineral acid but stable in alkali	• Anti-infective vitamin • Normal functioning of the immune system • Functions in limb development and formation of the heart, eyes, and ear
Vitamin D	Fish liver oil, egg, milk, and fortified foods	• Highly sensitive to light and oxidizing agents • Sensitive to heat, acids, and alkalis	• Efficient utilization of calcium by the body • Inhibits cell proliferation and stimulates the differentiation of cells • Decreases the risk of high blood pressure
Vitamin E	Vegetable oil, nuts, whole grains, and green leafy vegetables	• Loss of vitamin E during refining and deodorization of oil • Heating, frying, drying, and storage reduce the concentration of vitamin E in oil.	• Enhances vasodilation • Prevents chain reactions of lipid destruction • Acts as antioxidant
Vitamin K	Spinach, cabbage, oranges, fish, liver, meat, and eggs	• Vitamin K is destroyed by light • Slowly decomposed by atmospheric oxygen • Unstable against alkali but stable against heat	• Helps in blood clotting (coagulation) • Promotes bone calcification

(*Continued*)

Table 7.1 (Continued)

Vitamins	Source	Stability	Functions
Water-soluble vitamins			
Vitamin B1 (Thiamine)	Yeast, liver, outer layer of cereal grain, lean meat, and soybean	• Reduced by heat, oxygen, sulfur dioxide, or leaching neutral or alkaline pH	• Helps in glucose utilization by nerve tissue • Acts as a neurotransmitter • Aids in the production of RBCs
Vitamin B2 (Riboflavin)	Yeast, liver, green leafy vegetables, milk, wheat germ, eggs, and germinated legumes	• Acid-stable, thermostable, alkali-labile, and photo-labile	• Functions as antioxidants • Useful in carbohydrate, proteins, and fat metabolism
Vitamin B3 (Niacin)	Yeast, liver, legumes, peanuts, outer coat of cereals, meat, and fish	• Unaffected by light, heat, oxygen, acid, and alkali • May cause loss in blanching of vegetables	• Helps in the synthesis of cholesterol and fatty acids • Plays a role as a coenzyme
Vitamin B5 (Pantothenic acid)	• Yeast, liver, kidney, egg yolk, legumes, and nuts	• Stable in air, labile to dry heat • Stable in solution in pH range 5–7	• Important for cell metabolism • Important for the synthesis of cholesterol, fat, and steroid hormones
Vitamin B6 (Pyridoxine)	• Meat, fish, eggs, soybeans, and spinach	• Stable to heat and strong alkalis or acids but sensitive to light, especially ultraviolet light	• Helps in the generation of glucose from amino acids • Helps in the synthesis of neurotransmitters and heme
Vitamin B7 (Biotin)	• Egg yolk, liver, yeast, and intestinal bacteria	• Highly soluble in water. It is photo-stable and thermos-stable but acid-labile	• Helps in fatty acid synthesis • Metabolism of leucine

Table 7.1 (Continued)

Vitamins	Source	Stability	Functions
Vitamin B9 (Folic acid)	• Yeast, liver, spinach, amaranth, mint, and intestinal bacteria	• Easily destroyed by cooking, and fermentation of milk increases folate levels	• Metabolism for several important amino acids • Important role in DNA metabolism
Vitamin B12 (Cyanocobalamin)	• Meat, poultry, fish, and, to a lesser extent, milk	• Fairly thermostable in alkaline pH	• Cofactor for many enzymes • Synthesis of hemoglobin
Vitamin C (Ascorbic acid)	• Guava, cabbage, citrus fruits, milk, and liver	• Least stable of all vitamins • Light-sensitive • Loss of vitamin C in prolonged heating in the presence of O_2	• Strong antioxidant • Useful for the synthesis of collagen and neurotransmitters

the degradation percentage. It is interesting to note that the AA concentration of other experiments with tropical fruits, including mango, starfruit, muskmelon, watermelon, and papaya, did not change much after vacuum freeze drying. Pre-treatments have a significant impact on how well the chosen drying technique works. AA was shown to decrease in potatoes without a pretreatment of blanching by 7% following MD-VFD and VFD, although this difference was not statistically significant. However, when blanching was applied, a reduction of about 32% in AA was observed.

In the study conducted by Uscanga et al. (2021), after freeze drying, a rise in vitamin C concentration was seen; the higher the drying temperature, the greater the variation. This could be because a higher temperature shortened the freeze-drying process and because ascorbate peroxidase, an enzyme that uses vitamin C as a cofactor in its metabolic functions, is inactivated at certain temperatures.

In the study by Sarkar et al. (2022a,b), the amount of retinol (Vitamin A) is greatly reduced throughout the drying process. Raw mango pulp had the highest concentration of retinol, followed by microwave-dried mango pulp, sundried mango pulp, and freeze-dried mango pulp, with a 20%, 36%, and 86% decrease in retinol concentration compared to raw mango pulp, respectively. Heat-sensitive vitamins were significantly affected during sun or HAD due to the sample's prolonged exposure to the environment, where heat and oxidative degradation occur at a faster rate. Higher nutrient retention results from mild heating at lower temperatures and air pressure. Table 7.2 describes the effect of freeze drying on vitamins in various food materials.

Table 7.2 Effect of freeze drying (FD) on vitamins in various food materials.

Food materials	Sample preparation	FD conditions	Effect on vitamin	References
Mango	Mango pulp	Temp = −40 °C Air pressure = 0.1 mBar Time = 20 h	• Minimum destruction of vitamin C (29.94%) than other drying methods. • Vitamin B3 decreased most for freeze drying. • Vitamin B2 = 45.91% decreased. • Vitamin B6 = 81.67% decreased. • Vitamin A = 86.62% decrement	Sarkar et al. (2022a,b)
Brocolli, spinach, and kale	Broccoli florets, kale leaves, and spinach leaves frozen through IQF (individually quick freezing)	Condenser temperature = −64 °C Pressure = 0.5 mmHg Time = 24 h	• Higher retention of vitamin B1 and vitamin B5 • Riboflavin retention was more in case of broccoli • Vitamin B retention was significantly higher in brocolli than refractance window drying and hot air drying • In kale and spinach, vitamin B was found higher and insignificant difference was found between hot air drying and freeze drying	Vargas et al. (2022)
Tomato and ginger	$1 \times 1\,cm^2$	Temp = −50 °C Pressure = 0.133 mBar Time = 24 h	• Less loss of vitamin C than other thermal methods of drying	Gümüşay et al. (2015)

Table 7.2 (Continued)

Food materials	Sample preparation	FD conditions	Effect on vitamin	References
Human milk	Homogenized and frozen at −80 °C	Condensing plate temp = −46 °C Pressure = 10^{-3} mBar Time = 24 h	• Vitamin E content: Storage at 4 °C = no significant loss Storage at 40 °C = decreased significantly. • Vitamin C content gradually decreased at 40 °C and slight decrease at 4 °C was found	Lozano et al. (2014)
Guava and papaya	1 cm × 1 cm (cubes)	Shelf temp = 10 °C Pressure = <613.2 Pa Time = 24 h	Vitamin C retention: Papaya = 88% Guava = 63%	Hawlader et al. (2006)

7.3 Minerals

Minerals are the elements found in food that the body needs for healthy growth and development. Minerals that the body needs are referred to as essential minerals. They are micro, macro, or trace minerals. The quantity of minerals required by the body does not always reflect their importance. Carbon, hydrogen, and oxygen make up the majority of these minerals because they account for 96% of the weight of an adult human. Microminerals and macrominerals are two main categories that can be used to categorize the various forms of minerals.

7.3.1 Microminerals

Microminerals, commonly referred to as trace minerals, are needed in small amounts. Copper, iodine, manganese, iron, fluoride, cobalt, selenium, and zinc are among the trace elements on the list. Excessive use of the micro-minerals can result in mineral poisoning, which can cause a variety of health problems such as nausea, diarrhea, and discoloration.

7.3.2 Macro-minerals

The body requires vast amounts of macrominerals. Calcium, chloride, potassium, magnesium, phosphorus, and sulfur are a few examples of macrominerals. These minerals are necessary for the body's healthy metabolism and operation. The human body cannot synthesize these substances; hence, we must get these minerals from different dietary sources. Human physiology is adversely affected by the lack of macrominerals.

Freeze drying is a procedure that produces high-quality dried goods because it does not require liquid water and requires only very low temperatures. When water is frozen during the process, the primary structure is protected, and little to no changes in the product's shape occur with little loss of volume. Additionally, it helps to maintain the original flavor and aroma as well as elements like minerals and vitamins.

The study conducted by Ibrahim and Khalifa (2015) described the effect of freeze drying on the mineral content of camel milk. It was observed that all the major and trace element levels in the reconstituted freeze-dried camel's milk were somewhat greater than those found in fresh camel milk, with the exception of the Ca, K, and P contents. Table 7.3 shows the value showing the effect of freeze drying on camel's whole milk and skim milk as compared to the control sample.

Table 7.3 Effect of freeze drying on the mineral content of camel's milk.

	Freeze-dried milk	
Minerals	**Whole milk**	**Skim milk**
Major minerals		
Calcium	↑↑	↑↑↑
Magnesium	↑↑	↑↑
Phosphorous	↑↑↑	↑↑
Potassium	↑↑	↑↑
Sodium	↑↑	↑↑
Trace elements		
Zinc	↑↑	↑↑
Iron	↑↑	↑↑↑
Manganese	↑↑	↑↑↑
Copper	↑↑	↑↑↑

↑↑ = Slightly increased, ↑↑↑ = Fairly increased (Ibrahim and Khalifa 2015).

Considering that freeze drying keeps the samples at such a higher temperature and the moisture remaining allows for protein degradation, proteins were better maintained by freezing. Employing c-glutamyl transpeptidase as a measure of the functionality of proteins, no appreciable variations in protein content or enzymatic activity were discovered as an effect of freeze drying. Freeze drying has been found to protect protein functionality, whereas other high-temperature heating treatments destroy protein functionality. The involvement of this sugar in preserving protein functional integrity after freezing was validated by the simultaneous reduction of the enzyme and of glucose, which is more effective in freeze drying (micromolar ranges) as compared to freezing (molar ranges). On the other hand, while both freeze drying and freezing avoided oxidative stress as anticipated due to the cold, freeze drying under the combination treatment (FD + F) was not enhanced by subsequent freezing. This is significant because the oxidation of milk caused by superoxide results in the loss of nutrients and the exposure of young children to free radicals (Cortez and Soria 2016).

7.4 Bioactive Compounds

The most comprehensively studied dietary elements found in vegetables and fruits that act as antioxidants include polyphenols, fiber, conjugated isomers of linoleic acid, epigallocatechin, D-limolene, gallate, isoflavones, soy protein, vitamins A, B, C, and E (group of tocopherols), selenium, calcium, chlorophyllin, alipharin, sulfides, catechin, tetrahydrocurecumin, seasaminol, uric acid, glutathione, indoles, protease inhibitors, and thiocyanates. Antioxidants have a varied effect on the human body. They can lower blood pressure or blood sugar levels, or they can serve as agents with anti-inflammatory, anti-carcinogenic, immune-boosting, antibacterial, antifungal, and antiviral capabilities.

It is essential to keep in mind that factors such as degree of ripeness, harvesting techniques, processing temperature, and storage circumstances might have an impact on the antioxidant content of fresh tissues. Vegetables' and fruit's antioxidant capacity (AC) is also impacted by the types of pressures to which they are subjected and their exposure to ultraviolet-C irradiation. Therefore, even if bioactive components are present in high concentrations in fresh tissue, this does not guarantee that they will be bioavailable when ingested. To assess the strength of the correlation between the levels of bioactive chemicals found in fruits and vegetables, integral investigations are advised.

Freeze drying is typically a better technology than any other drying method to retain the bioactive compounds within foods. Asami et al. (2003) found that freeze drying preserved the total phenolic content in strawberries, corn, and marionberries more effectively than air drying. Similarly, Sablani et al. (2011) demonstrated that the processing method had an impact on the phytochemicals of blueberries and raspberries, whether they were conventional or organic. Freeze drying improved the retention of anthocyanins, phenolics, and antioxidant activity.

Reyes et al. (2011) found that freeze drying of blueberries showed a significant decrease in the AA content in all operating conditions, but that using a vacuum seemed to increase the total polyphenol content when compared to using atmospheric pressure. This increase was attributed to polyphenols' improved extractability. Vacuum freeze drying typically produces the greatest outcomes for phenolic content retention and vitamin C retention. Because autocatalytic oxidative reactions are accelerated at the extremely low water activities attained during freeze drying, the storage of those products should be done with great care in order to preserve lycopene, β-carotene, unsaturated oils, vitamin E, and other lipid-based oxidizable bio-compounds in vegetables and fruits.

Gutiérrez et al. (2008) showed that oils from the sea buckhorn berry pulp obtained from freeze-dried rather than air-dried berries had a much lower peroxide value, illustrating how the lipid-based biocompounds' quality could be hindered by the low water activity acquired during freeze drying.

Colucci et al. (2018) studied the effect of atmospheric freeze drying on the nutritional quality of eggplants. Total phenolic content degradation was observed to range from 58.45% for 8.8 mm cube size samples processed at −10 °C with 50 W US and a 2 m/s air flow to 76.28% for 17 mm cube-size samples handled at −10 °C without ultrasound application and a 2 m/s air flow. For a given sample size, processing temperature had a substantial impact on the degradation of total phenolic content, while neither air velocity nor ultrasound had a significant impact on the tested chemicals. However, sample size mattered; phenolic content reductions were higher in bigger samples. The average TPC loss values under this situation were 52% and 53% at 0 °C and −30 °C under 10 Pa and 64% and 67% at 0 °C and −30 °C at 10 Pa, respectively. These values are comparable to those that were attained by the air-freeze drying approach. Table 7.4 describes the effect of freeze drying on bioactive compounds in various food products.

The amount of phenolics in the freeze-dried residues was greater than that of the fresh residues, and the values after dehydration were higher than those of the fresh tomatoes. This activity might result from phenolic compounds being released as cellular components break down during the moisture removal process. The final phenolic content is higher during the slow freezing approach (Silva et al. 2017).

When onions were freeze-dried, the amount of flavonoids they contained after drying also considerably increased. Contrary to the findings for phenolic compounds, total flavonoid content (TFC) dropped after 48 hours of dehydration, which suggests that flavonoids were degraded as a result of excessive exposure to the freeze-drying process. Once more, samples that had been frozen in the freezer had marginally higher values for this bioactive substance (Silva et al. 2017).

Table 7.4 Effect of freeze drying on various food materials on bioactive compounds.

Food material	Sample prep	FD conditions	Effect on bioactive compounds	References
Oranges	Orange puree distributed in aluminum plates of 0.5 cm thickness	25 h at 30 °C, 7 h at 40 °C 6 h at 50 °C Pressure = 5 Pa and 100 Pa	• Total phenols were better preserved at 30 °C or 50 °C than 40 °C • β carotene is conserved quite well • Freeze drying at 30 °C leads to lower DPPH activity	Silva-Espinoza et al. (2019)
Yellow passion fruit	Passion fruit seeds, peels, and pulp	Pressure = 20 mmHg Condensing temp = −55 °C, Time = 24 h	• Phenolic content was higher in FD product • Flavonoid content ↑es • Citric acid content ↓es after 18 h of freeze drying • Ascorbic acid content ↑es in the first 6 h of freeze drying • Pectin content ↑es	Silva et al. (2017)
Orange	Orange juice with added gum arabic and bamboo fiber in the ratio of 100:5.1	Pressure = 0.05 mbar Condensing temp = −50 °C Shelf temp = 30 °C and 50 °C Time = 48 h and 18 h	• Antioxidant activity ↓es • More total phenols. • Less loss of total carotenoids	Uscanga et al. (2021)
Pomegranate	Pomegranate peel length = 20 ± 0.5 mm width = 20 ± 0.5 mm thickness = 5 ± 0.5 mm	Pressure = 7 mTorr Condensing temperature = −88.7 °C	• Highest concentration of rutin was found in freeze-dried peel. • On the dry basis, TPC, TTC, and TFC were between threefold and fivefold more in freeze dried peel than the oven dried peel	Mphahlele et al. (2016)

The citric acid content in the fresh residues for both freezing procedures was similar during the first 18 hours of freeze drying. After this time, the levels of citric acid dropped, indicating that this molecule had degraded (Silva et al. 2017).

The levels of amino acids were almost constant and somewhat lower than those present in the fresh residues at most processing times (Silva et al. 2017). Freeze-drying irradiation on grapefruit pulp powder had its AA concentration lowered by 15% from baseline levels. However, control fruit did not exhibit the decline. Vitamin C content in freeze-dried pulp from irradiated fruits may be lower than on the day of harvest due to oxidative stress brought on by irradiation and freeze drying. Vitamin C content in freeze-dried pulp from irradiated fruits may be lower than on the day of harvest due to oxidative stress brought on by irradiation and freeze drying (Vanamala et al. 2005). Pectin levels were higher after freeze drying than they were in the fresh leftovers, indicating that this fiber was released during dehydration. Depending on different freezing methods, samples frozen in liquid nitrogen showed the greatest pectin content value after 72 hours of freeze drying, ranging from 78.92 to 61.66%. Consequently, the lengthy freeze-drying procedure enhanced the preservation of this fiber in the finished product (Silva et al. 2017).

7.4.1 Antioxidant Activity

Numerous substances, such as phenols and vitamins, that can slow down material oxidation are referred to as antioxidants. It was hypothesized that antioxidants operate as hydrogen donors or acceptors for free radicals, which hinder chain processes. Antioxidants may be sensitive to high temperatures and may produce a variety of degradation products. As previously studied, various techniques can be used to evaluate AC. The 2,2-diphenyl-1-picrylhydrazyl method (DPPH), the ferric reducing antioxidant power (FRAP), the 2,2-azino-bis-3-ethylbenzothiazoline-6-sulfonic acid (ABTS) method, the oxygen radical absorption capacity (ORAC), the total radical absorption potentials (TRAP), and the cupric ion reducing antioxidant capacity (CUPRAC) are some of the common assays for AC. The final two approaches are based on hydrogen atom transfer, while the first three are dependent on single electron transfer.

Samples of tomato and ginger were assessed for their AC using the CUPRAC assay in research conducted to check the effect of VFD treatment. Figure 7.2 describes the effect of freeze drying on agricultural produce. The authors noted that the AC increased and speculated that it might be related to higher extraction effectiveness brought on by freezing and reduced component losses due to antioxidant activity (Chang et al. 2006). Different tests were used to determine the antioxidant content of various tomato types, and both VFD and hot air drying revealed a little increase or no change in the levels. The chelating power of ferrous ions, reducing power, and DPPH scavenging power of VFD-SN (Sheng-Neu variety)

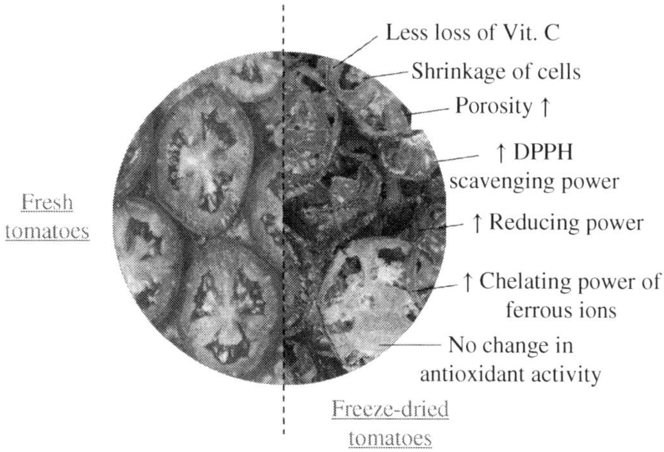

Figure 7.2 Effects of freeze drying on tomatoes.

tomatoes increased by 3.7%, 15.6%, and 1%, respectively. Similar differences were found in ITH (I-Tieng-Hung variety) tomatoes for the same tests in the same order: 3.7%, 2.4%, and 0% respectively. When compared to the combined VFD-MVD procedure, the FRAP values for the okra samples dried with the help of VFD were the highest. Following VFD and FD-MVD, the percentage changes for okra's AC were 2–9% decrease for AEAC, 6–19% decrease for FRAP, and lastly, 10–15% increase for the ABTS assay (Jiang et al. 2017b). Further, the FRAP assay for wheat-grass flour revealed an increase in antioxidant activity of above 50%, which is probably connected to the rise in TPC and flavonoids seen in this work. Additionally, DPPH assay results showed increases of about 10%. Numerous techniques were used to assess the pumpkin flour's antioxidant activity following hot air drying and vacuum freezing. While the total antioxidant activity in relation to BHA levels was greater for VFD samples than HAD, the reducing power in relation to BHT was greater for HAD. In comparison to BHT, samples dried by freeze drying also demonstrated greater DPPH free radical scavenging activity in various concentrations of methanol. Additionally, the HAD flour's ability to chelate metals was marginally greater than that of flour samples done by VFD. These findings suggested that the antioxidant chemicals produced by Maillard reactions may have formed in part as a result of drying's increased heating.

Sample size had a substantial detrimental impact on the freeze-dried blueberries DPPH-measured AC. In comparison to bigger samples cut in half, the smaller sample cut into quarters exhibited higher retention rates. The study found that the employment of IR and AFD in place of VFD had no discernible effects on qualitative features. Tropical fruits have also been used in numerous VFD research studies for antioxidant tests. After VFD, it was discovered that the DPPH levels for mango and starfruit decreased, but those for muskmelon, papaya, and watermelon did not show a significant effect (Harguindeguy and Fissore 2020).

7.4.2 Anthocyanin

Anthocyanins are water-soluble and natural pigments found in plants such as cereals, fruits, and vegetables. They exhibit purple, blue, and red colors. Due to their high pigment concentration and low toxicity, anthocyanins have attracted attention as potential replacements for synthetic colorants in food coloring. As a result, anthocyanins are now considered a powerful functional food. Freeze drying is considered a suitable method for the retention of anthocyanin, as it is suitable for the drying of thermally sensitive pigments.

In the study by Laokuldilok and Kanha (2015), Anthocyanin retention was, on average, $71.96 \pm 0.70\%$ greater in freeze-dried powders than in spray-dried powders. When BRM20 was employed as the wall material, the anthocyanin retention among the freeze-dried powders was determined to be the highest. Therefore, the freeze-drying process can trap a higher anthocyanin content while avoiding the negative effects of high temperatures on thermally sensitive anthocyanins. Despite this, the majority of the anthocyanin was maintained inside the freeze-dried microcapsules. However, anthocyanin degradation might continue due to the increased HCl content during freezing.

The study carried out by Sablani et al. (2011) showed that Red raspberries that had undergone freeze drying had higher levels of total anthocyanins, which were statistically similar to or higher than those that had undergone fresh raspberry processing. Red raspberry anthocyanin concentrations increased 7–26% after freeze drying, while blueberry anthocyanin concentrations decreased 6–32%, with the exception of conventional raspberry, where the amount of anthocyanin increased to 15%.

However, the color of anthocyanin varies with pH. The four structures flavylium cation (red), quinoidal base (blue), carbinol pseudobase (colorless), and chalcone (colorless or light yellow) coexist in equilibrium as a combination of anthocyanins in acidic aqueous solutions. Thus, a decrease in $H+$ concentration during freeze drying can alter the anthocyanin equilibrium in the production of flavylium cation, potentially intensifying the red color (Shishehgarha et al. 2002).

7.4.3 Carotenoids

Many fruits and vegetables naturally contain carotenoids, which are organic colors. They are produced by photosynthetic organisms, and they give plants their vivid red, orange, and yellow colors. Carotenoids are essential nutrients for human health and have a number of positive health effects. The type of carotenoid, processing circumstances, initial concentration, and storage circumstances all affect how carotenoids respond to freeze drying. While some studies have shown that freeze drying can aid in preserving the levels of carotenoids, others have claimed that freeze drying can lead to a drop in the quantities of carotenoids in foods (Eun et al. 2020).

In the study conducted by De Ancos et al. (2018), the freeze-drying method (frozen 24 hours at −80 °C, freeze-dried 48 hours at −50 °C, 4.00 Pa) demonstrated very slight variations in several component families (xanthones and flavonoids) but no appreciable effects on the total concentration of phenolic compounds in mango peel and paste. In contrast, freeze drying was better suited for dry mango paste.

Free radical production during freeze drying is assumed to be the cause of the carotenoids' deterioration because these molecules can be broken down by free radicals. However, by adopting low pressure and low temperature conditions and also varying the water activity of the product during freeze drying, the degree of degradation can be minimized (Eun et al. 2020).

The growth or reduction in these substances does not appear to follow a clear trend, as can be seen from the results obtained, and the scientists offered a variety of justifications for their work. As previously indicated, phenolic compounds can be broken down by enzymes and other processes, which may help explain why a drop in their concentration has been observed. Through various analytical techniques, polyphenols in an intermediate oxidation state can demonstrate stronger antioxidant effects, increasing their apparent content. Many of the authors who have gone before have also proposed improved extraction as the cause of the increases seen due to structural changes during freeze drying, and many metabolic pathways may result in the synthesis of phenolic compounds from other chemicals and the buildup of these compounds. According to Nayak et al. (2022), the increase in phenolic compounds could be a possible outcome of the degradation of high-molecular-weight phytochemicals into lower molecular-weight compounds.

7.5 Future Perspective

To ensure the best nutrient preservation, freeze drying is frequently used in conjunction with other processing techniques like blanching or high-pressure processing. The quality and nutritional content of food products may be enhanced in the future by novel processing method combinations (Ma et al. 2022). Overall, research and development with an emphasis on lowering energy consumption, increasing effectiveness, and maintaining the nutritional content and quality of freeze-dried products are expected to determine the future of freeze drying and its effect on nutrients.

7.6 Conclusion

It is challenging to completely comprehend the metabolic pathways and mechanisms causing variations in nutritional content after freeze drying due to the complexity of food matrices. Expanding our understanding of the variables influencing these pathways in relation to product attributes, process parameters, and the employment of assisted drying techniques could be beneficial for future studies. This would facilitate improved quality management and final product control. In addition, because the freezing process is very important, future research to enhance the effectiveness of the freeze drying process and the nutritional qualities of the result may be intriguing. Future research should also look at various combinations of new freezing and drying techniques from the viewpoint of the hurdle effect.

References

Asami, D.K., Hong, Y.J., Barrett, D.M., and Mitchell, A.E. (2003). Comparison of the total phenolic and ascorbic acid content of freeze-dried and air-dried marionberry, strawberry, and corn grown using conventional, organic, and sustainable agricultural practices. *Journal of Agricultural and Food Chemistry* 51 (5): 1237–1241.

Chang, C.-H., Lin, H.-Y., Chang, C.-Y., and Liu, Y.-C. (2006). Comparisons on the antioxidant properties of fresh, freeze-dried and hot-air-dried tomatoes. *Journal of Food Engineering* 77 (3): 478–485. https://doi.org/10.1016/j.jfoodeng.2005.06.061.

Colucci, D., Fissore, D., Rossello, C., and Carcel, J.A. (2018). On the effect of ultrasound-assisted atmospheric freeze-drying on the antioxidant properties of eggplant. *Food Research International* 106: 580–588. https://doi.org/10.1016/j.foodres.2018.01.022.

Cortez, M.V. and Soria, E.A. (2016). The effect of freeze-drying on the nutrient, polyphenol, and oxidant levels of breast Milk. *Breastfeeding Medicine* 11 (10): 551–554. https://doi.org/10.1089/bfm.2016.0102.

Darniadi, S., Ho, P., and Murray, B.S. (2018). Comparison of blueberry powder produced via foam-mat freeze-drying versus spray-drying: evaluation of foam and powder properties. *Journal of the Science of Food and Agriculture* 98 (5): 2002–2010. https://doi.org/10.1002/jsfa.8685.

De Ancos, B., Sánchez-Moreno, C., Zacarías, L. et al. (2018). Effects of two different drying methods (freeze-drying and hot air-drying) on the phenolic and carotenoid profile of 'Ataulfo' mango by-products. *Journal of Food Measurement and Characterization* 12 (3): 2145–2157. https://doi.org/10.1007/s11694-018-9830-4.

Eun, J.-B., Maruf, A., Das, P.R., and Nam, S.-H. (2020). A review of encapsulation of carotenoids using spray drying and freeze drying. *Critical Reviews in Food Science and Nutrition* 60 (21): 3547–3572. https://doi.org/10.1080/10408398.2019.1698511.

Gümüşay, Ö.A., Borazan, A.A., Ercal, N., and Demirkol, O. (2015). Drying effects on the antioxidant properties of tomatoes and ginger. *Food Chemistry* 173: 156–162. https://doi.org/10.1016/j.foodchem.2014.09.162.

Gutiérrez, L.-F., Ratti, C., and Belkacemi, K. (2008). Effects of drying method on the extraction yields and quality of oils from Quebec Sea buckthorn (Hippophaë rhamnoides L.) seeds and pulp. *Food Chemistry* 106 (3): 896–904. https://doi.org/10.1016/j.foodchem.2007.06.058.

Harguindeguy, M. and Fissore, D. (2020). On the effects of freeze-drying processes on the nutritional properties of foodstuff: a review. *Drying Technology* 38 (7): 846–868. https://doi.org/10.1080/07373937.2019.1599905.

Hawlader, M.N.A., Perera, C.O., Tian, M., and Yeo, K.L. (2006). Drying of guava and papaya: impact of different drying methods. *Drying Technology* 24 (1): 77–87. https://doi.org/10.1080/07373930500538725.

Herbig, A.-L. and Renard, C.M.G.C. (2017). Factors that impact the stability of vitamin C at intermediate temperatures in a food matrix. *Food Chemistry* 220: 444–451. https://doi.org/10.1016/j.foodchem.2016.10.012.

Ibrahim, A. and Khalifa, S. (2015). Effect of freeze-drying on camel's milk nutritional properties. *International Food Research Journal* 22: 1438–1445.

Jiang, N., Liu, C., Li, D. et al. (2017a). Evaluation of freeze drying combined with microwave vacuum drying for functional okra snacks: antioxidant properties, sensory quality, and energy consumption. *LWT - Food Science and Technology* 82: 216–226. https://doi.org/10.1016/j.lwt.2017.04.015.

Jiang, N., Liu, C., Li, D. et al. (2017b). Evaluation of freeze drying combined with microwave vacuum drying for functional okra snacks: antioxidant properties, sensory quality, and energy consumption. *LWT - Food Science and Technology* 82: 216–226. https://doi.org/10.1016/j.lwt.2017.04.015.

Laokuldilok, T. and Kanha, N. (2015). Effects of processing conditions on powder properties of black glutinous rice (Oryza sativa L.) bran anthocyanins produced by spray drying and freeze drying. *LWT - Food Science and Technology* 64 (1): 405–411. https://doi.org/10.1016/j.lwt.2015.05.015.

Liu, Y., Zhang, Z., and Hu, L. (2022). High efficient freeze-drying technology in food industry. *Critical Reviews in Food Science and Nutrition* 62 (12): 3370–3388. https://doi.org/10.1080/10408398.2020.1865261.

Lozano, B., Castellote, A.I., Montes, R., and López-Sabater, M.C. (2014). Vitamins, fatty acids, and antioxidant capacity stability during storage of freeze-dried human milk. *International Journal of Food Sciences and Nutrition* 65 (6): 703–707. https://doi.org/10.3109/09637486.2014.917154.

Ma, Y., Yi, J., Jin, X. et al. (2022). Freeze-drying of fruits and vegetables in food industry: effects on phytochemicals and bioactive properties attributes - a comprehensive review. *Food Reviews International* 1–19: https://doi.org/10.1080/87559129.2022.2122992.

Marques, L.G., Prado, M.M., and Freire, J.T. (2009). Rehydration characteristics of freeze-dried tropical fruits. *LWT - Food Science and Technology* 42 (7): 1232–1237. https://doi.org/10.1016/j.lwt.2009.02.012.

Mphahlele, R.R., Fawole, O.A., Makunga, N.P., and Opara, U.L. (2016). Effect of drying on the bioactive compounds, antioxidant, antibacterial and antityrosinase activities of pomegranate peel. *BMC Complementary and Alternative Medicine* 16 (1): 143. https://doi.org/10.1186/s12906-016-1132-y.

Nayak, P.K., Sundarsingh, A., and Kesavan, R.K. (2022). In vitro gastrointestinal digestion studies on total phenols, flavonoids, anti-oxidant activity and vitamin C in freeze-dried vegetable powders. *Journal of Food Science and Technology* 59 (11): 4253–4261.

Pei, F., Yang, W., Shi, Y. et al. (2014). Comparison of freeze-drying with three different combinations of drying methods and their influence on colour, texture, microstructure and nutrient retention of button mushroom (agaricus bisporus) slices. *Food and Bioprocess Technology* 7 (3): 702–710. https://doi.org/10.1007/s11947-013-1058-z.

Reyes, A., Evseev, A., Mahn, A. et al. (2011). Effect of operating conditions in freeze-drying on the nutritional properties of blueberries. *International Journal of Food Sciences and Nutrition* 62 (3): 303–306. https://doi.org/10.3109/09637486.2010.534078.

Sablani, S.S., Andrews, P.K., Davies, N.M. et al. (2011). Effects of air and freeze drying on phytochemical content of conventional and organic berries. *Drying Technology* 29 (2): 205–216. https://doi.org/10.1080/07373937.2010.483047.

Sarkar, T., Salauddin, M., Sheikh, H.I. et al. (2022a). Effect of drying on vitamin, carotene, organic acid, mineral composition, and microstructural properties of mango (*Mangifera indica*). *Journal of Food Processing and Preservation* 46 (2): https://doi.org/10.1111/jfpp.16237.

Sarkar, T., Salauddin, M., Sheikh, H.I. et al. (2022b). Effect of drying on vitamin, carotene, organic acid, mineral composition, and microstructural properties of mango (*Mangifera indica*). *Journal of Food Processing and Preservation* 46 (2): https://doi.org/10.1111/jfpp.16237.

Shishehgarha, F., Makhlouf, J., and Ratti, C. (2002). Freeze-drying characteristics of strawberries. *Drying Technology* 20 (1): 131–145. https://doi.org/10.1081/DRT-120001370.

Silva, N.C., Santana, R.C., Duarte, C.R., and Barrozo, M.A.S. (2017). Impact of freeze-drying on bioactive compounds of yellow passion fruit residues: Silva et al. *Journal of Food Process Engineering* 40 (4): e12514. https://doi.org/10.1111/jfpe.12514.

Silva-Espinoza, M.A., Ayed, C., Foster, T. et al. (2019). The impact of freeze-drying conditions on the Physico-chemical properties and bioactive compounds of a freeze-dried orange puree. *Foods* 9 (1): 32. https://doi.org/10.3390/foods9010032.

Tumbas-Saponjac, V., Cetkovic, G., Stajcic, S. et al. (2015). Optimization of the bioactive compounds content in raspberry during freeze-drying using response surface method. *Chemical Industry and Chemical Engineering Quarterly* 21 (1–1): 53–61. https://doi.org/10.2298/CICEQ131201005T.

Uscanga, M.A., Salvador, A., Camacho, M.d.M., and Martínez-Navarrete, N. (2021). Impact of freeze-drying shelf temperature on the bioactive compounds, physical properties and sensory evaluation of a product based on orange juice. *International Journal of Food Science & Technology* 56 (10): 5409–5416. https://doi.org/10.1111/ijfs.15086.

Vanamala, J., Cobb, G., Turner, N.D. et al. (2005). Bioactive compounds of grapefruit (*Citrus paradisi* cv. Rio red) respond differently to postharvest irradiation, storage, and freeze drying. *Journal of Agricultural and Food Chemistry* 53 (10): 3980–3985. https://doi.org/10.1021/jf048167p.

Vargas, L., Kapoor, R., Nemzer, B., and Feng, H. (2022). Application of different drying methods for evaluation of phytochemical content and physical properties of broccoli, kale, and spinach. *LWT* 155: 112892. https://doi.org/10.1016/j.lwt.2021.112892.

8

Packaging of Freeze-Dried Products

Tanuva Das, Donald Lyngdoh Nonglait, Devraj Rajput, and Shalini S. Arya

Department of Food Engineering and Technology, Institute of Chemical Technology, Mumbai, India

8.1 Introduction

By definition, packaging is a socio-scientific discipline that works in society to ensure that things are delivered to their intended consumers in the best possible condition (Lockhart, 1997). A package was defined as a device or container that fulfilled one or more of these requirements. Food packaging is a crucial component of the food system that upholds food safety, nutritional sufficiency, and crew acceptability. Food goods are shielded by packaging from microbial contamination, moisture, air, light, external foreign contamination, and many other degrading processes. The package needs to be light, have a strong gas and moisture barrier, excellent puncture resistance, heat sealability, and be approved for use with food. Packaging with insufficient moisture barriers can allow water to enter, which increases the inclination for microbial development. Water entry could make it more difficult for food products to rehydrate. Low oxygen barrier packaging increases the likelihood of food oxidation, which lowers food quality and nutritional value (Perchonok and Bourland, 2002). In general, the main packaging materials do not adequately protect natural and freeze-dried food products from moisture and oxygen. These goods are covered with an additional layer of packaging material, extending the food's shelf life by up to 18 months.

Food preservation techniques like freeze-drying were first applied to the processing of agricultural products, including peanuts (also known as groundnuts) and tobacco, in the early 1970s. The major goal of freeze-drying is to avoid heat in order to retain the structural and chemical integrity of the product without alteration because heat, which is frequently employed in crop and food

Freeze Drying of Food Products: Fundamentals, Processes and Applications, First Edition.
Edited by Roji Balaji Waghmare, Manoj Kumar, and Parmjit Singh Panesar.

processing, usually changes these properties. Thus, in terms of structure and chemistry, foods and crops that are freeze-dried are closest to their natural composition. When it was used to make freeze-dried ice cream, an example of astronaut cuisine, the method gained significant public attention. It is frequently employed to create flavors or essences that can be added to food. A variety of herbs, vegetables (such as vitamin-rich spinach and watercress), pre-boiled rice, and temperature-sensitive baker's yeast suspensions may also be freeze-dried. Herbs that have been air-dried are far more accessible and affordable. In Japan, freeze-dried tofu, also known as "koya-dofu" or "shimi-dofu," is a common meal.

Foods that have been freeze-dried are often sealed to stop the reabsorption of moisture, allowing them to be kept at room temperature for extended periods of time without refrigeration. Because there is far less water present, enzymes and microorganisms that would typically ruin or degrade the product are prevented from working. Compared with other dehydration techniques that employ higher temperatures, freeze-drying preserves the material better. However, the freeze-drying process is dependent on several environmental factors (Figure 8.1); thus, maintaining appropriate storage conditions results in a high-quality and storage-stable product. Additionally, in most cases, shrinkage or toughening of the material being dried does not result from freeze-drying. Hence, the technique is well-liked for food preservation because flavors, aromas, and nutritional value typically remain unaltered. Because the procedure leaves tiny pores, things that have been freeze-dried can be rehydrated considerably more rapidly and easily. Ice crystals that sublimate leave voids or pores in their wake, forming the pores. This is crucial when it comes to medicinal applications. This chapter features insights on the various aspects related to freeze-drying packaging.

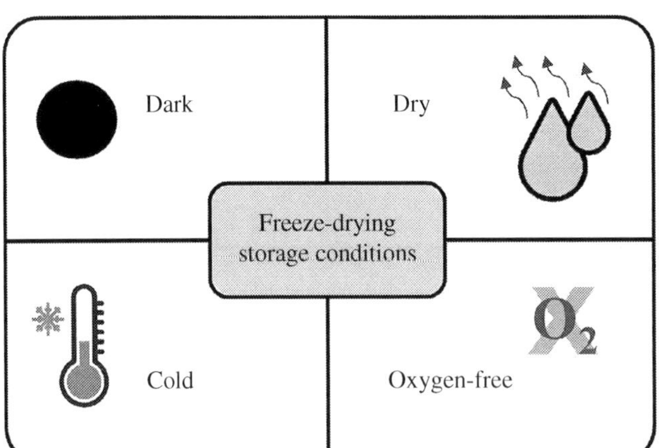

Figure 8.1 Freeze-drying storage conditions.

The parameters that are needed to be taken into consideration for the packaging of freeze-dried foods are as follows:

✓ *Moisture barrier*: Freeze-dried food is particularly vulnerable to moisture, which can cause spoilage or bacterial contamination. To prevent moisture from entering the box, its packaging should have a robust moisture barrier.
✓ *Barrier to oxygen*: Oxygen may also cause freeze-dried food to deteriorate or lose flavor and texture. To prevent air from entering the product, the packaging ought to include a robust oxygen barrier.
✓ *Durability*: Because freeze-dried food is delicate and quickly destroyed, the packaging must be strong enough to endure the pressures of shipping and handling.
✓ *Light protection*: Some freeze-dried foods might be light-sensitive, causing them to lose color or flavor. Exposure to light should be adequately protected by the packaging.
✓ *Size and weight*: Since freeze-dried food packaging is frequently used for outdoor activities or emergency supplies, it should be compact and lightweight.
✓ *Resealable*: Resealable packaging is especially advantageous for freeze-dried food since it allows consumers to utilize a portion of the food and then reseal the package to keep the remaining food fresh.
✓ *Labeling information*: It is critical to give accurate and unambiguous labeling on the packaging, including the type of food, expiration date, and any specific storage or preparation.

8.2 History of Freeze-Dried Foods

Jacques-Arsene d'Arsonval developed freeze-drying in 1906 at the College of France in Paris. Blood serum was preserved during World War II via freeze-drying. Since that time, freeze-drying food has become one of the most essential methods for preserving heat-sensitive foods. Industrial food freeze-drying gained popularity in the 1950s. The first freeze-dried product to be widely sold was coffee. Several foods are freeze-dried nowadays, including fruits, vegetables, meats, eggs, and culinary flavorings.

Moreover, freeze-dried foods have grown to constitute a significant part of astronaut and military supplies. Improved methods for rehydrating freeze-dried foods allowed astronaut crews to convert their initial tubed meals and difficult-to-reheat snacks into hot meals while in space.

8.3 Industrial Process of Freeze Drying

Freeze-drying, also known as lyophilization, is a method of eliminating water from a product (often food, medicines, or biological samples) by freezing it first and then submitting it to a low-pressure vacuum atmosphere. There are three

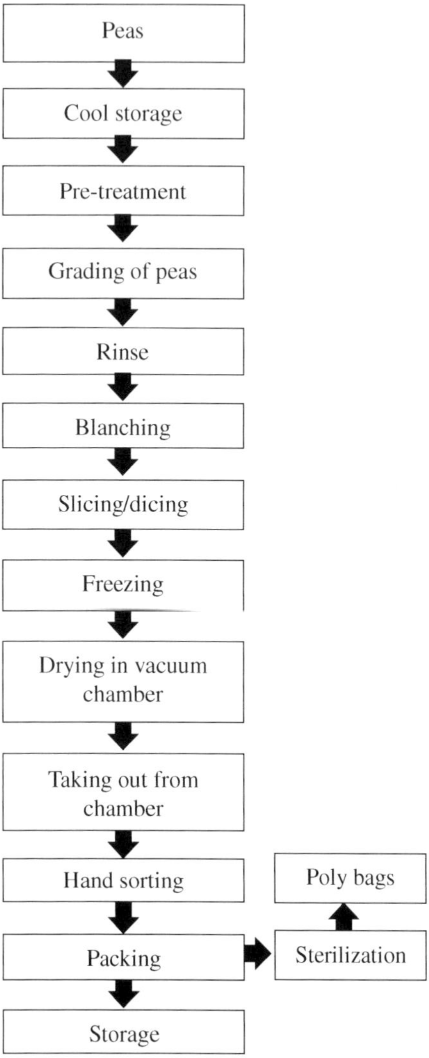

Figure 8.2 Process flow chart for industrial manufacturing of freeze-dried peas.

steps to the process: freezing, primary drying (sublimation), and secondary drying (desorption). In the first stage, the material is frozen to an extremely low temperature, often below −40 °C, to transform the water content into ice. The vacuum environment is applied in the second stage, forcing the ice to sublimate and remove the water from the material. Finally, heat is applied to the chemical in the third stage to eliminate any leftover water molecules. Microbiological spoilage and deteriorative chemical reactions are considerably reduced by lowering moisture levels in fruits and vegetables to a set level. Drying, in addition to preserving the product, reduces the cost of storage, packing, and transportation by lowering the finished product's weight and volume (Doymaz and Kocayigit, 2011). This section will be discussed using the example of freeze-dried pea processing. Peas are a highly perishable vegetable that must be preserved in order to be stored for an extended period of time. Because of the reduced weight and volume of the finished dried product, freeze-dried green peas have a longer shelf life, are easier to carry and handle, and have a lower cost for all procedures such as packing, storage, and transportation. Figure 8.2 depicts the process flow chart for the industrial manufacturing of freeze-dried peas. Fresh, sound, and green pea pods are properly rinsed in water before being separated and cleaned using pea powder. The peas are then punctured because pricking allows for speedy and uniform drying. They are then blanched and sulfited to preserve color, flavor, and texture in the finished product. Blanched peas are then frozen from −40 to −50 °C in a freeze-dryer, which reduces

Table 8.1 Different operating parameters for freeze-dried foods.

Food product	Freeze-drying operating condition	Results	References
Pineapple, mango, Barbados cherry	Total pressure and temperature in vacuum chamber are 1.3×10^{-1} mbar and −30 °C	Freeze-dried mango, Barbados cherry and pineapple conserved their color, flavor and taste	Marques et al. (2006)
Strawberry	Slow frozen at −40 °C	Structure, color and aroma remained unchanged but moisture-sensitive	Pääkkönen and Mattila (1991)
Green peas	Microwave-assisted freeze drying at temperatures of −6, −3, and 0 °C	In terms of product quality, drying time, and process control, 6 and 3 °C fared well	Eikevik et al. (2012)
Button mushroom	Pressure (0.04, 0.07, and 0.10 mbar), first drying temperatures (−2, −5, and −8 °C), second drying temperatures (25, 28, and 31 °C)	Primary and secondary drying temperatures had greater impacts than pressure and sample thickness	Tarafdar et al. (2017)

moisture to 7–8%. To produce a vacuum, frozen peas are placed in a vacuum chamber, and the pressure inside the chamber is reduced (Alves-Filho et al. 2004).

The temperature is then increased slightly within the chamber to allow the ice in the peas to sublimate without melting. When the procedure is complete, the peas are removed from the vacuum chamber and wrapped in polybags to prevent moisture from entering the peas again. The airtight container allows the freeze-dried peas to be stored at −23 to −18 °C for one year. Throughout the process, quality control methods are often employed to verify that the finished product meets the necessary criteria. Table 8.1 shows the different operating parameters for freeze dried foods.

8.3.1 Pre-Treatment Methods of Various Foods Before Freeze-Drying

8.3.1.1 Blanching

Blanching (scalding vegetables briefly in boiling water or live steam) is done to inactivate enzyme activity, which might result in flavor, color, and texture loss. Blanching also cleans the surface of dirt and germs, improves the color, and slows vitamin loss. Additionally, it reduces the difficulty of transporting vegetables by softening or wilting them. It is important to blanch vegetables at the appropriate time and according to their size. Numerous studies have looked into the possibilities of pre-treatment of food before FD using blanching. Suriya et al. (2016) discovered that blanching elephant foot yam (*Amorphophallus paeoniifolius*) slices with sodium metabisulfite solution before FD affected the physicochemical and functional aspects of freeze-dried and powdered yam. According to Ciurzyska et al. (2014), blanching pumpkin, *Cucurbita maxima*, influenced the process of dehydration and dried product characteristics after drying. The researchers blanched the osmotically dehydrated pumpkin in water and discovered that blanching aided in the elimination of water during FD. According to experts, pumpkin processing disrupted the cell walls, causing this effect. A further observation was that the freeze-dried pumpkin material had a lower moisture content than the unblanched pumpkin. Blanching impacts the drying course and quality of freeze-dried goods, depending on the raw material and the process parameters used.

8.3.1.2 Osmotic Dehydration (OD)

Dehydrating fruits and vegetables by osmosis (OD) increases their solute content by removing water. This is called "osmotic dehydration." Capillary flow and diffusion are the major means of extracting water from a substance during osmotic dehydration. This kind of pretreatment also allows for shorter drying times (Bialik et al., 2020) and lower drying costs, which has a favorable influence on the environment. Prosapio and Norton (2017) investigated how OD affects the FD process in strawberries. They compared four osmotic agents (maltodextrin, maltose, fructose, and sucrose) and chose fructose for more research. Researchers found that freeze-dried fruits were considerably rehydrated after OD treatment, and their drying time was reduced by roughly threefold. As well, they found that freeze-dried strawberries did not hold their structure as well as OD samples, which required a greater force to penetrate than unprocessed freeze-dried strawberries. Sette et al. (2016) observed that OD on raspberries using sugar resulted in a lower rehydration rate and a significant reduction in freeze-dried fruit firmness when compared with control samples.

8.3.1.3 Pulsed-Electric Field (PEF)

The PEF approach is based on the permeabilization (electroporation) of cell membranes with short electrical pulses (often in the range of 100–300 $\mathrm{V/cm^1}$ to 20–80 $\mathrm{kV/cm^1}$). The PEF approach is based on the permeabilization (electroporation) of cell membranes using brief electric pulses (often ranging from 100 to 300 $\mathrm{V/cm^1}$ to 20–80 $\mathrm{kV/cm^1}$). When PEF is used in food preparation, the electric pulses cause biological cells to plasmolyze. Lammerskitten et al. (2019) investigated the effects of PEF and FD processes on apple drying and found that this pre-treatment technique reduced drying time by about 25% and increased the rehydration capacity of freeze-dried apples. PEF was utilized as a pretreatment for red bell pepper and strawberries before FD in another investigation (Fauster et al., 2020).

8.3.1.4 Ultrasound (US)

US technology is widely used in the food sector to increase the speed of various operations, including cutting and slicing, filtering, freezing and crystallizing, thawing, extracting, pickling, and drying (Rybak et al., 2020). The use of the US for drying speeds up the process greatly (Kowalski et al., 2017). The action of the US is mostly mechanical rather than thermal. The application of US causes surface tension in capillaries, resulting in the formation of microchannels and an easier loss of water from the sample during drying. Before freeze drying (FD), the US can also be utilized as a pre-treatment procedure. Xu et al. (2021) employed a US-freeze–thawing pre-treatment to increase the freeze-drying efficiency of okra and discovered that okra pre-treated with this approach had the highest retention of bioactive chemicals and the shortest drying time. Colucci et al. (2018) investigated the effect of US intensity on eggplant antioxidant capacity and discovered that US had no negative effect on this parameter in freeze-dried eggplants. Schössler et al. (2012) used ultrasound throughout the freeze-drying process of red bell peppers and reported that it boosted the peppers temperature while decreasing drying time by around 12%.

8.4 Need for Good Packaging for Freeze-Dried Food Products

8.4.1 Microbial Quality

In terms of microbiological threats, dried food items are regarded as safe. There is a critical water activity (a_w) below which no microbes can grow. Pathogenic bacteria cannot grow in an a_w range of 0.85–0.86, although yeast and molds can tolerate an a_w range of 0.80. Because freeze-drying is a low-temperature dehydration technology, pathogens and spoilage organisms that are resistant to these conditions may still be present in the product. Low moisture levels inhibit bacterial growth but do not prevent them from surviving in food. If the product is not adequately maintained or wrapped, it may accumulate moisture and begin to proliferate previously inhibited bacteria.

Furthermore, considering freeze-dried samples have significant hygroscopic qualities, moisture during storage may have an adverse effect on them, reducing the quality of the dried fruit. Polyphenol stability during storage, as well as storage temperature, may have an influence on food quality. Browning, taste loss, and nutritional loss are the three basic ways that dried fruits deteriorate.

8.4.2 Chemical Quality

During drying and storage, lipids might become brownish, colors can fade, and flavors can change. Browning processes might be enzymatic or nonenzymatic (Salunkhe et al. 1991). The oxidation of polyphenol chemicals, both enzymatic and non-enzymatic, as well as the Maillard process, produces the color shift in dried fruits (McEvily et al. 1992). According to Gvozdenović et al. (2007), Piasecka et al. (2013), and Henríquez et al. (2013), adequate packaging material, as well as the right temperature and storage term, had a significant impact on the quality and sensory qualities of dried fruits. Intricate films, such as laminated materials and metalized films made of polyethylene and aluminum foils, give outstanding protection to dry foods due to their strong barrier qualities.

As a result, they are commonly used to wrap snacks and higher level meals (Zorić et al., 2016). Browning tends to occur primarily at the midpoint of the drying period. In carbohydrate foods, by eliminating or avoiding amines, browning can be minimized, and in protein meals, by removing reducing sugars. By producing inactive hydroxysulfonate sugar derivatives, sulfur treatment can avoid the first condensation process. Because of their decreased porosity, meals that have been air-dried are less susceptible to lipid oxidation than meals that have been freeze-dried (Sablani, 2006). The packaging of freeze-dried food products should take into account a number of factors to prevent product damage during handling and transit (Figure 8.2).

8.4.3 Nutritional Quality

Fruits, vegetables, and their dry derivatives are high in energy, minerals, and vitamins. However, nutritional quality varies throughout the dehydration phase (Sablani, 2006). Many vitamins, including A, C, and thiamine, are thermo-labile and susceptible to oxidative destruction. Sulfuring can destroy thiamine and riboflavin; however, pre-treatments like blanching and dipping in sulfite solutions can limit vitamin reduction during drying. Dried foods' sensory characteristics are equally significant in assessing quality. Color, scent, flavor, texture, and taste are examples of these. The most typical quality decline for dried items is the loss of volatile organic molecules, which causes aroma and flavor to alter. According to Singh et al. (2006), a low-temperature dryer produced less damage to green crops during drying and so maintained more nutrients than other dryers. Extreme temperatures or high humidity are two environmental variables that the packaging should be able to tolerate. In a nutshell, the packaging material should protect the freeze-dried meals' integrity and safety. Hence, packaging plays a crucial role in dictating the overall product quality and stability, as the former imparts barrier protection against external factors that may affect the efficacy of the freeze-drying process.

8.4.4 Influence of Storage Quality

During preservation, dried fruits and vegetables lose a substantial amount of nutrients. Storage temperature, pH, contact with oxygen, permeability, light, and the presence of naturally occurring acids all have an impact on this loss. The amount of loss is determined by the kind of vitamin and storage circumstances such as oxygen and light exposure (Sablani, 2006). As a result, airtight packing is necessary to prevent nutritious characteristics from deteriorating. For example, Kaminski et al. (1986) discovered that carotenoids degrade quickly in freeze-dried carrots. In general, freeze-dried items are more porous. This accelerates the fast oxidation of carotene and improves oxygen transport. Çinar (2004) found that carrots held at 40 °C had the maximum pigment loss (98.1%), whereas sweet potatoes stored at 4 °C had the lowest loss (11.3%) after 45 days of storage.

8.5 Selection of Packaging Material

It is important to consider factors such as product damage and fines. There are several factors to consider, including stickiness, bag opening size, and re-closure. A freeze-dried food packaging material should take into account the following factors.

8.5.1 Water Vapor Transmission Rate (WVTR)

Papers and aluminum foil exhibit considerable changes when folded, creased, or crumpled due to the WVTR of packaging material.

8.5.2 Gas Transmission Rate

Oxygen-sensitive fruit items have a shorter shelf life due to their permeability to gases such as oxygen. A freeze-dried candy's packaging should be permeable to volatiles and flavors in addition to oxygen transmission rate. A polyolefin is high in value; on the other hand, plastics such as polyester, nylon, and ethylene vinyl alcohol are good at preventing volatile contaminants from being transmitted.

8.5.3 Tensile Strength and Elongation

It is essential to study the tensile strength and elongation properties of materials before deciding whether they can run on high-speed machines.

8.5.4 Tear Strength

Tear strength plays an important role for fruit-processed products, since opening packages by hand can be difficult if the tear strength is too high.

8.5.5 Heat Seal Strength

If the package seal is not perfect, the resistance to water vapor, gases, and volatiles rises, affecting the end package's performance. Polyethylene and other thermoplastic films offer good heat seals.

8.5.6 Performance Properties

A material must operate well on machines in addition to the above-mentioned critical features; consequently, an understanding of physical attributes such as slip, stiffness, and blocking resistance is required. Twist retention is also important for twist wrap. The primary purpose of packing is to protect. However, the emotional function that packing plays is important as well, especially when the confection is a gift. A complex package made of expensive materials is frequently utilized to communicate sentiments. Packaging must also be tailored to certain demographics. For brief impulse purchases at gas stations and roadside stores, a different pack size is necessary than for supermarkets that sell mostly family-sized packs. Figure 8.3 illustrates the barrier properties of packaging against environmental factors during freeze-drying.

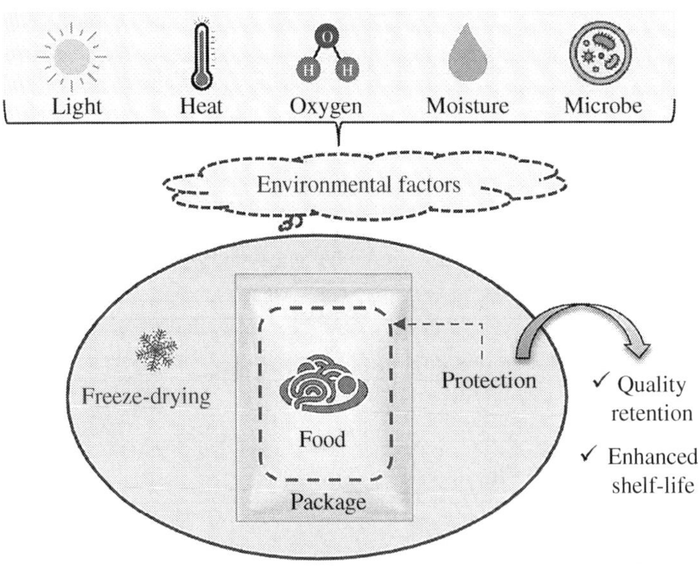

Figure 8.3 Barrier properties of packaging against environmental factors during freeze-drying.

8.6 Types of Packaging for Freeze-Dried Foods

There are numerous packaging types available depending on the requirements for storing the freeze-dried food product (Table 8.2). Figure 8.4 illustrates the packaging requirements of freeze-dried foods. To prevent freeze-dried foods from absorbing moisture from the air, they must be kept in airtight containers. For bulk packing, a variety of containers can be deployed, including plastic laminated foil pouches, metal and plastic cans, and metal and fiber drums. Certain freeze-dried foods are packaged using vacuum packing, in which the air is eliminated from the container before it is sealed. In order to replace the oxygen in the air and stop the food from oxidizing or decaying, other foods have an inert gas such as nitrogen injected into the container before sealing. The freeze-dry facility completes the packaging practically immediately after the items exit the drying chamber. The facility can shape, fill, and seal the packages to the final user's desired weight. Carton-packed, pallet-stacked packages that are meant to be sold directly to customers are delivered to the grocery warehouse.

8.6.1 Modified Atmosphere Packaging (MAP)

An effective strategy for maintaining the sensory quality and increasing the shelf life of several food commodities, including chicken meat, has been characterized in certain research as modified atmosphere packaging (MAP) (Chouliara et al., 2007). In order to prevent hardening and the loss of textural features, García-Esteban et al. (2004) found that modified-environment packaging was more effective than vacuum packaging at protecting meat (such as dry-cured ham). The principle of MAP is to replenish the environment around a product before closing it, with the most typical gases being carbon dioxide, oxygen, and nitrogen. Carbon dioxide has bacteriostatic properties (Nair et al., 2015). Nitrogen is necessary to prevent fat oxidation and pack collapse, whereas oxygen is necessary to preserve meat color. Depending on the exact requirements of the particular food products being preserved, these gases can be used singly or in combination to get the best outcome (Narasimha Rao and Sachindra, 2002).

Oxidative rancidity is the primary process of food degradation that affects dried foods with a high content of unsaturated fatty acids, including cereals, potato chips, almonds, cocoa powder, and dry milk. MAP inhibits this harmful reaction very efficiently. High gas and moisture barrier qualities are required for MAP materials. Films that have been metalized have these essential qualities. In order to prevent oxidative rancidity in certain dry foods, such as dried infant milk, residual oxygen levels should be kept below 0.2%. Oxygen scavengers may be added to MA packs to achieve very low residual O_2 levels. Moreover, these oxygen scavengers can be used on foods with little water activity, including bread goods. The use of oxygen scavengers extends the shelf life of freeze-dried foods, prevents

Table 8.2 Packaging materials for different types of freeze-dried food products.

Type of packaging	Packaging material composition	Freeze-dried food product	Function of package material	Comparison with control	Reference
Resealable pouch	Recyclable polyethylene films	Blueberry, strawberry, kiwi, banana, blackcurrant, sour cherry	High barrier properties, oxygen and water vapor resistant, sustainable	Resealable package for longer freshness	ePac Flexible Packaging (2023)
Edible coating	Whey protein isolate-Pullulan	Chinese chestnut	Inhibits microorganism growth and discoloration on surfaces	Chestnut uncoated in SEM revealed ruptured cells and void spaces unlike coated chestnut	Gounga et al. (2008)
Metalized polypropylene and aluminum packaging	Laminate PET 12 μm/PP 18 μm met/PE 100 μm	Sour cherry (*Prunus cerasus* var. Marasca)	Polyphenols are preserved for a longer period of time, almost all of them, after 12 months	PET/PPmet/PE foil more effective than PET/Al/PE in respect to color and sensory profile	Zorić et al. (2016)
Modified atmosphere packaging (MAP)	20%CO_2/80%N_2	Ozonated Chicken breast fillets	Color with texture and sensory quality preserved	Improved physicochemical and sensory quality for long time	Zouaghi and Cantalejo (2016)
Closed container and vacuum packaging	Vacuum bag with aluminum foil	*Moringa oleifera* seeds powder	Coagulation efficiency preserved in room and at refrigerated temperature (4 °C)	*M. oleifera* freeze-dried seeds provide better coagulation efficiency than non-freeze dried seeds	Katayon et al. (2006)
Chia mucilage films	Chia seeds oil	Mango fruit powder, instant coffee powder, pomegranate, apples, pineapple	Source of nutrients, edible sachets, biodegradability	Freeze-dried mucilage provided better mechanical performance than oven dried mucilage	Fernandes et al. (2020)
Arrowroot starch films	Microencapsulated blackberry pulp powder	Bioactive compound	Freeze-dried blackberry powder improved color, anthocyanin and antioxidant property to the films	Increment in water vapor permeability and excellent carrier of bioactive extract	Nogueira et al. (2019)

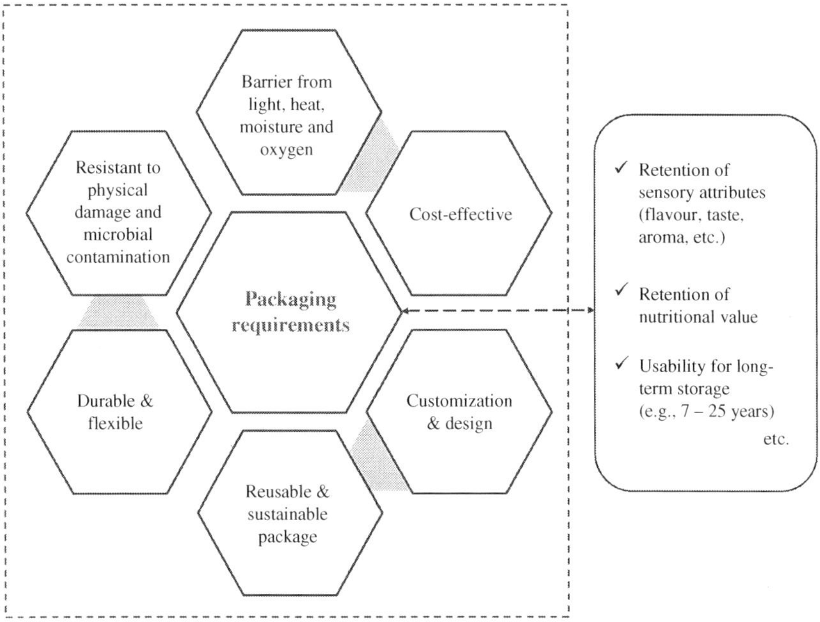

Figure 8.4 Freeze-drying packaging requirements.

bacterial and insect growth in food packets, maintains the freshness of food products, and eliminates the need for preservatives.

8.6.2 Flexible Packaging

Although still in development, the retortable pouch technology has certain practical uses and has just recently been given restricted authorization for use by the appropriate regulatory bodies. The marketing system as a whole is projected to be significantly impacted by further use of this technology, particularly in the areas of energy, food storage, transportation, labor, and retailing. The retortable pouch technology is therefore a major priority for evaluation. The pouch is a thermos processing temperature-resistant multilayer (plastic laminate with a center layer of aluminum foil), adhesively linked bag. It combines a number of the benefits of both the plastic boil-in-the-bag and the metal can. Whether it comes to packaging dehydrated, air-dried, or freeze-dried (lyophilized) goods, MylarFoil strikes a wonderful balance between affordability and high barrier qualities.

This pouch is flexible, light, and UV-resistant, thanks to the biaxially oriented polyethylene terephthalate film that is used to cover a thin layer of foil. Its strong tensile characteristics and barrier qualities separate foods, components, and supplements from its surroundings. Without worrying about damaging or puncturing them, you can vacuum seal them as well. Ingredients stored in packaging with oxygen have a much shorter shelf life, and dehydrated foods must naturally be kept dry until you are ready to rehydrate them. The quality and longevity of the product are significantly improved when oxygen absorbers or desiccant are used in conjunction with high barrier packaging.

8.6.3 Shrink Film

A large number of polymeric materials, in addition to cellulosics and aluminum foil, are utilized for packaging. For some purposes, paperboard and metal containers are also utilized. Although a variety of packaging materials are available, the overall wrapper choice is determined by the obligatory shelf life, wrapping machine efficiency, and other factors and costs, which are solely determined by the manufacturer's target market segment. Plastic (usually flexible) is the most often used packaging medium because it offers adequate protection and preservation, as well as friction resistance, durability, productivity, and printability. Because plastic is lighter in weight, it is suitable for candy packaging. Candy packaging is undergoing many changes. Because of their superior properties, plastic films and laminates are gradually replacing waxed papers, while aluminum foil laminates are replacing waxed papers due to their lower cost and superior flex fracture resistance. Packaging films made of plastic differ depending on the form of package, such as twist wrap, pillow pack, vertical flow pack, or roll pack.

8.6.4 Cardboard Packaging

Chipboard, often known as cardboard, is widely used in the packaging of freeze-dried goods. Chipboard boxes are frequently used to package pies, vegetables, pizza, cakes, and a variety of other frozen food goods. Chipboard boxes are strong, long-lasting, and simple to shrink wrap.

8.6.5 Wax-Coated Cardboard

Another common type of frozen food packaging is wax-coated cardboard, specifically for frozen items that are going to be cooked in the container.

8.6.6 Edible Films

To stop oxidative deterioration, mold growth, and moisture penetration, the majority of food products can be coated; food must be removed from the coating to achieve these goals, as these coatings are usually not edible. Freeze-dried foods, such as those produced through a well-known process, pose a food safety concern. It is not enough to coat dehydrated food with edible coatings to prevent oxidative deterioration, mold attack, and moisture penetration; it is also crucial that the coating should not negatively influence the food's ability to rehydrate and its flavor after rehydration. Furthermore, dehydrated food tends to be brittle and crumble when subjected to mechanical stress or shock. Dehydrated food products are therefore highly desirable to protect against fragmentation by providing edible coating compositions. A variety of food products have been coated with edible cellulose ethers. It should be noted that edible cellulose ethers have poor moisture barriers and are extremely brittle, resulting in coatings that crack easily, making them unsuitable for food protection. The edible cellulose ethers are modified by adding plasticizers to improve their toughness and coherence. Food-safe plasticizers that accomplish these improvements are, however, either not available or, if they are, result in a substantial reduction in edible cellulose ethers' moisture barrier properties.

There is little research done in the edible coating of freeze-dried food products, which is commercially unavailable due to consumer acceptance of the product in the market and limited investments made in this area. According to certain studies, covering whey protein isolate (WPI) films with a low water vapor permeability (WVP) material can help reduce the rate of rehydration. Because of the presence of covalent disulfide linkages, WPI films were water-insoluble, as opposed to caseinate films. Glycerol is added to films, which decreases internal hydrogen bonding and raises WVP. It has been demonstrated that the WVP properties of whey protein films are exponentially affected by relative humidity. For instance, under 11% and 65% humidity levels, the WVP of whey protein–glycerol film was 6.6 and 119.8 g/cm^2, respectively (McHugh and Krochta 1994). Greater humidity and greater glycerol content in the films promote flexibility. At lower humidity levels, higher glycerol content films are more flexible than lower glycerol content films (Coupland et al., 2000). Researchers created a whey protein coating solution to lower the rehydration ratios of freeze-dried strawberry pieces due to their physicochemical properties (Huang et al., 2009).

8.6.7 Cans and Glass Jars

Depending on the particular needs and specifications of the product, cans can be a useful packaging choice for freeze-dried goods. Because they are strong and resilient, cans can survive rigorous handling while being transported and stored. Moreover, they can shield the product from being crushed and pierced. Cans are a practical choice for both retailers and customers because they are simple to stack and store. Cans may cost more than other forms of packaging, particularly if they are made of premium materials. If they offer the required durability and protection for the goods, they might be worth the expense. Combining the anti-corrosion, 100% recyclable, and pliable qualities of tin and aluminum results in food

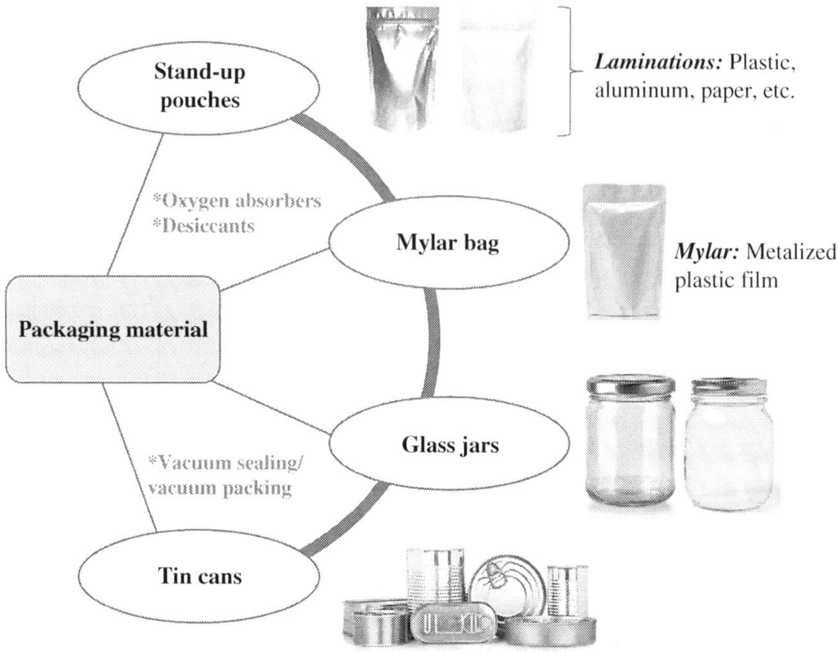

Figure 8.5 Common types of packaging material for freeze-dried food products.

packaging that keeps food fresher for longer. Tin cans are used in a variety of ways to package food. There are rectangular cans and the typical cylindrical cans. Tin cans of every type are used by manufacturers for packaging. Tin cans are divided into two primary varieties, however, in addition to shape, size, and form: two-piece cans and three-piece cans. High-end freeze-dried food products are a fantastic fit for glass jars because they offer exceptional protection from light and oxygen. They are more delicate and heavier than alternative packing choices, though. Figure 8.5 shows the commonly used types of packaging material for freeze-dried foods.

8.7 Quality Consideration During Packaging

Quality control of packed products is checked before reaching the customer:

1) Package weight
2) Freeze-dried food product weight
3) Arrangement of the freeze-dried product in the boxes
4) Damage to the product
5) Moisture content

8.8 Case Studies of the Packaging of Freeze-Dried Food Products

8.8.1 Packaging of Strawberries

Pääkkönen and Mattila investigated the processing and packaging of freeze-dried strawberries in 1991. It has been discovered that freeze-dried strawberries are responsive to changes in storage conditions. The distinctive scent ingredient in farmed strawberries is *Fragaria ananassa* var. *Senga Sengana* was 2,5-dimethyl-4-methoxy-3(2*H*)-furanone, which is a major volatile component in wild strawberries. They discovered that freeze-dried reference strawberries maintained at −40 °C preserved their fresh flavor for a year but were susceptible to moisture in their investigation. The odor and flavor intensity of the samples held in laminate bags in a vacuum were determined to be the same as those of the reference sample.

8.8.2 Packaging of Sour Cherries

Zorić and his colleagues researched freeze-dried sour cherry packing. Sour cherries are high in polyphenols, which impact the fruit's quality and nutritional worth. The researchers freeze-dried the sour cherry to maintain this feature. Despite having significant hygroscopic characteristics, these freeze-dried samples are impacted by moisture during storage. As a result, the sour cherries were packaged for up to 12 months in high-hedge metalized polypropylene and aluminum packaging (PET/PPmet/PE and PET/Al/PE) at 4, 20, and 37 °C. The experimenters set up that regardless of the type of packaging used, the cherries stored at a lower temperature for 12 months retained a higher content of flavonol glycosides (FGs) (quercetin-3-glucoside, kaempferol-3-glucoside, and kaempferol-3-rutinoside) and hydroxycinnamic acids (HCAs) (neochlorogenic, chlorogenic, p-coumaric, caffeic, and ferulic acid) than anthocyanins (ANTs) (cyanidin-3-glucosylrutinoside, cyanidin-3-rutinoside, cyanidin-3-glucoside, and cyanidin-3-sophoroside). Furthermore, the food preserved its unique dark red color. This study found that laminated packaging improved product quality (Zorić et al., 2016).

8.8.2.1 Packaging of Accelerated Freeze-Dried Fish Products

Packaging of accelerated freeze-dried fish goods comprises fish, prawns, or any other fishery product that is in great demand in the export sector. The moisture percentage of AFD products is typically about 2%. These foods are particularly hygroscopic due to their low moisture content and vast surface area. Because fish consists of fat, it is plausible that it will absorb taint from the packing material. Mechanical damage is caused by the structural features of freeze-dried items. Because of their light, porous nature, they are extremely delicate and quickly breakable.

Rigid containers, both glass and plastic, are utilized to package these foods. However, metalized polyester laminated with polyethylene or aluminum foil/paper/polyethylene and metallized polyester/linear low-density polyethylene (LLDPE) or LLDPE–polyamide (PA)–low-density polyethylene (LD) co-extruded film bags with nitrogen flushing, 2 packs in a 5 ply corrugated fibreboard (CFB) shipping is just starting to be employed due to their low oxygen and water vapor transmission rates. The majority of the containers contain an inert gas. The product can also be vacuum-packed for added protection against damage. Simon et al. (1971) investigated the durability of freeze-dried shrimp bars packed in various materials. They created an algorithm for forecasting dry food storage durability in semipermeable containers. They detected pigment loss and color deviation from the original color, although the organoleptic criteria were acceptable. This color loss was discovered to be connected to oxygen transport inside the prawn bars. It was determined that more viable packaging for freeze-dried shrimp bars should be devised.

8.9 Market Scenario

The market for freeze-dried food products is ever-growing, with different categories of food products available globally (Figure 8.6).

8.9.1 Asia-Pacific

The market for freeze-dried food in the Asia-Pacific area is expanding quickly due to rising demand for processed food and ready-to-eat meals as well as new lifestyle trends such as on-the-go among customers. These variables develop Asia as a prospective market for freeze-dried meals; therefore, manufacturers are looking into this trend to cater to the youthful consumers in the region. A growing trend of supermarket shopping through online distribution channels and rising internet usage in the region are both anticipated to increase demand for freeze-dried ready meals. Due to Chinese consumers' enthusiasm for novel, wholesome products that fit into their hectic schedules and are manufactured using Chinese ingredients, foreign retail manufacturers can grow their businesses in China. Tyson Foods Inc., for instance, introduced frozen Bites, Nuggets, and Strips in 2021 in a few store outlets and on e-commerce platforms in the Asia-Pacific under the brand name First Pride (Report, 2022).

8.9.2 European Market

The freeze-dried food industry in Europe is made up of a number of different groups, including fruits, vegetables, beverages, meat and fish, milk products, and

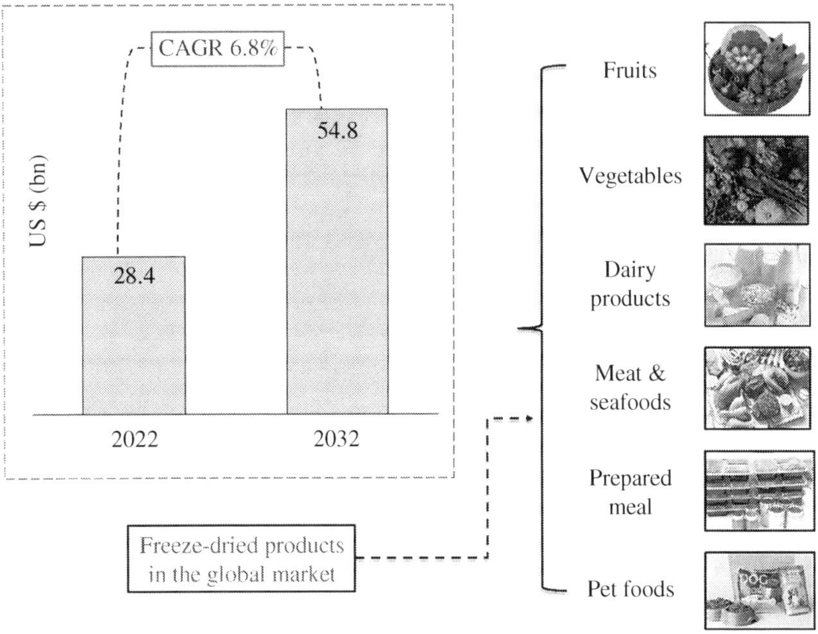

Figure 8.6 Market freeze-dried products available globally. Sourced from: https://www.futuremarketinsights.com/reports/freeze-dried-food-market.

prepared dishes. The category for freeze-dried beverages includes sections for coffee, tea, and other frozen drinks. Each product type is retailed via supermarkets and hypermarkets, convenience stores, department stores, online retailing, and other distribution methods. The European freeze-dried food business is extremely competitive due to the large number of companies. Because of their market position, the top corporations concentrate on product development, mergers and acquisitions, alliances, and expansion. Some of the leading participants include AMG Group Ltd., European Freeze Dry, Katadyn Products Inc., LYO FOOD Sp. Z o.o., Harmony House Foods Inc., and other businesses. Due to rising income levels and consumers' expanded choice of varied ready-to-eat food products, major producers in the freeze-dried business are concentrating on seizing opportunities provided by emerging countries like Germany, Russia, and Spain to boost their income levels.

8.9.3 Freeze-Dried Food Market: Recent Developments

In June 2022, Conagra Brands unveiled a substantial summer portfolio of upcoming product advancements. These inventions incorporated well-liked flavors, ingredients, and practical preparation techniques in order to satisfy desires and solve mealtime problems. The frozen snack items were added to Prasuma's menu in April 2022, an Indian company that sells frozen cuisine; some of the most recent products introduced were Bacon, Frozen Veg and Chicken Spring Rolls, Frozen Chicken Nuggets, Chicken and Veg Mini Samosas, Chicken and Mutton Shammi Kababs, and Seekh Kababs. Also, in January 2022, the New Jersey-based company called "Crispy Green" debuted Pia Picante, a brand of dry pineapple snacks, as its newest flavor. The product is produced using simple components, has a clean label, and is shelf-stable.

8.10 Sustainability Aspects

Sustainable management is now gaining popularity since it is viewed as a method intended to use all of the ingredients used to make food with the least amount of waste possible thereafter. The food sector still faces a sizable and expensive problem with waste, despite the fact that current technology is increasingly useful in reducing waste from homes and restaurants. Food loss and landfill waste can be greatly reduced by using edible fruits and vegetables that are labeled as "green trash" because they do not meet grocery criteria for color, uniformity of surface, or calibrated shape and size. By using freeze-drying, it is possible to create a useful, long-lasting product that eliminates the issues of rotting produce and the requirement for cold storage. Sustainability is the main concern in the food industry with respect to the environment (Janiszewska-Turak et al., 2022).

The majority of businesses that produce freeze-dried food products for the market adhere to sustainability standards. The corporations have set a target of becoming climate-neutral, or with zero carbon emissions, by 2025; however, they are aware that they will still leave some carbon footprints, which will be steadily decreased. Reusing packaging minimizes carbon emissions, and using recycled printed paper is one of the steps taken by food firms. Packaging materials are 98% recyclable, and all cardboard boxes and trays are Forest Stewardship Council (FSC)-certified. Table 8.3 displays the advantages and disadvantages of packaging freeze-dried food products.

8.11 Safety Aspects of Freeze-Dried Product Packaging

Freeze-dried food products retained their quality parameters such as color, flavor, chemical composition, and texture; therefore, the freeze-drying method is considered the best dehydration technique for food preservation. However, it has a negligible or minor damaging effect on microorganisms (Xiao, 2018). At the same time, improper handling during packaging and the type of packaging material can cause deterioration of the freeze-dried product. Hence, packaging is a crucial

Table 8.3 Advantages and disadvantages of packaging freeze-dried food products.

Advantages	Disadvantages
If packaging is lightweight, shipping and storage costs are even further reduced	Packaging materials and equipment for freeze-dried products are costly
Proper packaging enhances the durability of the product	Can have an environmental impact if not biodegradable or recyclable
Proper packaging protects the food by retaining much of its color, shape, texture, and flavor	If not airtight or of low quality, it can compromise the safety of the product
Good packaging designs can make the product attractive to consumers	
It is simple to freeze-dry a solution, thin slices of meat, and perishable fruits and vegetables	

stage in the freeze-drying process, and when done properly, the product is shielded from physical damage as well as microbial and chemical contamination. Moreover, the maximum allowed limits for microbial contamination in freeze-dried food can vary depending on the type of microbe. For instance, the maximum permitted level for *Escherichia coli* is 10 CFU/g, while the maximum allowable limit for total coliforms in dried fruits is 1000 CFU/g. In terms of food safety, it is imperative to follow the maximum permissible level for microbiological contamination in freeze-dried foods. Depending on the food and the intended application, the limit may change. For instance, the limits for microbiological contamination of food, including freeze-dried food, as defined by the U.S. Food and Drug Administration (FDA) allow 10 000 CFU/g as the upper permissible limit for yeast and mold in dried fruits. Likewise, the European Food Safety Authority (EFSA) establishes limitations for microbiological contamination in food. These restrictions guarantee the safety of freeze-dried food for consumption.

Another aspect of freeze-dried food safety is the assurance of using the right packaging type and material. The integrity of the packaging material must be validated so that the freeze-dried food retains its quality and safety. The packaging is usually done in a controlled environment, and numerous testing techniques (depending on the packaging type) are used to ensure that the packaging material is not affected during freeze-drying. For instance, the pressure decay method is a commonly used leak integrity test, while seal strength integrity is extensively used to detect seal strength by detecting properties like burst testing and creep testing, to name a few. Also, the packaging type must be able to withstand the storage conditions (shelf temperature level, freezing rate, total moisture, total gas pressure, etc.) without changing the quality-determining properties of the dried product.

However, different packages are used for different products, and testing methods vary from product to product; therefore, selecting the appropriate method is crucial in detecting package integrity, although consideration of factors like packaging material and design, process line, and packaging environment is needed. Hence, packaging testing and validation are significant aspects of freeze-drying packaging because the detection of packaging issues dictates the efficacy of the freeze-drying process, thereby ensuring that the product's shelf life is not compromised.

8.12 Principal Instrumental Techniques Employed for Packaging Controls

8.12.1 Spectrophotometry

It is used to assess the color and opacity of packaging materials, including films, sheets, and inks. This can assist producers in ensuring that their packaging materials fulfill color and opacity specifications. It also tracks the breakdown of packing materials over time, allowing it to estimate the shelf life and stability of packed goods.

8.12.2 Thermal Analysis Techniques

The thermal characteristics of packing materials are evaluated using this method. These include differential scanning calorimetry (package material heat capacity), thermogravimetric analysis (package material breakdown behavior), and dynamic mechanical analysis (mechanical characteristics).

8.12.3 Gas Transmission Analysis

It is a method for determining the permeability of packing materials to different gases. Under controlled conditions, the quantity of gas traveling through a film of packaging material is measured. The material is sealed over a tiny aperture in a diffusion cell, the cell is filled with a known concentration of gas, and the amount of gas that diffuses through the film over time is measured.

8.13 Conclusion and Future Scope

Long-term freeze-dried fruits contain a variety of fragrant and therapeutic plant parts, such as blooms, seeds, stems, leaves, and roots, which can deteriorate with time, making them susceptible to contamination by toxigenic fungus and bacteria. One of the key forces influencing the market is this. Moreover, these fruits contain a powdered blend of many medicinal plants, which could produce poisons. Fruit that has been frozen-dried naturally, without the use of artificial additives or preservatives, has a long shelf life. Therefore, it has been challenging in the market to develop packaging material accordingly.

There is a lot of intriguing research and development currently underway in the field of packaging freeze-dried goods. The demand for environmentally friendly packaging options is rising as consumers become more conscious of their environmental impact. Although some companies have already begun to use it, more environmentally friendly packaging alternatives are still something we may anticipate in the future. The market has not yet witnessed the development of intelligent packaging for freeze-dried items, which might be a game-changer for further extending their shelf life. We can anticipate seeing more creative package options that offer outstanding protection while being simple to use, store, and transfer. Overall, a focus on sustainability, functionality, and innovation is expected to characterize packaging for freeze-dried food products in the future.

References

Alves-Filho, O., García-Pascual, P., Eikevik, T.M., and Strømmen, I. (2004, August). Dehydration of green peas under atmospheric freeze-drying conditions. In: *XIV Simposio Internacional de Secado, São Paulo (Brasil)*, 1521–1528.

Bialik, M., Wiktor, A., Witrowa-Rajchert, D., and Gondek, E. (2020). The influence of osmotic dehydration conditions on drying kinetics and total carotenoid content of kiwiberry (*Actinidia arguta*). *International Journal of Food Engineering* 16.

Chouliara, E., Karatapanis, A., Savvaidis, I.N., and Kontominas, M.G. (2007). Combined effect of oregano essential oil and modified atmosphere packaging on shelf-life extension of fresh chicken breast meat, stored at 4 C. *Food Microbiology* 24 (6): 607–617.

Çinar, I. (2004). Carotenoid pigment loss of freeze-dried plant samples under different storage conditions. *LWT - Food Science and Technology* 37 (3): 363–367.

Ciurzyńska, A., Lenart, A., and Greda, K.J. (2014). Effect of pre-treatment conditions on content and activity of water and colour of freeze-dried pumpkin. *LWT - Food Science and Technology* 59: 1075–1081.

Colucci, D., Fissore, D., Rossello, C., and Carcel, J.A. (2018). On the effect of ultrasound-assisted atmospheric freeze-drying on the antioxidant properties of eggplant. *Food Research International* 106: 580–588.

Coupland, J.N., Shaw, N.B., Monahan, F.J. et al. (2000). Modeling the effect of glycerol on the moisture sorption behavior of whey protein edible films. *Journal of Food Engineering* 43 (1): 25–30.

Doymaz, İ. and Kocayigit, F. (2011). Drying and rehydration behaviors of convection drying of green peas. *Drying Technology* 29 (11): 1273–1282.

Eikevik, T.M., Alves-Filho, O., and Bantle, M. (2012). Microwave-assisted atmospheric freeze drying of green peas: a case study. *Drying Technology* 30 (14): 1592–1599.

ePac Flexible Packaging. (2023, February 17). ePac Flexibles—Custom Flexible Packaging Company—Digital Printing for Custom Mylar Bags, Stand Up Pouches & More. https://epacflexibles.com/

Fauster, T., Giancaterino, M., Pittia, P., and Jaeger, H. (2020). Effect of pulsed electric field pretreatment on shrinkage, rehydration capacity and texture of freeze-dried plant materials. *LWT* 121: 108937.

Fernandes, S.S., Romani, V.P., da Silva Filipini, G. et al. (2020). Chia seeds to develop new biodegradable polymers for food packaging: properties and biodegradability. *Polymer Engineering & Science* 60 (9): 2214–2223.

García-Esteban, M., Ansorena, D., and Astiasarán, I. (2004). Comparison of modified atmosphere packaging and vacuum packaging for long period storage of dry-cured ham: effects on colour, texture and microbiological quality. *Meat Science* 67 (1): 57–63.

Gounga, M.E., Xu, S.Y., Wang, Z., and Yang, W.G. (2008). Effect of whey protein isolate–pullulan edible coatings on the quality and shelf life of freshly roasted and freeze-dried Chinese chestnut. *Journal of Food Science* 73 (4): E155–E161.

Gvozdenović, J., Aljilji, A., Lazić, V. et al. (2007). Influence of packaging materials protective characteristics on the quality changes of dried apples. *Časopis za procesnu tehniku i energetiku u poljoprivredi/PTEP* 11 (3): 112–115.

Henríquez, C., Córdova, A., Lutz, M., and Saavedra, J. (2013). Storage stability test of apple peel powder using two packaging materials: high-density polyethylene and metalized films of high barrier. *Industrial Crops and Products* 45: 121–127.

Huang, L., Zhang, M., Yan, W. et al. (2009). Effect of coating on post-drying of freeze-dried strawberry pieces. *Journal of Food Engineering* 92 (1): 107–111. https://doi.org/10.1016/j.jfoodeng.2008.10.031.

Janiszewska-Turak, E., Rybak, K., Pobiega, K. et al. (2022). Sustainable production and characteristics of dried fermented vegetables. *Fermentation* 8 (11): 659.

Kaminski, E., Wasowicz, E., Zawirska, R., & Wower, M. (1986). Effect of drying and storage of dried carrots on sensory characteristics and volatile constituents. *Die Nahrung= Food*.30 819–828.

Katayon, S., Noor, M.M.M., Asma, M. et al. (2006). Effects of storage conditions of *Moringa oleifera* seeds on its performance in coagulation. *Bioresource Technology* 97 (13): 1455–1460.

Kowalski, S.J., Mierzwa, D., and Stasiak, M. (2017). Ultrasound-assisted convective drying of apples at different process conditions. *Drying Technology* 35: 939–947.

Lammerskitten, A., Mykhailyk, V., Wiktor, A. et al. (2019). Impact of pulsed electric fields on physical properties of freeze-dried apple tissue. *Innovative Food Science and Emerging Technologies* 57: 102211.

Lockhart, H.E. (1997). A paradigm for packaging. *Packaging Technology and Science: An International Journal* 10 (5): 237–252.

Marques, L.G., Silveira, A.M., and Freire, J.T. (2006). Freeze-drying characteristics of tropical fruits. *Drying Technology* 24 (4): 457–463.

McEvily, A.J., Iyengar, R., and Otwell, W.S. (1992). Inhibition of enzymatic browning in foods and beverages. *Critical Reviews in Food Science and Nutrition* 32 (3): 253–273.

McHugh, T.H. and Krochta, J.M. (1994). Water vapor permeability properties of edible whey protein-lipid emulsion films. *Journal of the American Oil Chemists' Society* 71: 307–312.

Nair, D.V., Kiess, A., Nannapaneni, R. et al. (2015). The combined efficacy of carvacrol and modified atmosphere packaging on the survival of Salmonella, *Campylobacter jejuni* and lactic acid bacteria on turkey breast cutlets. *Food Microbiology* 49: 134–141.

Narasimha Rao, D. and Sachindra, N.M. (2002). Modified atmosphere and vacuum packaging of meat and poultry products. *Food Reviews International* 18 (4): 263–293.

Nogueira, G.F., Fakhouri, F.M., Velasco, J.I., and de Oliveira, R.A. (2019). Active edible films based on arrowroot starch with microparticles of blackberry pulp obtained by freeze-drying for food packaging. *Polymers* 11 (9): 1382.

Pääkkönen, K. and Mattila, M. (1991). Processing, packaging, and storage effects on quality of freeze-dried strawberries. *Journal of Food Science* 56 (5): 1388–1392.

Perchonok, M. and Bourland, C. (2002). NASA food systems: past, present, and future. *Nutrition* 18 (10): 913–920.

Piasecka, E., Uczciwek, M., Klewicki, R. et al. (2013). Effect of long-time storage on the content of polyphenols and ascorbic acid in osmo-convectively dried and osmo-freeze-dried fruits. *Journal of Food Processing and Preservation* 37 (3): 198–209.

Prosapio, V. and Norton, I. (2017). Influence of osmotic dehydration pre-treatment on oven drying and freeze drying performance. *LWT - Food Science and Technology* 80: 401–408.

Report (2022). *Freeze Dried Food Market Outlook (2022–2032)*. Future Market Insights Retrieved March 11, 2023, from https://www.futuremarketinsights.com/reports/freeze-dried-food-market.

Rybak, K., Samborska, K., Jedlinska, A. et al. (2020). The impact of pulsed electric field pretreatment of bell pepper on the selected properties of spray dried juice. *Innovative Food Science and Emerging Technologies* 65: 102446.

Sablani, S.S. (2006). Drying of fruits and vegetables: retention of nutritional/functional quality. *Drying Technology* 24 (2): 123–135.

Salunkhe, D.K., Bolin, H.R., and Reddy, N.R. (1991). Storage, processing, and nutritional quality of fruits and vegetables. In: *Fresh Fruits and Vegetables*, 2e, vol. vol. I. CRC Press.

Schössler, K., Jäger, H., and Knorr, D. (2012). Novel contact ultrasound system for the accelerated freeze-drying of vegetables. *Innovative Food Science and Emerging Technologies* 16: 113–120.

Sette, P., Salvatori, D., and Schebor, C. (2016). Physical and mechanical properties of raspberries subjected to osmotic dehydration and further dehydration by air- and freeze-drying. *Food and Bioproducts Processing* 100: 156–171.

Simon, I.B., Labuza, T.P., and Karel, M. (1971). Computer-aided predictions of food storage stability: oxidative deterioration of a shrimp product. *Journal of Food Science* 36 (2): 280–286.

Singh, U., Sagar, V.R., Behera, T.K., and Suresh, K.P. (2006). Effect of drying conditions on the quality of dehydrated selected leafy vegetables. *Journal of Food Science and Technology-Mysore* 43 (6): 579–582.

Suriya, M., Baranwal, G., Bashir, M. et al. (2016). Influence of blanching and drying methods on molecular structure and functional properties of elephant foot yam (*Amorphophallus paeoniifolius*) flour. *LWT - Food Science and Technology* 68: 235–243.

Tarafdar, A., Shahi, N.C., Singh, A., and Sirohi, R. (2017). Optimization of freeze-drying process parameters for qualitative evaluation of button mushroom (*Agaricus bisporus*) using response surface methodology. *Journal of Food Quality* 2017: 5043612.

Xiao, H.-W. (2018). Handbook of Drying of Vegetables and Vegetable Products edited by Min Zhang, Bhesh Bhandari, and Zhong Xiang Fang: CRC Press, Boca Raton. ISBN: 978-1-4987-5386-9 (Hardback), 2017, 517 pp. *Drying Technology* 36 (9): 1142. https://doi.org/10.1080/07373937.2017.1370769.

Xu, X., Zhang, L., Feng, Y. et al. (2021). Ultrasound freeze-thawing style pretreatment to improve the efficiency of the vacuum freeze-drying of okra (*Abelmoschus esculentus* [L.] Moench) and the quality characteristics of the dried product. *Ultrasonics Sonochemistry* 70: 105300.

Zorić, Z., Pedisić, S., Kovačević, D.B. et al. (2016). Impact of packaging material and storage conditions on polyphenol stability, colour and sensory characteristics of freeze-dried sour cherry (*Prunus cerasus* var. Marasca). *Journal of Food Science and Technology* 53: 1247–1258.

Zouaghi, F. and Cantalejo, M.J. (2016). Study of modified atmosphere packaging on the quality of ozonated freeze-dried chicken meat. *Meat Science* 119: 123–131. https://doi.org/10.1016/J.MEATSCI.2016.04.032.

9

Advances in Freeze Drying to Improve Efficiency and Maintain Quality of Dehydrated Fruit and Vegetable Products

*Swati Sharma[1], Kalyan Barman[2], Hare Krishna[1], Surya N. Chaurasia[1], and Arun S. Mujumdar[3]**

[1] Division of Vegetable Production, ICAR-Indian Institute of Vegetable Research, Varanasi, Uttar Pradesh, India
[2] Department of Horticulture, Institute of Agricultural Sciences, Banaras Hindu University, Varanasi, Uttar Pradesh, India
[3] Department of Bioresource Engineering, Macdonald College, McGill University, Montreal, Quebec, Canada

9.1 Introduction

Drying is an age-old technique practiced in every corner of the world. Sunshine has been used to preserve and store food products by drying them for convenient usage during off-season. The drying process is a major consumer of energy, about 20–25% in the food processing industry (Waghmare et al., 2023). Several techniques of drying are currently used, like hot air-controlled drying, spray drying, freeze drying, infrared drying, superheated steam drying, osmotic dehydration, microwave drying, and hybrid drying methods, etc. (Sagar and Kumar, 2010). Among different drying methods, freeze drying has always been recognized as the technique that can yield dried produce of the highest quality, which is quite close to the fresh state, albeit at a higher cost (Liu et al., 2022; Rahman and Mujumdar, 2008; Zhang et al., 2017). It is rightly pointed out that freeze drying is appropriate for high-value and heat-sensitive products (Liapis and Bruttini, 2014; Rahman and Mujumdar, 2008; Stratta et al., 2020). The running cost of a lyophilizer comprises expenses incurred for refrigeration, vacuum pumping, and a skilled personal operator, along with a heavy initial investment. Even though the tissue structure, color, bioactive compound retention, rehydration ratio, rehydration time, lesser shrinkage, and visual appearance of the freeze-dried produce are greatly valued, the technique is marred by persistent questions on its cost, time, and energy effectiveness. Tray, multi-batch,

*Corresponding Author: arunmujumdar123@gmail.com

Freeze Drying of Food Products: Fundamentals, Processes and Applications, First Edition.
Edited by Roji Balaji Waghmare, Manoj Kumar, and Parmjit Singh Panesar.

tunnel, vacuum spray, and continuous freeze dryers are available for lyophilizing different kinds of samples on an industrial scale (Liapis and Bruttini, 2014). An industrial plant for freeze drying vegetables at atmospheric pressure was started in 2005 in Hungary (Rahman and Mujumdar, 2008). Atmospheric (AFD) and vacuum (VFD) freeze-drying techniques remove moisture by sublimating ice crystals from the frozen produce (Rahman and Mujumdar, 2012; Stratta et al., 2020). Microbial spoilage and other biochemical and enzymatic changes are almost eliminated due to the low temperatures (−40 to −96 °C) used during the freeze-drying process and the presence of water in the form of ice crystals. This also helps in preserving tissue integrity and cell structure, causing minimal damage, and also preserving the bioactive compounds. Yet, despite obtaining the highest quality freeze-dried produce, the concerns about the high energy consumption, longer drying times, consequent high cost per unit, operational and maintenance complexities, and the environmental impact due to a higher carbon footprint are worthy of attention (Ratti, 2001; Merone et al., 2020). These have resulted in trials and studies of different hybrid drying methodologies by several researchers on different products to optimize the process conditions (Khampakool et al., 2019; Ekezie et al., 2017; Waghmare et al., 2021). Among these, the combinatorial use of microwave, infrared, and ultrasound energy with freeze drying has also been evaluated. The effect on quality parameters like color, visual appearance, texture, rehydration ratio, shrinkage, glass transition temperature, and drying kinetics has been evaluated.

Infrared (IR 2.5–200 μm) radiation is used in food drying by getting absorbed on the surface or penetrating to a depth to heat the surface. The strong absorption wavelengths of water and organic materials present in food are in the range of 3–6 μm (Pan and Atungulu, 2010). Different parameters like IR power, wavelength, intensity, distance of penetration, and temperature. The benefits of using IR with FD include rapid heating, reduced unit costs, reduced energy consumption, and high rates of moisture removal, leading to increased drying efficiency. The wavelength coinciding with the absorption wavelength of food constituents enhances the molecules movement, generating heat and shortening drying times. However, there is a risk of overheating, and the quality of the dried food product is inconsistent (Pan and Atungulu, 2010). Microwaves are electromagnetic waves that are used for drying, baking, roasting, and tempering purposes in the food industry. The most commonly used frequencies for microwaves, used generally in industry and for domestic purposes, are 915 and 2450 MHz, respectively. Recently, microwaves have also been used in conjunction with freeze drying to shorten process times, save time and energy, and produce a dried produce of comparable quality. Yet, some issues such as corona discharge, non-uniform heating, tissue structure damage, high electrical power consumption, etc. can impede industrial implementation (Ratti, 2001; Ekezie et al., 2017; Rattanadecho and Makul, 2016). It also helps to maintain antioxidant capacity and bactericidal effects. Ultrasound is used as an effective, non-thermal method to advance the drying intensity by increasing heat and mass transfer and, hence, saving time and energy, resulting in reduced costs. Due to the cavitation and sponge effect created by the acoustic effects of ultrasound in food, quick migration of moisture to the surface and its rapid removal increase drying rates (Schossler et al., 2012; Waghmare et al., 2023, 2021). Besides accelerating the drying rates, it preserves the quality of dried produce by improving color and maintaining nutritional content and texture. This book chapter is an attempt to compile information about the recent accomplishments on the uses of infrared, ultrasound, and microwaves coupled with freeze drying, their benefits and drawbacks, the challenges involved, and their future scope.

9.2 Challenges in Freeze Drying

9.2.1 Enhancement of Drying Kinetics

The main challenge in freeze drying is the slow drying rate due to the longer times required to obtain the required lower moisture activity in the dried produce. The sublimation of ice crystals from fresh produce is a major step that takes up the major share of the energy (45%) consumed in the lyophilizing process (Ratti, 2001). The drying rates are often variable and dependent on the internal tissue structure and compactness of the commodity. The whole drying kinetics, the cost of the final produce, and the final quality of freeze-dried produce are influenced by the drying rate. It would be a great boon for freeze drying if the drying rates could be accelerated by coupling with other drying technologies.

9.2.2 Energy Intensity Reduction

Freeze drying is a highly energy-intensive process, often discouraging potential entrepreneurs from upscaling the process to industrial levels. In general, the production of freeze-dried produce like dried okra, strawberry slices, any vegetable bar, or any dried snack or soup particles exhibits the highest quality in terms of color retention, tissue porosity, and rehydration characteristics (Karwacka et al., 2022). However, the energy costs and longer time requirements are major deterrents to its adoption. It has been estimated that almost double the energy is required for the removal of one kilogram of water in a freeze-drying process in comparison to a conventional drying process. A study on energy consumed in different steps of freeze drying by Ratti (2001) concluded that the sublimation step required the maximum share (46% of total energy consumed), while vacuum and condensation required about 25% each, and the minimum energy was utilized for freezing the produce (4%). Merone et al. (2020) reported that energy consumption varies with the internal structure of the horticultural crop. It was observed that the energy requirement (kWh/kg of dried product) was maximum for carrots with a dense and compact internal structure, followed by apples and eggplants with a comparatively porous tissue arrangement. This influences the moisture migration rate, and hence the time required for drying is affected. Notably, assisting the freeze-drying unit with operations like sublimation and vacuum necessities is the crucial point that can help accelerate drying rates and reduce energy consumption, carbon footprint, and final costs. There is a need to optimize the variables in assisting drying technologies for increasing energy efficiency when projecting freeze drying at industrial levels.

9.2.3 Scale-Up and Technical Transfer

Currently, the main challenges in scaling up freeze drying are its huge energy consumption, long drying times, increased cost per unit, and difficulties in maintaining day-to-day operations at a large scale, particularly in vacuum freeze drying (VFD). There is a dearth of technical personnel skilled enough to operate the freeze dryers. More so, any technical difficulties, notwithstanding their minuteness, can impede the process and also affect the quality of lyophilized produce. Maintaining the batch-to-batch product uniformity and quality of freeze-dried commodities produced at industrial scale is quite a task, especially in vacuum freeze drying. The environmental impact was found to be reduced by a whopping 60–70% for various factors like climate change, fine particulate matter formation, fossil depletion, freshwater and marine ecotoxicity and eutrophication, human toxicity, metal depletion, photochemical ozone formation ecosystems and human health, stratospheric ozone depletion, terrestrial acidification, and ecotoxicity when conducting an assessment for freeze drying of different products like eggplant, apple, and carrot at atmospheric pressures using ultrasound with FD (Merone et al., 2020). Thus, the scale-up of the technology should take care of all variables for judicious industrial establishment (Figure 9.1).

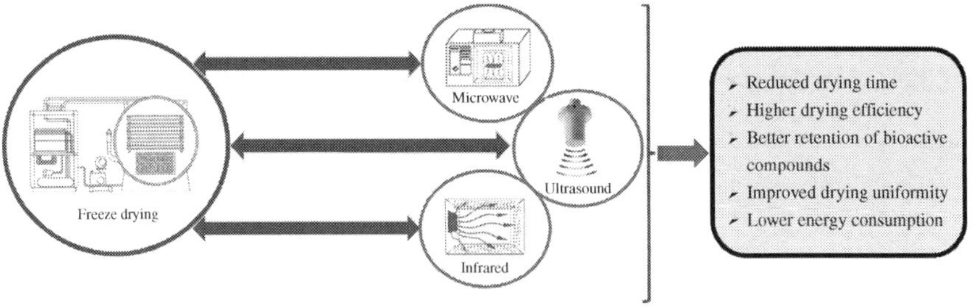

Figure 9.1 Advantages of microwave/ultrasound/infrared-assisted freeze drying.

9.3 Novel Freeze Drying Systems

9.3.1 Infrared-Assisted Freeze Drying

Infrared radiation belongs to the electromagnetic spectrum in the wavelength regions of near-infrared (0.75–3.00 μm), mid-infrared (3.00–25.0 μm) and far-infrared (25–100 μm). The major infrared radiation absorption bands of water lie in the 3, 4.7, 6, and 15.3 μm wavelength ranges. Due to the wider absorption spectrum of water in its liquid state, moisture removal from food material is rapid (Sakare et al., 2020). Infrared radiation has been used for various operations like drying, roasting, peeling, etc. in food processing. Various process factors like power, intensity, wavelength, sample penetration depth, temperature, exposure time, time of application, and thickness of the sample influence drying rates, energy consumption, and quality (Huang et al., 2021). The depth of infrared radiation can be zero, i.e., falling on the product surface to generate heat, or at a certain depth of the sample, which increases the movement of molecules heating the product from inside and aiding moisture removal. Generally, it has been noted that drying rates can be enhanced by increasing IR power, intensity, and temperature. The lowering of infrared distance has also been found to increase drying. The falling of IR radiation results in increased molecular vibration, which causes friction and generates rapid heating. Infrared radiation can be generated by passing electricity through an element or using gas. However, the conversion efficiency is higher for electrical emitters (78–85%). The incandescent vacuum lamp, tungsten halogen lamp, quartz tube, and ceramic IR emitters are common examples of electrical sources of infrared radiation. It is vital to consider the absorptivity, transmissivity, and reflectivity of parts used for producing infrared radiation and drying food products (Sakare et al., 2020; Pan and Atungulu, 2010). High transmissive rates are useful for drying food products to prevent uneven and extreme heating, which can injure and burn the tissue structure. It is desirable that the reflectivity of the reflector component of equipment in infrared radiation be high, while the product should be low to maximize the utilization of the generated heat. However, the effect of different process variables on the final dried produce quality has not been found to be consistent. Yet, the majority of the studies have observed that using infrared radiation with freeze drying is useful to minimize the water activity, increase the stability of the produce, better preserve the bioactive components like ascorbic acid, phenolics, etc., maintain a better texture, and lower the changes in the hue and chroma of the final dried product (Baeghbali et al., 2020; Chen et al., 2022a, b; Rashid et al., 2022; Wang et al., 2015; Ren et al., 2022; Oliveira et al., 2021; Carames et al., 2021; Liu et al., 2007, 2020, 2021a, 2021b). Several workers have advocated the use of infrared drying with freeze drying in different crops like okra, quinoa, mushrooms, and ginger, etc. to shorten drying time while maintaining the quality of the freeze-dried produce. The application of infrared drying in combination with other drying methods, particularly freeze drying, has the potential to improve mass and heat transfer rates and simultaneously lower drying time and energy consumption. Mainly, it has been observed that infrared radiation is useful to bring down the energy consumption in the sublimation step by increasing the heat transfer rates and moisture migration during the drying process. There is a need to develop advanced intelligent drying systems to read the dynamic drying curve and identify timings for the application of infrared radiation at the appropriate time for pulsed durations to increase heat transfer rate and help in moisture removal. Some of the recent studies conducted to evaluate the effectiveness of infrared drying combined with freeze-drying process and the consequent key findings are presented in Table 9.1.

9.3.2 Microwave-Assisted Freeze Drying

The microwaves (2450 Hz) are often used domestically for various activities like drying, tempering, sterilization, and baking, etc. It also causes non-uniform moisture distribution during drying, which renders the control of the drying process difficult and complicated to obtain an appreciable final dried quality. It may cause under-processing, over-heating, and burning of the sample, reducing the final quality of dried produce. The inconsistent heat distribution is influenced by the type of material, internal tissue compactness, dielectric properties, and design of the microwave instrument. The major benefits include enhanced drying rates, particularly when supplemented with other drying techniques due to vibrational and cavitation effects; efficient drying in the falling rate period; specific process control; quick attainment of energy dispersion in the whole sample; higher retention of volatile compounds; and shortened drying time (Ekezie et al., 2017; Menon et al., 2020). The foremost disadvantages of microwave drying are uneven heating distribution, which may cause burning and tissue damage; restricted depth of penetration, variable product quality; slow drying at lower power levels; higher investment costs; and higher electricity utilization. High drying efficiency with shortened drying time and negligible changes in dried product quality and retention of bioactive compounds with sensory acceptability have been reported by several workers using microwave drying with freeze drying. Due to the higher penetration of the microwave field, it is suitable for drying food materials with high moisture content, high viscosity, and heat sensitivity, like raspberries, blueberries, and other berries (Ozcelik et al., 2020). In real time, microwaves create an

Table 9.1 Some recent studies conducted on vegetables and fruits using infrared radiation for freeze drying.

Plant sample	Drying methods	Key results	References
Okra	US- and infrared-assisted conductive hydro-drying (US frequency – 28 kHz, US power – 166 W, temperature ~ 82.5 °C, IR power – 139 W, lateral airflow – 0.95 m/s, time – 45 min)	• US- and IR-assisted conductive hydrodrying is a new patented drying method. • UIACHD was significantly better than oven drying. • Quality of the products by UIACHD was comparable to FD. • Microstructure of dried okra was maintained. • Minimized energy consumption of the process.	Baeghbali et al. (2020)
Pepper, carrot, and cantaloupe	IR-assisted thawing (20 s), microwave-assisted thawing (10 s), ambient thawing at 25 °C.	• Thawing time and loss by using combined microwave and IR were reduced by 65–75% and 10–40%, respectively. • Infrared heating prior to microwave was found better to lower excessive heating, cell damage, and thawing loss, and maintain quality.	Chen et al. (2022a)
Sweet potato	Ultrasound (US) (20, 40, 60 kHz, time – 30 min, acoustic power 300 W/cm^3, 50 W/l ultrasonic power density) Convective HAD (hot-air drier – 60, 70, and 80 °C; 25% RH and 1.5 m/s air velocity). Catalytic infrared drying (CIR) 60, 70, and 80 °C	• US (40 kHz) + IRD and HAD (70 °C) significantly reduced drying time and water content. • No effect on bioactive and other quality-related attributes was found. • US-HAD (60 kHz, 80 °C) resulted in poor quality and high hardness. • Improved texture in sweet potato slices treated with 40 kHz US frequency in a HAD over IRD. • US pretreatments with HAD and IRD are energy-efficient and preserve quality of dried sweet potatoes.	Rashid et al. (2022)

(Continued)

Table 9.1 (Continued)

Plant sample	Drying methods	Key results	References
Yam	FIR (freeze-dried-far infrared radiation) power – 200 W) FD (under vacuum at absolute pressure – 0.3 mmHg)	• FIR with FD lowered the time taken for drying and the total color difference. • Optimum distance between sample and FIR heater was found to be 50 mm.	Lin et al. (2007)
Blueberry	Microwave vacuum drying (MVD) (MV density – 3, 5, 7 W/g). Pulse-spouted microwave freeze drying (PS-MFD)	• LF-NMR and NIR combination was found to be apt for predicting TPS and drying endpoints.	Liu et al. (2021a)
Banana	IRAFD IR lamp installed in to FD. Continuous IRAFD (2.7 kW/m^2), IRAFD (2.7 kW/m^2 at 20% weight reduction), and IRAFD (2.7 kW/m^2 at 20% WR to 4.0 kW/m^2 at 0 °C); sample distance – 150 mm from IR lamp	• Continuous IRAFD significantly reduced drying time (213 min) than FD (696 min), which cutbacks 70% of the needed time. • FD used maximum electrical energy (27.0×10^3 kJ). • Continuous IRAFD saved electrical energy consumption up to 8.4×10^3 kJ worth of rapid drying. • Superior crispness in IRAFD dried bananas over FD is due to the dense surface layer with large pores in the center.	Khampakool et al. (2019)
Acai	FD, FD + continuous NIR (NIRFD) NIRFD after 20% weight reduction ($NIRFD_{20\%WR}$) FD + continuous far IR (FARFD), FARFD after 20% weight reduction, NIR-150 W, FAR-150 W	• NIRFD and FARFD took 49.4% and 33.4% less time than FD, respectively. • Significant reduction in total drying time presents direct implications for energy savings. • Powder prepared using IRFD showed larger pores and maintained quality. • NIRFD was found to be better than FARFD.	Oliveira et al. (2021)

Acai	Frozen at −28 °C and freeze dried (pressure – 2000 μHg)	• NIR-PLS (near-infrared spectroscopy with partial least squares regression) model was useful to predict the bioactive content of freeze-dried acai pulp.	Carames et al. (2021)
Quinoa	Pressure (PCIRFD), MV – (MCIRFD) atmospheric pressure cooked – (APCIRFD), frozen – 80 °C, IRFD (wavelength – 3.0 μm, flux density – 0.703 W/cm^2, temperature – 50 °C) or FD	• Significant reduction in drying time by IRFD over FD. • Maintenance of the gelatinized state of cooked quinoa, rehydration ability, texture quality, and color by IRFD. • Maximum flavor retention ability in IRFD over FD. • Pressure cooking and IRFD can be used for the production of instant quinoa products of high quality.	Chen et al. (2022b)
Shiitake mushroom	FD with mid-infrared (MIRD) drying before and after FD, mid-IR-assisted freeze drying (FD-MIRD), FD (FD-2 h – MIRD, FD-4 h – MIRD, FD-6 h – MIRD)	• FD (4 h) – MIRD (15 min) lowered 48% drying time, highest retention of aroma over FD. • FD–MIRD was superior over MIRD-FD as collapsed tissue and poor rehydration ability were noted in the latter.	Wang et al. (2015)
Mushroom	FD (−40 °C) and IRFD (optimal drying temperature – 70 °C) microwave pre-gelatinization (MPG) with infrared freeze drying (IRFD)	• IRFD lowered drying time (23.81%). • IRFD reduced energy consumption (28.43%) • Improved starch stability • IRFD did not improve the quality of FD products • MPG (time – 120 s at 8 W/g energy) was found to be most appropriate for wheat flour.	Liu et al. (2020)
Ginger	US (40 kHz, 300 W, 25 ± 1 °C for 15 min). Catalytic infrared drying (temperature – 60 ± 2 °C, distance between sample and catalytic infrared plate – 200 ± 1 mm)	• Ethanol + US reduced CIR drying time by 58%. • Ethanol + US showed the highest drying efficiency, hardness, total phenolic content, and total flavonoid content retention. • Slight decrease in rehydration capacity and gingerol content.	Ren et al. (2022)

oscillating electrical field, increasing molecular movement, causing friction, producing heat, and rapidly removing free water during the initial stages as well as bound moisture during later stages to attain quicker overall drying rates (Moreno et al., 2020; Ozcelik et al., 2020). The dipole rotation, ionic conductance, pulsing time, and pulse interval of the application of an electromagnetic field influence the microwave freeze-drying characteristics. The haphazard heating of high-moisture products like fruits and vegetables due to microwave application can be avoided by using it in vacuum conditions (Zhou et al., 2021). Moreover, microwave drying has been found to retain strong antioxidant capacity.

It has been recorded in several recent studies that the use of microwave drying in combination with freeze drying has reduced drying time significantly by increasing the efficiency of moisture migration to the surface and loss in different fruits and vegetables. Table 9.2 presents the studies ascertaining the influence of microwave drying in combination with freeze drying on some vegetables and fruits like bottle gourd, beet, carrot, Chinese yam, mushroom, quinoa, blueberry, strawberry, cranberry, raspberry, and dragon fruit, etc.

9.3.3 Ultrasound-Assisted Freeze Drying

It is well established that the dried products obtained by freeze drying are of the highest quality in terms of the maintenance of internal tissue structure, retention of color and bioactive compounds, rehydration ratio, long shelf stability, and overall acceptability. Nevertheless, the longer drying times required and the consequently higher energy costs translate to expensive final dried produce. Moreover, there may be a requirement for a skilled person at various critical points, more so if any unit component becomes dysfunctional, during the operation of freeze dryers, particularly while using a vacuum pump for drying. In freeze drying, the final product quality is best due to the drying carried out after freezing fresh samples, which results in inhibition of all biochemical changes, enzymatic activities, and microbial deterioration. Moreover, the moisture is directly lost from ice to water vapor at pressure below the triple point of water by sublimation in vacuum conditions. This results in the preservation of porosity and internal tissue structure with minimal cell damage, thus giving the best rehydration characteristics (Waghmare et al., 2023). However, this sublimation step takes up the major chunk of the total energy consumption, which implies that the additional use of energy to increase heat and mass transfer during this phase can accelerate drying rates. Several researchers are working to evaluate the effectiveness of using ultrasound waves by changing the process variables (wavelength, amplitude, frequency, probe length, simultaneous or sequential

Table 9.2 Some recent studies conducted on some vegetables and fruits using microwave drying in conjunction with freeze drying.

Plant sample	Drying treatment	Major findings	References
Bottle gourd	Ultrasound-assisted extraction (UAE) – (temperature – 40, 50, 60 °C, power – 100 W, 150 W) Microwave-assisted extraction (MAE) (time – 40, 60, 80 s, power – 360, 480, 600 W; solvent-to-solid ratio – 20 ml/g; time of extraction – 30 min)	• UAE and MAE were more powerful than SE for polyphenolic compound extraction. • MAE was the most efficient technique. • Maximum % yield by UAE, MAE, and SE was 5.08%, 4.00%, and 3.20%, respectively. • Percent yield, TPC, and TFC were higher in the FD sample.	Abbas et al. (2021)
Beet	Oven drying (45 °C, 4 h) Gravitational and microwave-assisted thawing	• Efficiency of concentration and product features were not influenced by the thawing method. • MW-assisted thawing was faster (1–5 min). • Due to the low temperature used, the bioactive compounds were maintained.	Arend et al. (2022)
Carrot	MWFD (frequency – 2.4 GHz, power – 100, 200, and 300 W)	• Efficiency of MFD was improved by up to 62.4% by shortening drying time and having more efficiency than FD. • Quality of dried carrots by MFD was equivalent to FD.	Sujinda et al. (2021)
Chinese yam	Low frequency MW (915 MHz) assisted FD (temperature – 40 °C, pressure – 80 Pa, spouting time – 0.3 s, pressure recovery time – 2 min, spouting interval – 10, 20, 30 min) High-frequency MW (2450 MHz) assisted FD (temperature – 40 °C, pressure – 80 Pa, power – 600 W, spouting interval – 20 min, time – 0.2 s) FD (cold trap temperature: –40 °C, pressure – 80 Pa)	• Drying time was reduced by 41.4%–48.3% in MFD-915 Hz over FD. • Longer interval caused a shorter drying time and improved drying uniformity. • MFD-915 Hz lowered total energy consumption (34.4% lower) over FD. • Yam product dried using MFD-915 showed the same quality (color, rehydration ratio, TPC, and TFC) as FD and MFD-2450.	Li et al. (2019)

(Continued)

Table 9.2 (Continued)

Plant sample	Drying treatment	Major findings	References
Cordyceps militaris, mushroom	FD (−40 °C, pressure – 80 Pa, maximum temperature – 50 °C, power – 100, 300, and 500 W) MPFFD (microwave-assisted pulse fluidized bed freeze drying) (cold trap temperature – −40 °C, absolute pressure – 80 Pa, MW power – 660, 760 and 860 W) HD (hot air dryer) (temperature – 40, 50 and 60 °C, air flow rate – 1 m/s)	• MPFFD reduced rehydration time and lowered shrinkage in dried mushrooms. • Rehydration ratio, color, crispness, volatile compounds, and antioxidant activities in dried mushrooms were maintained better over HD. • Final water content, water activity, rehydration time, and rehydration ratio of mushrooms dried by MPFFD were very close to FD. • Quality of dried mushrooms prepared using MPFFD was similar to FD samples. • Drying time in MPFFD was lowered by 71.9% over FD.	Wu et al. (2019)
Mushroom	MW pre-gelatinization (MPG + IRFD) Frozen (−50 °C for 10 h) IRFD (IR temperature – 70 °C)	• IRFD reduced drying time and energy consumption by 23.81% and 28.43%, respectively, over FD. • No improvement in the quality of IRFD products was noted. • MPG did not affect drying time or energy consumption in the IRFD process. • MPG significantly enhanced the flavor and viscosity of soup after rehydrating. • MPG (120 s, 8 W/g) was most suitable. • MPG and IRFD combined to improve the drying efficiency and quality of mushroom soup.	Liu et al. (2020)

Quinoa	Frozen at −80 °C for 12 h. Dried by IRFD or FD. IR (wavelength – 3.0 μm, maximum heat flux density – 0.703 W/cm^2, temperature – 50 °C) Quinoa prepared using pressure cooking (PCIRFD), microwave cooking (MCIRFD), and atmospheric pressure cooking (APCIRFD) Freeze-dried quinoas made by pressure cooking (PCFD), microwave cooking (MCFD), and atmospheric pressure cooking (APCFD)	• IRFD significantly reduced drying time with higher process efficiency and quality than FD. • IRFD maintained the gelatinized state of quinoa samples. • Pressure cooking and microwave cooking resulted in a softer and thicker consistency after rehydration. • Pressure cooking and IRFD can be used for making instant quinoa products of high quality.	Chen et al. (2022b)
Blueberry	Microwave-assisted foam-mat drying (MFD) using a microwave oven dryer (microwave intensity of 3 W/g)	• Microwave foam drying of foamed blueberry pulp using egg white powder, maltodextrin, and carboxymethyl cellulose (16 : 20 : 1) yielded powder with a high drying rate, good physicochemical quality, high solubility, and porosity.	Gao et al. (2022)
Blueberry	Microwave vacuum drying (MVD) (3, 5, 7 W/g) Pulse-spouted (PS-MFD) frozen blueberries (−80 °C for 10 h). 3 W/g MW loading density, 20 min spouting interval, and 0.3 s spouting holding time	• Models based on LF-NMR and NIR spectra accurately forecasted the moisture and drying endpoints during the PS-MFD process.	Liu et al. (2021a)
Strawberries	Ultrasound-assisted osmotic dehydration (USOD) (15 min at 25 ± 5 °C, 45 Hz, power – 180, 240, 360 W, power intensity – 0.36, 0.48, 0.6 W/g) Pulse fluidized bed microwave freeze drying (PFBMFD) (pressure – 80 ± 5 Pa, pulse interval – 10 min, pulse period – 0.1 s)	• PFBMFD reduced drying time by 45% over FD, while USOD pretreatment further increased the drying rate and lowered drying time by 10%. • PFBD and MV can greatly improve drying efficiency and maintain quality with the retention of total polyphenols and flavonoids.	Jiang et al. (2021)

(Continued)

Table 9.2 (Continued)

Plant sample	Drying treatment	Major findings	References
Dragon fruit	Frozen (−20 °C, 12 h). Thermo-ultrasound-assisted freeze–thaw (US-FT) (power – 500 W, frequency – 40 kHz, temperature – 25, 50, 70 °C, time – 30, 8, 4 min) Microwave-assisted freeze–thaw (MW-FT) (frequency – 2450 MHz, power – 100, 300, 500 W, time – 7, 4, 2 min) Infrared drying (temperature – 60 °C, rotation speed – 40 rpm, irradiation distance – 20 cm)	• Thermo-US and MW-assisted freeze–thaw (US-FT and MW-FT) pretreatments caused ice formation and affected the IR drying of dragon fruit slices. • Treatments facilitated moisture migration and reduced drying time. • Low-temperature US-FT improved physicochemical and bioactive properties, while high-temperature USFT preserved TFC, FRAP, and CUPRAC better. • High-power MW-FT (500 W) showed better physicochemical attributes. • Low-power MW-FT (100 W, 300 W) improved phenolics, flavonoids, vitamin C, betalain pigment, and antioxidant capacity.	Bassey et al. (2022)
Raspberry	FD MWFD microwave freeze drying MW power (1.0, 1.5 and 2.0 W/g)	• MWFD was found to be more efficient with comparable product quality to FD. • Uniform drying was obtained at 1.0 W/g. • Drying time was drastically shortened by 70% by using MW energy. • Drying time in MWFD was eightfold less than in FD. • Energy demand lowered slightly due to a shorter drying time with more power. • However, a linear correlation was not seen between MW power input and energy requirement.	Ozcelik et al. (2019)
Raspberry	FD, MWFD (frequency – 2.45 GHz, chamber pressure – 0.1 mbar, maximum temperature – 30 °C)	• MWFD samples showed equivalent or slightly better storage stability characteristics over conventional FD methods.	Ozcelik et al. (2020)

Raspberry	PSMFD (temperature – −15 °C, spouting time interval – 20 min, spouting time – 0.2 s).	• Drying rate using PSMFD presents the same trend under different microwave powers. • Complete drying process included an increasing and falling rate with no obvious constant-rate period. • PSMFD significantly improved moisture uniformity in dried raspberries over MFD.	Sun et al. (2021)
Cranberries	Hot air convective drying (HACD) (airflow velocity – 1.5 m/s, temperature – 80 °C, sample mass: 0.15 kg) Two-stage hot air convective drying and microwave vacuum drying (HACD + MWVD). (MW power – 150 W; absolute pressure – 5 ± 1 kPa; rotational speed – 6 rpm)	• MWVD reduced drying time by 54% over HACD. • MW application in the final drying stages resulted in a drying rate 17 times higher than over using convection in initial stage. • HACD + MWVD dried cranberries presented higher polyphenols, flavonoids, and anthocyanin content, antioxidant activity, and a more pleasing overall appearance over fruits dried using HACD. • MWVD of dehydrated samples without osmosis gave a healthier product with more polyphenols and stronger antioxidant capacity in less time. • HACD + MWVD treatment presented high-quality dried cranberries and reduced drying time.	Zhou et al. (2021)

application, power, temperature, pulsing duration, medium of application, and other pretreatments) in conjunction with freeze drying to optimize methodology for producing freeze-dried produce with lower time requirements, lesser energy utilization, and, consequently, lower environmental degradation and expenses incurred in processing.

Ultrasound waves can be used to pass mechanical energy into food samples, generating heat and shortening drying times by aiding moisture migration. The frequencies ranging from 16 to 100 kHz have been found useful in the food industry (Schossler et al., 2012). The sponge effect and cavitation created by these pressure waves due to the creation of temporary, quick, alternating high and low compression regions result in increased tissue amenability for the removal of water during primary drying as well as bound water loss during secondary drying. Higher drying efficiency is due to lowered air vapor pressure in rarefaction cycle and the creation of localized pockets with a higher air temperature, resulting in a change in surface tension and viscosity formed by friction and compression forces. The creation and bursting of bubbles and the formation of micro-channels in tissue help to accelerate water removal while maintaining the quality of freeze-dried produce. Additionally, the application of ultrasound waves at intervals prevents overheating of the sample. It has been noted that USFD hybrid drying technology has shortened drying times while maintaining comparable dried product quality. An additional benefit is the destruction of microbes and pathogens due to US treatment. The frequency and power need to be standardized for drying different fresh fruits and vegetables to prevent overheating and tissue damage. In using ultrasound drying, the product color was mostly found to be better than samples obtained from other traditional drying techniques, which might be due to reduced drying time, lower pigment deterioration, and the inactivation of color-degrading enzymes like polyphenol oxidase and peroxidase. The color of dried products is influenced with the medium, particular fruits and vegetables, and the different ultrasound parameters. It was rightly suggested by Caglayan and Mazi (2018) that the uniformity in the final dried product for hue and chroma values cannot be maintained every time due to the involvement of several factors. Ultrasound waves increase the sublimation rate during freeze drying when applied intermittently, which prevents the rise in temperature of the sample; otherwise, the sample may show overheating and deterioration in quality. Several workers have found and advocated that ultrasound-assisted freeze drying increases moisture removal rates and maintains quality in crops like blueberries, strawberries, apples, and quince, and vegetables like onion, yam, pumpkin, okra, etc. (Table 9.3).

Table 9.3 Some recent studies conducted on some fruits and vegetables using ultrasound drying as a pretreatment followed by freeze drying.

Plant sample	Drying treatment	Major findings	References
Strawberry	US – 20 and 40 kHz, alternate 20/40 kHz (SeDM), 20 and 40 kHz frequency at the same time (SiDM), power density – 30 W/l, duration – 30 min, temperature (25 °C)	• Drying time was lowered by 15–50%. • Minimum drying time, lowest moisture ratio, and best rehydration (3.47 g/g) in SeDM. • Highest redness (a^* value) in SeDM. • Lower color change was seen in pretreated samples.	Xu et al. (2021a)
Strawberry	US-assisted OD (frequency – 40 kHz, power – 130 W, temperature – 50 °C, time – 15 to 180 min) FD	• USOD followed by FD increased rehydration capacity, shortened drying time, and saved energy costs. • Retained color closest to fresh strawberries, showing the lowest change.	Prosapio and Norton (2018)
Strawberry	US (temperature – 25 °C, power – 300 W, frequency – 40 kHz, time – 30 min) US + UHP (200 MPa for 5 min, temperature – 20 °C) FD	• Maximum anthocyanin content was found in US-UHP-pretreated samples. • Maximum drying speed and minimum process time were noted in samples pretreated with UHP-US followed by FD.	Zhang et al. (2020)
Blueberry	US [water, 30 min], [10% sodium bicarbonate, 30 min], [1.5% chitosan, 30 min], [10% sodium bicarbonate, 30 min, then 1.5% chitosan, 30 min], [10% sodium bicarbonate, 30 min, then 1.5% chitosan, 30 min], power – 180 W, frequency – 20 kHz	• US (sodium bicarbonate and chitosan) pretreatment followed by FD. • Maximum rehydration rate. • Best internal porous structure. • Highest antioxidant capacity. • Highest drying rates.	Liu et al. (2021b)
Apple	US-CD (1 kW, 40 kHz, temperature – 26–40 °C)	• Minimum moisture content (5% w.b.). • Higher rehydration ratio. • Better color retention. • Shortened drying time (45%) in US pretreated samples.	Kahraman et al. (2021)

(*Continued*)

Table 9.3 (Continued)

Plant sample	Drying treatment	Major findings	References
Apple	US (temperature – 15, 25 °C, frequency – 50, 60, 80, 90, 100 kHz, time – 5, 10, 15 min)	• Drying rate was significantly higher. • Increased rehydration ratio.	Ren and Bai (2018)
Pear	US (power – 360, 600, 960 W, frequency – 20 kHz, temperature – 20–25 °C)	• Higher US input power reduced a_w and moisture content faster.	Islam et al. (2014)
Peach	US (power – 180, 240, 300 W, frequency – 20, 28, 40 kHz)	• Higher pigment retention. • Lower color degradation. • Increased retention of phenolic content. • Lower browning enzyme activity. • Reduced water activity (0.26–0.31). • US (240 W) presented the best final product quality. • Significant change in dehydration at varying power levels was not seen.	Chu et al. (2021)
Quince	US (frequency – 28 kHz, power – 50 W, time – 10, 20, and 30 min, intensity – 100 kW/m^3)	• USFD quince samples were lighter than those with only FD. • US pretreatment (20 min) followed by FD retained better color, lower water activity, and a higher rehydration ratio over control samples.	Yildiz and Izli (2019)
Pumpkin	Ultrasound-assisted osmotic dehydration (UAOD) (40% and 60% sucrose, frequency – 53 kHz, temperature – 30 °C)	• UAOD lowered the rehydration ratio in dried pumpkin, negatively affecting rehydration ability. • UAOD samples were darker in color. • UAOD pretreatment was not recommended for drying pumpkin.	Caglayan and Mazi (2018)
Okra	US and IR-assisted conductive hydro-drying (US: frequency – 28 kHz, power – 166 W, temperature – 82.5 °C, IR power – 139 W, time – 45 min)	• Lowest moisture and maximum drying speed were found in US and IR pretreatment over FD. • Dried samples showed higher L^*, a^* and b^* values.	Baeghbali et al. (2020)

Okra	US (frequency – 40 kHz, power – 25 W/l, time – 30 min)	• Maximum retention of green color after drying in USFD over others. • Lower chlorophyll degradation • Drying rate accelerated. • Energy costs reduced by 25–62%.	Xu et al. (2021b)
Beet	US (frequency – 40 kHz, power – 180 W, time – 5 and 10 min)	• Minimum water activity in US pretreated beet.	Ciurzynska et al. (2021)
Carrot	US (1.5 kW, 20 kHz, amplitude – 24.4, 42.7, and 61.0 μm, time – 3, 10 min, temperature – 25 °C)	• Highest carotenoids were retained with the smallest amplitude in the US. • US pretreated carrots retained better color than others.	Rawson et al. (2011)
Bell pepper	US (power – 76, 90, and 110 W)	• Intermittent US treatment retained the color of freeze-dried bell peppers.	Schossler et al. (2012)
Onion	US (20 kHz, 70 °C, power – 40%, 60%, 80%, amplitude – 24.4, 42.7, 61 μm), time [1, 3, 5 min], pulse duration – 5 s on, 5 s off]	• Longer US duration resulted in higher color change and was not affected by amplitude. • USFD retained the maximum bioactive compounds and showed the strongest antioxidant capacity.	Ren et al. (2018)
Chinese yam	USOD (frequency – 20 kHz, time – 10 min, pulse – 10s on and 2 s off, power – 40%, 60%, 80% of 2000 W)	• The color of the USOD pretreated yam was acceptable and good. • USOD retained higher phenolic content and enhanced antioxidant capacity over control.	Li et al. (2020)
Sweet potatoes	US (temperature – 25 °C, frequency – 30 kHz, time – 30 min, power – 200, 400, and 600 W)	• Minimum moisture content and maximum drying rate were observed. • Higher input power shortened drying time significantly.	Wu et al. (2020)

9.4 Conclusion and Future Prospects

Drying is undoubtedly the most effective technique to lower food loss and waste, reduce volume, increase shelf stability, and facilitate easy storage. Yet, the raw material, possible cost escalation of the drying process, ease of operation, time required, and final product quality, particularly rehydration characteristics, color and appearance, tissue structure, and flavor preservation, weigh in for the appropriate selection of the drying method. It is necessary that the drying process be optimized for maximum preservation of quality at the lowest possible cost. Freeze drying in combination with infrared, microwave, and ultrasound drying can be utilized to significantly shorten the drying time and obtain dried produce of comparable quality.

Future studies should focus on evaluating the energy consumption in the making of dried food products to assess their impact and carbon footprint. Attempts should be made to reduce the energy needed for drying, reduce complications in operation and maintenance, increase ease of handling, and increase the possibilities of scaling up for industrial applications while maintaining the high quality of dried produce. Numerical modeling studies to understand the drying kinetics and get insights for useful optimization to obtain high-quality dried produce at lower costs are required. Studies on the development of new products, proper utilization of the by-products to reduce loss and wastage, and their sensory acceptability, nutritional, and functional characteristics should be taken up. There is a need to optimize variables that can increase the efficacy of the freeze-drying process by coupling it with newer, upcoming hybrid drying technologies for industrial applications. Studies to integrate the drying technologies, especially ultrasound, microwave, and infrared radiation, for complementing the freeze-drying process, monitoring real-time data, and integrating them into the instrument for easy up-scaling of the drying process at industrial levels need to be taken up.

References

Abbas, M., Ahmed, D., Qamar, M.T. et al. (2021). Optimization of ultrasound-assisted, microwave-assisted and Soxhlet extraction of bioactive compounds from *Lagenaria siceraria*: a comparative analysis. *Bioresource Technology Reports* 15: 100746.

Arend, G.D., Almeida, E.S., Byruchko, R.T. et al. (2022). Gravitational and microwave-assisted multi- stages block freeze concentration process to obtain enriched concentrated beet (*Beta vulgaris* L.) by-products extract: bioactive compounds and simulated gastrointestinal profile. *Food and Bioproducts Processing* 133: 77–86.

Baeghbali, V., Ngadi, M., and Niakousari, M. (2020). Effects of ultrasound and infrared assisted conductive hydro drying, freeze drying and oven drying on physicochemical properties of okra slices. *Innovative Food Science & Emerging Technologies* 63: 102313.

Bassey, E.J., Cheng, J.H., and Sun, D.W. (2022). Thermoultrasound and microwave-assisted freeze-thaw pretreatments for improving infrared drying and quality characteristics of red dragon fruit slices. *Ultrasonics Sonochemistry* 91: 106225.

Caglayan, D. and Mazi, I.B. (2018). Effects of ultrasound-assisted osmotic dehydration as a pretreatment and finish drying methods on the quality of pumpkin slices. *Journal of Food Processing & Preservation* 42: e13679.

Carames, E.T.S., Baqueta, M.R., Conceiçao, D.A., and Pallone, J.A.L. (2021). Near infrared spectroscopy and smartphone-based imaging as fast alternatives for the evaluation of the bioactive potential of freeze-dried açai. *Food Research International* 140: 109792.

Chen, B., Zhang, M., Wang, Y. et al. (2022a). Comparative study of conventional and novel combined modes of microwave- and infrared-assisted thawing on quality of frozen green pepper, carrot and cantaloupe. *LWT - Food Science and Technology* 154: 112842.

Chen, K., Zhang, M., Bhandari, B., and Chen, J. (2022b). Instant quinoa prepared by different cooking methods and infrared assisted freeze drying: effects of variables on the physicochemical properties. *Food Chemistry* 370: 131091.

Chu, Y., Wei, S., Ding, Z. et al. (2021). Application of ultrasound and curing agent during osmotic dehydration to improve the quality properties of freeze-dried yellow peach (*Amygdalus persica*) slices. *Agriculture* 11: 1069.

Ciurzynska, A., Falacinska, J., Kowalska, H. et al. (2021). The effect of pre-treatment (blanching, ultrasound and freezing) on quality of freeze-dried red beets. *Foods* 10: 132.

Ekezie, F.G.C., Sun, D.W., Han, Z., and Cheng, J.H. (2017). Microwave-assisted food processing technologies for enhancing product quality and process efficiency: a review of recent developments. *Trends in Food Science & Technology* 67: 58–69.

Gao, R., Xue, L., Zhang, Y. et al. (2022). Production of blueberry pulp powder by microwave-assisted foam-mat drying: effects of formulations of foaming agents on drying characteristics and physicochemical properties. *LWT - Food Science and Technology* 154: 112811.

Huang, D., Yang, P., Tang, X. et al. (2021). Application of infrared radiation in the drying of food products. *Trends in Food Science & Technology* 110: 765–777.

Islam, M.N., Zhang, M., Liu, H., and Xinfeng, C. (2014). Effects of ultrasound on glass transition temperature of freeze-dried pear (*Pyrus pyrifolia*) using DMA thermal analysis. *Food and Bioproducts Processing* https://doi.org/10.1016/j.fbp.2014.02.004.

Jiang, J., Zhang, M., Devahastin, S., and Yu, D. (2021). Effect of ultrasound-assisted osmotic dehydration pretreatments on drying and quality characteristics of pulsed fluidized bed microwave freeze-dried strawberries. *LWT - Food Science and Technology* 145: 111300.

Kahraman, O., Malvandi, A., Vargas, L., and Feng, H. (2021). Drying characteristics and quality attributes of apple slices dried by a non-thermal ultrasonic contact drying method. *Ultrasonics Sonochemistry* 73: 105510.

Karwacka, M., Ciurzynska, A., Galus, S., and Janowicz, M. (2022). Freeze-dried snacks obtained from frozen vegetable by-products and apple pomace-selected properties, energy consumption and carbon footprint. *Innovative Food Science and Emerging Technologies* 77: 102949.

Khampakool, A., Soisungwan, S., and Park, S.H. (2019). Potential application of infrared assisted freeze drying (IRAFD) for banana snacks: drying kinetics, energy consumption, and texture. *LWT - Food Science and Technology* 99: 355–363.

Li, L., Zhang, M., and Wang, W. (2019). A novel low-frequency microwave assisted pulse-spouted bed freeze-drying of Chinese yam. *Food and Bioproducts Processing* 118: 217–226.

Li, L., Zhang, M., and Wang, W. (2020). Ultrasound-assisted osmotic dehydration pretreatment before pulsed fluidized bed microwave freeze-drying (PFBMFD) of Chinese yam. *Food Bioscience* 35: 100548.

Liapis, A.I. and Bruttini, B. (2014). Freeze drying. In: *Handbook of Industrial Drying* (ed. A.S. Majumdar). Boca Raton: CRC Press. ISBN: 9780429169762 https://doi.org/10.1201/b17208.

Lin, Y.P., Lee, T.Y., Tsen, J.H., and King, V.A.E. (2007). Dehydration of yam slices using FIR-assisted freeze drying. *Journal of Food Engineering* 79: 1295–1301.

Liu, W., Zhang, M., Adhikari, B., and Chen, J. (2020). A novel strategy for improving drying efficiency and quality of cream mushroom soup based on microwave pre-gelatinization and infrared freeze drying. *Innovative Food Science and Emerging Technologies* 66: 102516.

Liu, W., Zhang, M., Bhandari, B., and Yu, D. (2021a). A novel combination of LF-NMR and NIR to intelligent control in pulse-spouted microwave freeze drying of blueberry. *LWT - Food Science and Technology* 137: 110455.

Liu, W., Zhang, M., Mujumdar, A.S. et al. (2021b). Effects of chitosan coating on freeze-drying of blueberry enhanced by ultrasound pre-treatment in sodium bicarbonate medium. *International Journal of Biological Macromolecules* 181: 631–643.

Liu, Y., Zhang, Z., and Hu, L. (2022). High efficient freeze-drying technology in food industry. *Critical Reviews in Food Science and Nutrition* 62 (12): 3370–3388.

Menon, A., Stojceska, V., and Tassou, S.A. (2020). A systematic review on the recent advances of the energy efficiency improvements in non-conventional food drying technologies. *Trends in Food Science & Technology* 100: 67–76.

Merone, D., Colucci, D., Fissore, D. et al. (2020). Energy and environmental analysis of ultrasound-assisted atmospheric freeze-drying of food. *Journal of Food Engineering* 283: 110031.

Oliveira, N.L., Silva, S.H., Figueiredo, J.A. et al. (2021). Infrared assisted freeze drying (IRFD) of acai puree: effects on the drying kinetics, microstructure and bioactive compounds. *Innovative Food Science & Emerging Technologies* 74: 102843.

Ozcelik, M., Heigl, A., Kulozik, U., and Ambros, S. (2019). Effect of hydrocolloid addition and microwave-assisted freeze drying on the characteristics of foamed raspberry puree. *Innovative Food Science and Emerging Technologies* 56: 102183.

Ozcelik, M., Ambros, S., Morais, S.I.F., and Kulozik, U. (2020). Storage stability of dried raspberry foam as a snack product: effect of foam structure and microwave-assisted freeze drying on the stability of plant bioactives and ascorbic acid. *Journal of Food Engineering* 270: 109779.

Pan, Z. and Atungulu, G.G. (2010). *Infrared Heating for Food and Agricultural Processing*. Boca Raton: CRC Press https://doi.org/10.1201/9781420090994.

Prosapio, V. and Norton, I. (2018). Simultaneous application of ultrasounds and firming agents to improve the quality properties of osmotic + freeze-dried foods. *LWT - Food Science and Technology* 96: 402–410.

Rahman, S.M.A. and Mujumdar, A.S. (2008). A novel atmospheric freeze-dryer using vortex tube and multimode heat input: simulation and experiments. *Asia-Pacific Journal of Chemical Engineering* 3: 408–416. https://doi.org/10.1002/apj.135.

Rahman, S. M. A. and Mujumdar, A. S. (2012). Atmospheric freeze drying. In: Progress in Food Preservation. Eds. Bhat, R., Karim A. A., and Paliyath G. Wiley. ISBN: 9780470655856 https://doi.org/10.1002/9781119962045.

Rashid, M.T., Liu, K., Jatoi, M.A. et al. (2022). Developing ultrasound-assisted hot-air and infrared drying technology for sweet potatoes. *Ultrasonics Sonochemistry* 86: 106047.

Rattanadecho, P. and Makul, N. (2016). Microwave assisted drying: a review of the state of the art. *Drying Technology* 34 (1): 1–38.

Ratti, C. (2001). Hot air and freeze drying of high value foods: a review. *Journal of Food Engineering* 49: 311–319.

Rawson, A., Tiwari, B.K., Tuohy, M.G. et al. (2011). Effect of ultrasound and blanching pretreatments on polyacetyleneand carotenoid content of hot air and freeze dried carrot discs. *Ultrasonics Sonochemistry* 18: 1172–1179.

Ren, Z. and Bai, Y. (2018). Ultrasound pretreatment of apple slice prior to vacuum freeze drying. In: *Advances in Engineering Research (AER), 169, 2nd International Conference on Material Science, Energy and Environmental Engineering*.

Ren, F., Perussello, C.A., Zhang, Z. et al. (2018). Impact of ultrasound and blanching on functional properties of hot-air dried and freeze-dried onions. *LWT - Food Science and Technology* 87: 102–111.

Ren, M., Ren, Z., Chen, L. et al. (2022). Comparison of ultrasound and ethanol pretreatments before catalytic infrared drying on physicochemical properties, drying, and contamination of Chinese ginger (*Zingiber officinale* Roscoe). *Food Chemistry* 386: 132759.

Sagar, V.R. and Kumar, S.P. (2010). Recent advances in drying and dehydration of fruits and vegetables: a review. *Journal of Food Science and Technology* 47 (1): 15–26.

Sakare, P., Prasad, N., Thombare, N. et al. (2020). Infrared drying of food materials: recent advances. *Food Engineering Reviews* 12: 381–398.

Schossler, K., Jager, H., and Knorr, D. (2012). Novel contact ultrasound system for the accelerated freeze-drying of vegetables. *Innovative Food Science & Emerging Technologies* 16: 113–120.

Stratta, L., Capozzi, L.C., Franzino, S., and Pisano, R. (2020). Economic analysis of a freeze-drying cycle. *Processes* 8: 1399.

Sujinda, N., Varith, J., Shamsudin, R. et al. (2021). Development of a closed-loop control system for microwave freeze-drying of carrot slices using a dynamic microwave logic control. *Journal of Food Engineering* 302: 110559.

Sun, Y., Zhang, M., Mujumdar, A.S., and Yu, D. (2021). Pulse-spouted microwave freeze drying of raspberry: control of moisture using ANN model aided by LF-NMR. *Journal of Food Engineering* 292: 110354.

Waghmare, R.B., Choudhary, P., Moses, J.A. et al. (2021). Trends in approaches to assist freeze drying of food: a cohort study on innovations. *Food Reviews International* https://doi.org/10.1080/87559129.2021.1875232.

Waghmare, R., Kumar, M., Yadav, R. et al. (2023). Application of ultrasonication as pre-treatment for freeze drying: an innovative approach for the retention of nutraceutical quality in foods. *Food Chemistry* 404: 134571.

Wang, H., Zhang, M., and Adhikari, B. (2015). Drying of shiitake mushroom by combining freeze-drying and mid-infrared radiation. *Food and Bioproducts Processing* 94: 507–517.

Wu, X., Zhang, M., Bhandari, B., and Li, Z. (2019). Effects of microwave assisted pulse fluidized bed freeze-drying (MPFFD) on quality attributes of *Cordyceps militaris*. *Food Bioscience* 28: 7–14.

Wu, X.F., Zhang, M., Ye, Y., and Yu, D. (2020). Influence of ultrasonic pretreatments on drying kinetics and quality attributes of sweet potato slices in infrared freeze drying (IRFD). *LWT – Food Science and Technology* https://doi.org/10.1016/j.lwt.2020.109801.

Xu, B., Chen, J., Tiliwa, E.S. et al. (2021a). Effect of multi-mode dual-frequency ultrasound pretreatment on the vacuum freeze-drying process and quality attributes of the strawberry slices. *Ultrasonics Sonochemistry* 78: 105714.

Xu, X., Zhang, L., Feng, Y. et al. (2021b). Ultrasound freeze-thawing style pretreatment to improve the efficiency of the vacuum freeze-drying of okra (*Abelmoschus esculentus* [L.] Moench) and the quality characteristics of the dried product. *Ultrasonics Sonochemistry* 70: 105300.

Yildiz, G. and Izli, G. (2019). The effect of ultrasound pretreatment on quality attributes of freeze-dried quince slices: physical properties and bioactive compounds. *Journal of Food Process Engineering* 42: e13223.

Zhang, M., Bhandari, B., and Fang, G. (2017). *Handbook of Drying of Vegetables and Vegetable Products*. Boca Raton: CRC Press. ISBN: 9781315152677 https://doi.org/10.4324/9781315152677.

Zhang, L., Liao, L., Qiao, Y. et al. (2020). Effects of ultrahigh pressure and ultrasound pretreatments on properties of strawberry chips prepared by vacuum-freeze drying. *Food Chemistry* 303: 125386.

Zhou, Y.H., Staniszewska, I., Liu, Z.L. et al. (2021). Microwave-vacuum-assisted drying of pretreated cranberries: drying kinetics, bioactive compounds and antioxidant activity. *LWT - Food Science and Technology* 146: 111464.

10

Commercial Applications of Freeze Drying in Food Processing

Shiv Kumar[1], *Rahul Mehra*[1], *Mouandhelmamou Hassani*[2], *Anuradha Mishra*[1], *and Arun Kumar Pandey*[2]

[1] *MMICT&BM (HM), Maharishi Markandeshwar (Deemed to be a University), Mullana, Ambala, Haryana, India*
[2] *Amity Institute of Food Technology, Amity University Rajasthan, Jaipur, India*

10.1 Introduction

Inevitably, the majority of raw food items are perishable by nature and need to be processed to increase shelf life if not used for consumption after harvesting. Moreover, nowadays, the demand for food items like processed meals, convenience foods, ready-to-eat foods, etc. is increasing enormously due to busy schedules and changing human lifestyles (Lavelle et al. 2023; Scholliers 2015). There are several techniques like fermentation, smoking, salting, pickling, and drying that are frequently applied traditionally to preserve foods and their by-products. However, sun drying is one of the most often used traditional techniques for the preservation of perishable food items at the household level. Later, different mechanical drying systems were developed and used at the commercial level to overcome the limitations of the traditional sun drying method of food preservation. Traditional sun drying or mechanical drying methods are suitable for drying or processing food items that are less susceptible to high processing temperatures, air, and sun light. The products obtained through traditional or mechanical drying methods are of low quality, especially when they are applied to the processing or drying of perishable food items. However, freeze drying is one of the most innovative alternatives to traditional drying methods as it works at a low temperature under vacuum and is suitable for food items that are sensitive to high thermal processing (Mehra et al. 2022).

The industrialization of freeze-drying technology (lyophilization) started during World War II, when the preservation of human blood plasma became a critical

Freeze Drying of Food Products: Fundamentals, Processes and Applications, First Edition.
Edited by Roji Balaji Waghmare, Manoj Kumar, and Parmjit Singh Panesar.

requirement (Sailliol et al. 2013). Though the effect of freezing on concentration had been known for a long time, the lyophilization and sublimation of ice remained a laboratory mystery. However, in the early seventh century, the Uighurs of Turfan invented "frozen wine," where the freeze-drying procedure was performed by combining the freezing and drying unit operations (Franks 1998). Freeze drying preserves the maximum quality attributes of food, including nutrients, volatiles, color, aroma, flavor, and freshness (Assegehegn et al. 2019). Nevertheless, sublimation of ice during freeze drying causes a small structural deformation (Mehra et al. 2021). Technically, the application of latent heat to a frozen object causes sublimation of ice, in which water evaporates from the food by transforming from its solid state to a gaseous state without converting into the liquid state (Nowak and Jakubczyk 2020). The sublimation of ice occurs at a certain pressure and temperature below the triple point of water, i.e., 4.579 mmHg at 32.01782 °F (Yamamoto et al. 2021). In freeze-drying operations, the material to be dried is first frozen at a temperature below −18 °C, and then, under high vacuum conditions, it is subjected to heat radiation, heat conduction, or both, which leads to a condition where frozen liquid transforms to its gaseous state without converting into liquid and leaving behind just dry particles (Figure 10.1). As the temperature of the dehydrated product does not rise above 40 °C, the natural quality attributes of the product remain marginally affected in freeze-drying operations (Ganguly and Alexeenko 2012). Despite the lengthy drying durations and costly techniques, freeze dryers have been used in a variety of sectors, including food, dairy, and pharmaceutical (Bhatta et al. 2020). Freeze drying is often used to

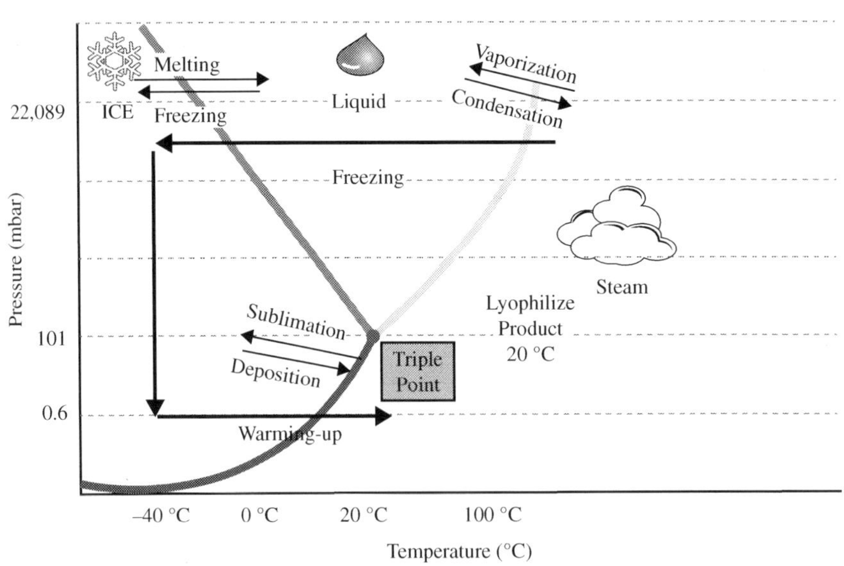

Figure 10.1 Schematic of the phase for triple points.

stabilize food, biological materials, and medicines such as proteins, vaccines, bacteria, and dairy products (Nowak and Jakubczyk 2020).

The lyophilization process has evolved over time to meet different industrial demands, including those for specialized drying. Today, freeze-drying techniques are most frequently utilized for the manufacturing of spices, coffee, dried fruits and vegetables, food powders, and nutritional microencapsulation of food for military or aerospace usage (Bhatta et al. 2020). It is also used for the production of high-quality products from entire plants or their parts without the addition of color stabilizers. The main feature of freeze drying is that it protects the functional properties of the thermally unstable material during dehydration, including colors, fibers, minerals, and vitamins (Liapis and Bruttini 2020; Różyło 2020). Lyophilization can be used to change the properties of active compounds and create novel receptors for oral medication delivery systems. The lyophilization process parameters are essential for obtaining the required product quality characteristics with excellent drying efficiency (Liapis and Bruttini 2020). However, to meet industrial demands, many variants of the lyophilization technique have emerged over time.

Today's market offers a vast selection of lyophilized products, and consumer interest in these items is also on the rise. The high expense of freeze drying restricts its widespread application in the food business. However, by updating the equipment based on the nature of raw materials, the time and effort required for this activity can be reduced. In recent years, as the market shifted toward "natural" and "organic" products, consumer demand for both low- and high-quality healthy ingredients has also expanded considerably (Tsinontides et al. 2004). In the near future, the commercial production of freeze-dried powders from "organic" fruits, vegetables, and other food items, including soy and algae, will increase significantly. Moreover, lyophilization is the ideal drying method to create dust microorganisms as it results in the least degree of cell viability loss in comparison to other drying methods (Bhatnagar and Tchessalov 2020; Bhatta et al. 2020). The present book chapter gives a deep insight into various aspects of freeze-drying technology, its mechanism of action, and its application in food processing industries, including various food products. The chapter also discusses various factors affecting the quality of freeze-dried products and different freezing techniques that are currently being used as a major part of freeze-drying operations.

10.2 Commercial Applications of Freeze-Dried Products

10.2.1 Food Industry

The increasing popularity of freeze-dried items is due to their quality attributes, increased shelf-life, and marketability (Tsinontides et al. 2004). Reducing water activity to an appropriate level permits the transformation of perishable resources into a sustainable commodity, minimizing microbial spoilage and quality deterioration owing to unwanted biochemical processes (Ciurzynska and Lenart 2011; Liapis and Bruttini 2020). The freezing process is beneficial for reducing postharvest losses, packing, and storing food products. Freeze-dried food products retain their sensory attributes as much as possible and are convenient ready-to-eat foods, which promotes their sale as value-added products (Duan et al. 2016; Nowak and Jakubczyk 2020). However, processing procedures affect the appearance, water content, antioxidant components, rehydration, and overall phytochemical concentration of a dehydrated product (Karami et al. 2021; Tarafdar et al. 2018; Lopez-Quiroga et al. 2016).

Despite being rich in vitamins, minerals, and amino acids, fruits and vegetables have a very limited shelf life. However, the shelf-life of fruits and vegetables can be extended by the application of combination technologies, including freeze drying (Silva-Espinoza et al. 2019). Ma et al. (2022) provide an overview of freeze-drying processes and the challenges associated with manufacturing freeze-dried fruits and vegetables. According to their findings, freeze-dried fruits and vegetables with quick freezing rates, higher drying temperatures, and shorter drying times can retain more phytochemicals and bioactive activities. Carotene and other oxygen-sensitive chemicals degrade during storage; however, freeze drying better preserves these molecules due to the tiny pore size and low processing temperature. Additionally, during the freezing phase, phenolic compounds bound to the cell wall degrade and are liberated.

Coffee is the most commonly traded tropical agricultural commodity globally, and processing techniques have evolved over time (Witik et al. 2019). Recent studies investigated the effects of different drying methods, such as sun drying, hot air drying, vacuum drying, and freeze drying, on proximate composition, dry matter, water activity, energy consumption, total phenolic compounds, chlorogenic acid, and 2,2-caffeic acid (Bauer et al. 2018). Using vacuum and freeze drying, they claim to have created coffee pulp with a potent antioxidant capability. To create the finest quality dry coffee pulp, freeze drying has been proposed as an effective method due to its exceptional ability to preserve bioactive components. There are several investigations on the application of vacuum freezing (VF) as a novel step in freeze-drying processing of coffee extract (Silva and Schmidt 2022). According to studies, the initial phase of the lyophilization process, i.e., VF, accelerates the rate of coffee extract drying. Also, compared to conventional freeze-dried coffee, vacuum-frozen coffee has a greater retention of volatile components, suggesting it is suitable for aromatic solutions. Spray freeze drying is another kind of drying technique that may produce a porous microstructure and free-flowing consistency in coffee powder. Spray lyophilization is superior to traditional lyophilization in preserving fragrance and flavor (Padma Ishwarya and Anandharamakrishnan, 2015).

10.2.2 Dairy Industry

Apart from its application in other food items, freeze drying is a prevalent dehydration method in the dairy industry, where its primary uses include the preservation of probiotics, starter cultures, and biological materials (Anandharamakrishnan 2017). Low-temperature freeze drying has a number of advantages for dairy products, including greater preservation of thermally sensitive vitamins, suppression of milk protein degradation and Mallard reaction, extended shelf life, decreased moisture content, firmness, and porosity. Owing to the costly and time-consuming nature of freeze-dried specialty goods, they are utilized in dairy manufacturing to concentrate useful components of probiotics and preserve starter cultures (Bhushani and Anandharamakrishnan 2017; Gaidhani et al. 2015). In addition, freeze drying offers various advantages in terms of viability, quality, and shelf-life of lactic acid bacteria cultures (Morgan et al. 2006; Vesterlund et al. 2012). Deshwal et al. (2020) examined the influence of spray freeze-drying technology on the reconstitution, bioactive characteristics, distribution of particle size, and morphology of camel milk samples, as well as the thermal coagulation of reconstituted camel powder. Freeze-dried camel milk powder has been shown to have the best dispersibility and solubility. Zhang et al. (2022) investigated the lipid profile and absorption of freeze-dried and spray-dried goat milk. Compared with sprayed goat milk, freeze-dried goat milk fat has a longer digesting time and a lower lipolysis rate. Moreover, a variety of dairy products, including yogurt, cheese, and butter, are freeze-dried (Bhushani and Anandharamakrishnan 2017; Hnin et al. 2021; Silva et al. 2021).

10.2.3 Other Industry

In addition to the food and dairy industries, freeze drying is widely used in the pharmaceutical, cosmetic, medical device, flower, microorganism, meat, and poultry industries. Freeze drying is an effective method for the manufacturing of pharmaceutical compositions, including chemically active components and carriers for drug delivery systems (Izutsu 2018). The processing temperature does not result in considerable protein degradation, as is the case with other drying procedures. First, freeze drying is employed to isolate solutes from an unstable liquid composition (Assegehegn et al. 2019). It significantly extends the shelflife of unstable liquid formulations by removing solvents that encourage the physicochemical breakdown of materials. In addition to safeguarding the medicine and guaranteeing the formulation's stability, the sublimation of water during freeze-drying operations forms a wide surface area that may be utilized to improve the bioavailability of the drug tablets (Kobayashi et al. 2020; Siow et al. 2019). Moreover, it has been observed that freeze-dried tablets may have both liquid and solid formulations (Izutsu 2018).

Owing to their similarity to fresh meat products and lengthy shelf lives, cured meat products are gaining popularity across the world. Traditional meat drying processes, such as hot air drying or the sun, have been shown to be sluggish, which can result in unfavorable side effects such as browning, shrinkage, loss of nutrients and aromatic compounds, oxidation, and lipid degradation (Mediani et al. 2022). Several researchers evaluated different drying methods to determine the impacts of drying on meat quality characteristics. In recent years, substantial research has been conducted on the application of freeze drying to meat and poultry products. Meat that has been freeze-dried preserves its natural color, taste, and nutritional content. In this method, meat protein as well as fat-soluble vitamins, i.e., vitamin A and D, remain unaffected. This method is found to be very effective for the dehydration of meat and meat products due to its ability to protect heat- and oxygen-sensitive components of food (Lee et al. 2021; Nowak and Jakubczyk 2020; Oyinloye and Yoon 2020).

Fresh and liquid eggs have been substituted with powdered egg whites due to their resistance to microbiological and oxidative deterioration and cheap packing costs (Vargas-del-Río et al. 2022). Due to the quick loss of water from the matrix, freeze-dried eggs are primarily amorphous solids with an uneven structure. The advantages of freeze drying include its appropriateness for drying heat-sensitive substances and its capacity to dry liquefied food items (Nowak and Jakubczyk 2020). Nevertheless, freeze drying is an energy-intensive process with high product prices, which restricts its commercial use for high-value items. Vargas-del-Río et al. (2022) evaluated the effects of spray drying at various outlet temperatures and freeze drying on foam potential and associated physicochemical parameters, interfacial adsorption behavior, and spray-dried egg white structure. According to them, spray drying at a low outlet temperature is appropriate for preserving the qualities of egg white foam, but the strength of foam and spray-dried egg white was inferior to that of freeze-dried egg white (Vargas-del-Río et al. 2022; Yang et al. 2022).

10.3 Cost Analysis of Freeze-Dried Food Products

The freeze-drying process is superior to the other drying methods, but it has the drawback of requiring high-priced equipment and significant amounts of energy (Dziki 2020). The application of freeze drying continues to be the method of choice for preserving the quality of dried food products. According to research findings, it was observed that the most expensive part of the process is not the lyophilizer but the initial expenditure and high processing and handling costs. Recent regulations concerning the use of process fluids containing greenhouse gases in cooling systems are another important aspect. It was hypothesized that the reduction in initial drying time could be an alternative way of lowering operating costs by increasing the number of drying cycles of the lyophilizer (Stratta et al. 2020). The longer lifespan of lyophilizers will ultimately result in high productivity. However, the overall processing cost is determined by factors such as the type of raw material used, the product itself, the packaging, the capacity of the plant, the duration of the cycle, and so on. Therefore, depending on the expenses involved, freeze-drying techniques are not utilized very frequently in the food business. In the past 10 years, new cost-cutting innovations like adsorption, liquefaction, and microwaves have been used; however, vacuum freeze drying is still the sole method that is utilized at industrial scale to dry coffee, spices, meat, food additives, and other food materials to achieve high-quality products (Stratta et al. 2020). Microwave freeze drying (MFD) and air freeze drying (AFD) are two methods that have been developed in the recent past to cut down on the amount of energy that is required for freeze-drying operations. These two drying processes produce products with a high quality that is comparable to that of freeze-dried products. However, the price of processing differs based on the kind of raw material, the products, the packaging, the capacity of the facility, and the length of the cycle, among several other factors (Duan et al. 2010).

10.4 Industries Freeze-Drying Operation

10.4.1 Pretreatments

Pre-freezing of a product depends on cycle time, product needs, or lyophilization theory. It includes cooling before freezing and may also require formulation adjustments, high vapor pressure to decrease the amount of solvent/liquid in food, or a surface area increase (Liu et al. 2005). Concentrating and purifying the samples prior to freeze drying using ultrasound, liquid–liquid extraction, quenchers, and solid-phase extraction are also typical types of pretreatments (Dziki 2020; Zhang et al. 2021). The freeze-drying procedure includes four different phases, as shown in Figure 10.2.

10.4.2 Freezing

The moisture content of food is the major component for freezing, where rapid freezing leads to the formation of tiny ice crystals that help maintain the microscopic structure of the product but make it more difficult to be lyophilized (Figure 10.3). The material that needs to be freeze-dried must first be adequately pre-frozen, as the process involves a sharp transition of water from its solid state to the vapor phase. The effectiveness of freeze-drying operations is influenced by

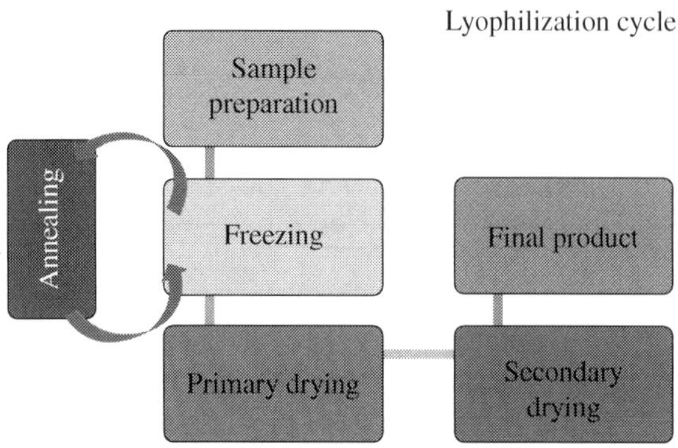

Figure 10.2 Steps involved in freeze drying, from sample preparation to final product development.

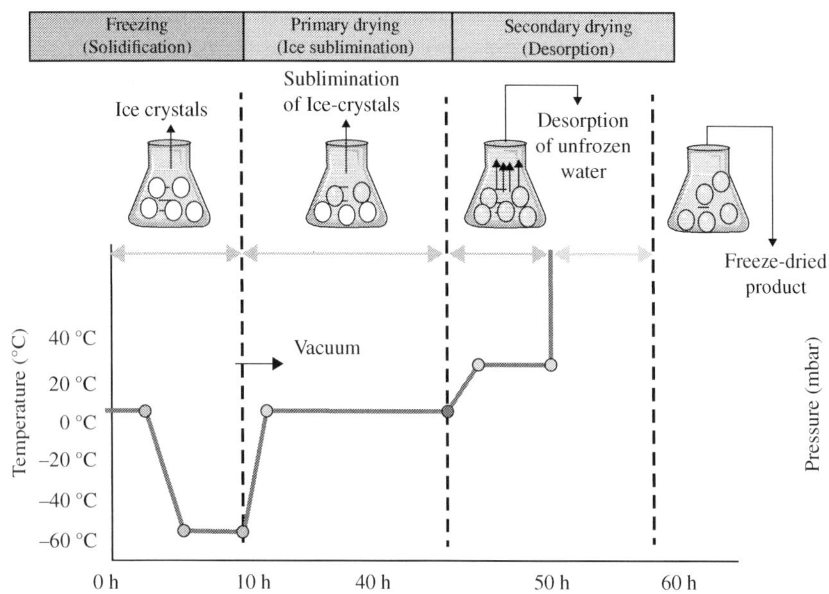

Figure 10.3 Step involved in the industrial freezing process.

both the freezing method and the temperature of the finished product. Rapid cooling produces small ice crystals that preserve the microscopic features of the product while making it more difficult to freeze-dry. Slow cooling produces larger ice crystals and fewer constricted channels in the matrix during drying (Tan et al. 2021). Based on the nature of the product, there are two possible freezing methods (Marshall et al. 2003). The bulk of goods exposed to freeze drying consist of water/solvent, and the solute components are dissolved or suspended in the water/solvent (I and An 2022). Most of the samples that require lyophilization are eutectic in nature, which refers to a mix of various substances that freeze at a temperature lower than the surrounding water/solvent. When the aqueous solution cools, the concentration of solutes in the product matrix also changes. In addition, with the progress of the cooling phase, water/solvent separates from solutes, resulting in zones of progressively concentrated solutes. These regions have a comparatively lower freezing point as compared to the surrounding water (Bano et al. 2020). In a suspension, eutectic is produced when various solute concentrations and the solvent are mixed together. Until the entire eutectic mixture is frozen, the suspension cannot be frozen thoroughly; that is why this particular thermal point is known as the eutectic temperature (Berk 2018). Prior to beginning the freeze-drying procedure, it is essential that the product be pre-frozen to a temperature below the eutectic temperature. The structural integrity of a freeze-dried product is compromised due to the expansion of tiny pockets of unfrozen material that remain in the product (Nowak and Jakubczyk 2020). However, suspensions are the second type of frozen product that crystallizes when frozen. Besides the eutectics, the thickening of the entire suspension increases with the decrease in temperature. Eventually, a vitreous solid is generated when the product freezes at its glass transition temperature. The freeze drying of such type of products is incredibly challenging.

10.4.3 Primary Drying

Once the freezing process is over, a vacuum is applied to the lyophilizer to reduce the inside pressure. Based on the product target temperature and the features of the system container, the pressure of a standard pharmaceutical lyophilization chamber ranges between 30 and 300 mTorr. To initiate the sublimation of ice and transit of water vapor to the ice storage condenser, where it is deposited as ice, the pressure of the chamber needs to be reduced below the ice vapor pressure at the sublimation interface of the product (Bhushani and Anandharamakrishnan 2017; Kawasaki et al. 2019). Exceedingly high chamber pressures reduce the pressure differential between the sublimation contact and the chamber, slowing the sublimation rate while continuing to remove ice. There will be no mass transfer if the chamber pressure is higher than the vapor pressure at the sublimation contact (Brülls and Rasmuson 2002). Yet, extremely low pressures (50 mTorr) are also harmful to rapid sublimation rates since they dramatically limit the rate of heat transmission to the product (Nowak and Jakubczyk 2020). At the sublimation interface, the vapor pressure of ice and the product temperature become inversely proportional to each other. Once the chamber pressure falls below the vapor pressure of the ice in the product, sublimation occurs. This is the process in which frozen water from the product is removed from the top layer by directly moving it to the vapor phase without going to the liquid state. Transported water vapor is deposited on the coils or plates of the ice condenser, where it is continually cooled to a temperature corresponding to the extremely low vapor pressure of the condensed ice. The product cools due to the energy required to sublimate the water (temperature-dependent, around 670 cal/g) (Feistel and Wagner 2007). At a prescribed temperature, the heated shelves must provide the energy necessary for the ice to continue sublimating. In freeze-drying operations, the temperature of the product, particularly at the sublimation interface at the primary drying stage, is often the most important product characteristic. Low product processing temperatures result in low vapor pressure in ice, resulting in lengthy primary drying times. By utilizing more aggressive product temperatures, it has been discovered that a 1 °C increase in product temperature can reduce the time required for primary drying by up to 13% (Bano et al. 2020; Jeeraruangrattana et al. 2020; Patel et al. 2010).

10.4.4 Secondary Drying

When the initial lyophilization process has been completed and all the ice has melted, the product still contains bound water. Water desorption from the cake takes place after the complete removal of ice, and the secondary drying process begins once the primary drying process is over (Nowak and Jakubczyk 2020). Once the complete ice is removed from the product container, the shelf temperature is typically raised and maintained between 20 and 40 °C for several hours (Patel and Pikal 2011). It is not essential to further lower the chamber pressure since diffusion inside the solute phase as well as desorption from the surface govern the rate of desorption to attain a particular moisture level (Andrieu and Vessot 2011). Typically, the objective of secondary drying is to reduce the cake's moisture level to less than 1%. It is desirable to eliminate as much leftover moisture as possible since, for the vast majority of lyophilized pharmaceutical active ingredients (APIs), the stability increases as the amount of moisture decreases. Yet, it is essential to consider the thermal strains imposed on the API by the high product temperature. It is vital, particularly for proteins, to determine the optimal secondary drying conditions that yield the correct moisture content without being affected by heat. The optimal stability for several protein formulations has been identified at moisture values between 1% and 3% (residual moisture) (Trelea et al. 2016). Determining and monitoring the desired moisture content for each vial in a batch might be difficult. Even if the product seems to be dry, it must continue to dry at a higher temperature to achieve the specified residual moisture in the finished product. In this process, the term "isothermal desorption" is used to describe the phenomenon of desorption of the bound water from the product.

Secondary drying happens often at a temperature above the product's ambient temperature but is consistent with product sensitivity (Nowak and Jakubczyk 2020). Secondary drying occurs at a higher temperature than the first drying step (Kawasaki et al. 2019; Nowak and Jakubczyk 2020).

10.4.5 Factors Affecting the Quality and Uniformity of Freeze-Dried Products

Numerous variables can impact the durability and quality of freeze-dried substances, of which moisture and oxygen are the two most influential factors (Czarnowska and Gujska 2012; Oyinloye and Yoon 2020). All freeze-dried items include residual moisture, which varies depending on the kind of product and the duration of secondary drying. Among other methods, moisture content can be detected chemically, chromatographically, monometrically, and gravimetrically. It is stated as a weight percentage of the product's total weight. Most items have residual moisture levels ranging from 1% to 3% (Nowak and Jakubczyk 2020). Being hygroscopic in nature, freeze-dried items can become unstable when exposed to moisture during storage (Cal-Vidal and Falcone 1985). For goods intended for use as freeze-dried products, packaging must be resistant to atmospheric moisture. To reduce the likelihood of items decaying due to exposure to moisture and atmospheric oxygen, they are often kept in places with low humidity in airtight containers. The rate of freeze-dried product degradation is often higher at higher storage temperatures; therefore, a temperature between 4 and 8 °C, or a refrigeration temperature, is preferred to extend the shelf life of freeze-dried products during storage (Nowak and Jakubczyk 2020). The keeping quality of a freeze-dried product may be determined by observing how rapidly it degrades at high temperatures. By selecting the appropriate time and temperature correlations at elevated temperatures, it is possible to predict the rate of product deterioration at lower storage temperatures. Initial and end product temperatures, the quantity of heat removed, the product's size (particularly thickness), its form, the heat transfer method, and the temperature all impact the freeze-dried product's quality. The International Institute of Refrigeration (1972) set several freezing time requirements for frozen products and freezing equipment (Czarnowska and Gujska 2012). Below is a list of some essential components:

a) Distribution of ice crystals
b) Concentration of the solution
c) Concentration gradient
d) Formation of an amorphous, glassy product

The list of the major industries producing freeze-dried products is given in Table 10.1.

Table 10.1 Leading Industries Producing Freeze-Dried Products.

Product	Industry	Application	References
Dairy product	Happy family brand (New York, US), Nutra vita (England, UK), Mountain House (California, US), Fresh way (Florida, US)	Freeze drying dairy products reduces their moisture content, making them easier to transport and reducing their storage costs by extending their shelf life. It also improves the re-dispersibility of milk powder in aqueous systems.	Baldelli et al. (2022); Jouppila and Roos (1997)
Instant food (noodles, soup, purees)	Slurrp farm (Haryana, India), Nestle (Vevey, Switzerland), Knorr (Heilbronn, Germany)	Instant foods are in great demand among consumers since they take less time to make and are healthy.	Du et al. (2022); Omah et al. (2015)
Fruit (mango, sapodilla, strawberry, kiwi, banana) **Fruit juice and fruit powder**	Moon (Delhi, India), Crunchy-munchy (Noida, India), Yum earth (Mumbai, India)	Freeze-dried fruits can be added to morning cereals, candies, icecream, and other foods to enhance their flavor, texture, appearance, and nutritional content.	Nualkaekul et al. (2012); Różyło (2020)
Vegetable (peas, carrot, beetroot, beans) **Vegetable soup, Vegetable powder**	Electric herb (west coast), Everyday essentials (Pune, India), Mother Earth (Bangalore, India)	Nowadays, freeze-dried veggies are utilized globally in the preparation of meals. According to WHO and FAO, this helps to reduce illness incidence.	Bhatta et al. (2020)
Space food	Astronaut (Colorado, USA)	To prevent astronauts from starving and provide them with palatable sustenance, freeze-dried foods are manufactured.	Cooper et al. (2011); Gronwald et al. (2022)
Beef products (diced, steaks, sliced, strewn and gravy)	Augason farm (Salt Lake city, US), Hormel Foods (Austin, US)	Freeze-dried beef is found to be low in moisture, to have a longer storage life, to have a superior flavor, and to be rehydratable.	Messina et al. (2015); Reidy and Heldman (1972)

10.5 Industrial Freeze-Drying Device

Freeze drying is an excellent technique for preserving a wide variety of drugs and biological materials (Merivaara et al. 2021). Freezing is the most common technique used in microbiology to store and maintain culture conditions. Lyophilizing microorganisms is a multi-step procedure that includes increasing microbial culture resistance to fast freezing, collecting organisms at the optimal moment, and optimizing cryoprotectant, freezing, primary drying, and secondary drying conditions (Prosapio and Lopez-Quiroga 2020). Numerous experts in the field have advocated for the use of lyophilization in the pharmaceutical industry (Nowak and Jakubczyk 2020). Lyophilization needs either a high vacuum or low pressure to get the best results. It is common practice for commercial freeze dryers to have the freezer and dryer in the same setup. Freeze drying can alter the size, pore structure, and form of a product, which can have a negative impact on its durability (Liapis and Bruttini 2020). The limited heat transfer rate of conventional freeze drying necessitated the first attempts at creating interior heat with microwave radiation (Liapis and Bruttini 2020). MFD is a fast-dehydrating procedure used on different types of foods, such as fish, hard soup, fruit, and vegetables (Duan et al. 2007). Given the multitude of different freezer options, most things can be frozen using more than one technique. Hence, it is vital to do a cost–benefit analysis taking into account the economics, functionality, and practicality of various freezing systems before making a final decision. Although their relative relevance may shift from one technique to another, these considerations can nevertheless help you narrow down your freezer options. Due to the reduced processing rate and the high capital investment, the cost factor becomes more relevant than any other variable for developing nations where the freezing application is still relatively new (Liapis and Bruttini 2020; Nowak and Jakubczyk 2020; Prosapio and Lopez-Quiroga 2020).

10.5.1 Air Blast Freezer

It is one of the oldest and most popular methods for freezing foods because it keeps the temperature steady and can be used for many different types of food (Dempsey and Bansal 2012). Either pushed air or still air is used as the freezing medium in the air blast freezer. Food is frozen by putting it in special freezers called "sharp freezers" (Dempsey and Bansal 2012). Air blast freezing is usually considered as the cheapest way of freezing commodities; it also keeps the temperature stable during the freezing operation, which makes it good for large-scale freezing of raw materials like beef quarters and fish. However, it is the slowest freezing method as the air circulates inside the compartment, which has a low surface heat transfer coefficient. The freezing time of sharp freezers is mostly affected by the size, type, and initial temperature of the raw material and the temperature inside the freezing chamber (Dempsey and Bansal 2012). The forced-air freezer is a better version of the still-air freezer. It uses convection to move air around inside the room where the food is frozen. Even if more refrigerator space and blowers are added, the airflow over the food in the sharp freezer cannot be controlled (Dempsey and Bansal 2012). There are many different designs and combinations for air blast freezers, depending on whether the process is in-line or batch (Chang and Hartel 2002). Continuous freezers, in which the goods are moved by trucks or conveyors, are the best way to make a lot of packaged items with consistent freezing times. When trucks are used, the system functions in semi-batch mode because the trucks stay still until a new truck comes to the entry point of the tunnel and pushes the others along until a finished vehicle comes out from the exit point. Batch freezers are more versatile because they can freeze different things at the same time on different carts. These freezers need to be checked more often than continuous systems because they could get too full (James and James 2014).

10.5.2 Fluidized Bed Freezers

The fluidized bed freezer is a modified form of the conventional air blast freezer for specific freezing purposes (Liapis and Bruttini 2020). In this type of freezer, cold air is forced upward vertically through the perforated bottom of the sample holder. The unpackaged food samples that need to be frozen are placed inside the sample holder, where they are completely surrounded by the moving cold air, which freezes them effectively in a very short time (Garcia-Amezquita et al. 2016). Fluidization is better than traditional freezing methods in a number of ways since individual pieces of product can be frozen quickly with this method. The process is also known as individual quick freezing (IQF), which is good for foods that tend to stick together. The goal of IQF freezing is to provide a reliable way to freeze small pieces that might stick together while they are being frozen. Today, different small-size vegetables, prawns, shrimp, french-fried potatoes, diced meat, and fruits can be frozen individually using this freezing technique (Fikiin and Fikiin 1998; Joy and Lada 2006).

10.5.3 Contact Freezer

This type of freezer is most efficient for freezing food samples due to its high heat transfer rate. During the freezing process, there may be either direct or indirect contact between the product and the freezing medium (Visek et al. 2014). In a contact freezer, heat transfer efficiency is at its maximum as the material to be frozen is completely surrounded by the freezing medium/refrigerant. By employing a belt or plate in an indirect contact freezer, the product does not come into direct contact with the freezing medium. Despite the technique's significant limitations, heat transmission occurs efficiently during indirect freezing, especially when utilized for packed objects due to the package's barrier to heat transfer. Corrosion may also result from the interactions between metal packing and the heat transfer surfaces (Gaidhani et al. 2015; Nowak and Jakubczyk 2020).

10.5.4 Immersion Freezer

The chilled freezing medium, such as glycol, glycerol, sodium chloride, calcium chloride, or a mixture of salt and sugar, is used for immersion freezing (Zarkadas and Mitrakas 1999). A high rate of temperature reduction can be achieved through direct heat exchange between the product and the freezing medium as the product is either immersed in or sprayed with this freezing solution in this freezing method. Putting an item directly into direct contact with the liquid refrigerant is the quickest way to freeze, as the rate of heat conduction is much higher through liquids as compared to air. For products to freeze effectively, the solution must be denser than the product itself, and non-toxic as well as tasteless solutes must be used in the freezing system. Immersion freezing machines are often applied for shell freezing of larger food pieces as they minimize moisutre loss from the product due to rapid freezing of the outside layer. A common problem in these freezing systems is dilution of the freezing medium due to leaching of product moisture content, which can change the concentration and process parameters. Therefore, flexible membranes can be used to shield food from the liquid refrigerant (George 1993; Resch et al. 2011).

10.5.5 Plate Freezer

It is also one of the most common types of contact freezers that you could find. In this method, the product is compressed between the cold metal plates, and a refrigerant is pumped inside the hollow spaces of the plates (Liapis and Bruttini 2020). The only kinds of materials or blocks that can be frozen using this technique are things like beef patties and other pre-packaged commodities that come in the shape of blocks (Kyu 2018; Persson and Londahl 1993).

10.5.6 Cryogenic Freezers

In this method, food directly comes into contact with liquefied forms of gases, i.e., liquid nitrogen or liquid carbon dioxide, to expose it to a sub-60 °C environment (Fennema 1978). This is a relatively new technology for freezing food. This technique differs from others in that it is not connected to a refrigeration plant; rather, the liquid refrigerants are used to pre-freeze food before transferring it to the freezing vessel, especially at large industrial facilities. The compact nature of cryogenic freezers offers flexibility in plant design and efficient freezing applications. Typically, cryogenic freezers have cheap initial prices but significant operating expenses (Khadatkar et al. 2004).

Nitrogen in its liquid form boils at −196 °C under standard atmospheric pressure. The refrigerant evaporates when it contacts the freezer's contents and spray nozzles. The transfer efficiency is optimized by having the refrigerant flow in the opposite direction of the items' movement on the conveyor belt (Fountain et al. 1997). It takes around 1.2 kilograms of product to equal 1 kilogram of refrigerant. Fillets of fish, seafood, fruit, and berries are frequently used in this method (Dawson 2019). Liquid carbon dioxide may be held as either a solid or a gas if the pressure is maintained at the same level as that of the surrounding air. When the gas is discharged into the atmosphere at a temperature of about −70 °C, half of it becomes dry-ice snow, and the other half continues to exist as a vapor state (Liapis and Bruttini 2020). Before the application of nitrogen spray, the product is subjected to liquid carbon dioxide due to its unique property of preliminary freezing (Dawson 2019).

10.6 Future Prospects

Freeze drying is among the highest-quality drying options. Since before World War II, it has been used in a variety of food sectors to keep nutrients stable throughout the storage and transport of perishable items, including blood plasma and organs. Vitamins and polyphenols, which degrade at high temperatures, are preserved better in freeze-dried foods. A wide variety of products, including fruits and vegetables, dairy products, pharmaceuticals, nutraceuticals, meats, quick foods, and more, can benefit from the freeze-drying process. Bakery goods, cereals, fruit snacks, and veggie snacks are just a few examples of categories where freeze-dried goods are now in great demand. The freeze-dried food business is expanding all the time. Although the market for freeze-dried goods is far larger in the West and other wealthy countries than it is in developing countries, the latter have been expanding their use of freeze-dried goods in recent decades. Although the market for freeze-dried goods grew during the Covid-19 pandemic, this method will continue to be useful in the future as growing demand for nutritious products and nutraceuticals would necessitate longer storage times. Although freeze drying is an expensive process, several pretreatment methods are utilized to manage the price of the freeze-dried product; moreover, other methods, such as microwave and atmospheric pressure, are employed to increase yield while decreasing the price of the freeze-dried product.

10.7 Conclusions

The increasing demand for ready-to-eat items has led to a greater use of freeze-drying techniques in the food industry to guarantee both quality and longevity in this category. The freeze-drying technique has found a wide range of applications in the food and pharmaceutical industries. The recent advancements in the hybrid freeze-drying process have led to shorter drying times, lower processing costs, and higher quality end products, all of which are important factors in ensuring the long-term sustainability of the food industry. Low storage costs and portability are added advantages of freeze-dried products that meet the current needs of the canned food industry. Lyophilization has also been found useful in creating a stable injectable dosage of pharmaceutical solutions by reducing the formulation moisture content, increasing stability, easing handling, and increasing dissolution properties. The freeze-drying process increases the durability of the product to a greater extent by reducing the rate of physico-chemical changes. However, successful lyophilization requires deep knowledge of the complexities of the process and operating conditions to maximize product quality and process performance.

References

Anandharamakrishnan, C. (2017). *Handbook of Drying for Dairy Products*, 1ste. Wiley https://doi.org/10.1002/9781118930526.

Andrieu, J. and Vessot, S. (2011). Characterization and control of physical quality factors during freeze-drying of pharmaceuticals in vials. In: *Modern Drying Technology* (ed. E. Tsotsas and A.S. Mujumdar), 51–90. John Wiley & Sons, Ltd. https://doi.org/10.1002/9783527631667.ch3.

Assegehegn, G., Brito-de la Fuente, E., Franco, J.M., and Gallegos, C. (2019). The importance of understanding the freezing step and its impact on freeze-drying process performance. *Journal of Pharmaceutical Sciences* 108 (4): 1378–1395. https://doi.org/10.1016/j.xphs.2018.11.039.

Baldelli, A., Oguzlu, H., Liang, D.Y. et al. (2022). Spray freeze drying of dairy products: effect of formulation on dispersibility. *Journal of Food Engineering* 335: 111191. https://doi.org/10.1016/j.jfoodeng.2022.111191.

Bano, G., De-Luca, R., Tomba, E. et al. (2020). Primary drying optimization in pharmaceutical freeze-drying: a multivial stochastic modeling framework. *Industrial & Engineering Chemistry Research* 59 (11): 5056–5071. https://doi.org/10.1021/acs.iecr.9b06402.

Bauer, D., Abreu, J., Jordão, N. et al. (2018). Effect of roasting levels and drying process of coffea canephora on the quality of bioactive compounds and cytotoxicity.

International Journal of Molecular Sciences 19 (11): 3407. https://doi.org/10.3390/ijms19113407.

Berk, Z. (2018). Freeze drying (lyophilization) and freeze concentration. In: *Food Process Engineering and Technology* (ed. Z. Berk), 567–581. Elsevier https://doi.org/10.1016/B978-0-12-812018-7.00023-3.

Bhatnagar, B. and Tchessalov, S. (2020). Advances in freeze drying of biologics and future challenges and opportunities. In: *Drying Technologies for Biotechnology and Pharmaceutical Applications*, 1ste (ed. S. Ohtake, K. Izutsu, and D. Lechuga-Ballesteros), 137–177. Wiley https://doi.org/10.1002/9783527802104.ch6.

Bhatta, S., StevanovicJanezic, T., and Ratti, C. (2020). Freeze-drying of plant-based foods. *Foods* 9 (1): 87. https://doi.org/10.3390/foods9010087.

Bhushani, A. and Anandharamakrishnan, C. (2017). Freeze drying. In: *Handbook of Drying for Dairy Products* (ed. C. Anandharamakrishnan), 95–121. John Wiley & Sons, Ltd. https://doi.org/10.1002/9781118930526.ch6.

Brülls, M. and Rasmuson, A. (2002). Heat transfer in vial lyophilization. *International Journal of Pharmaceutics* 246 (1–2): 1–16. https://doi.org/10.1016/S0378-5173(02)00353-8.

Cal-Vidal, J. and Falcone, M. (1985). Processing conditions affecting the hygroscopic behavior of freeze-dried passion fruit juice. *Journal of Food Science* 50 (5): 1238–1241. https://doi.org/10.1111/j.1365-2621.1985.tb10451.x.

Chang, Y. and Hartel, R.W. (2002). Development of air cells in a batch ice cream freezer. *Journal of Food Engineering* 55 (1): 71–78.

Ciurzynska, A. and Lenart, A. (2011). Freeze-drying-application in food processing and biotechnology-a review. *Polish Journal of Food and Nutrition Sciences* 61 (3).

Cooper, M., Douglas, G., and Perchonok, M. (2011). Developing the NASA food system for long-duration missions. *Journal of Food Science* 76 (2): R40–R48. https://doi.org/10.1111/j.1750-3841.2010.01982.x.

Czarnowska, M. and Gujska, E. (2012). Effect of freezing technology and storage conditions on folate content in selected vegetables. *Plant Foods for Human Nutrition* 67 (4): 401–406. https://doi.org/10.1007/s11130-012-0312-2.

Dawson, P.L. (2019). Effects of freezing, frozen storage, and thawing on eggs and egg products. In: *Freezing Effects on Food Quality* (ed. Jeremiah), 337–366. CRC Press.

Dempsey, P. and Bansal, P. (2012). The art of air blast freezing: design and efficiency considerations. *Applied Thermal Engineering* 41: 71–83. https://doi.org/10.1016/j.applthermaleng.2011.12.013.

Deshwal, G.K., Singh, A.K., Kumar, D., and Sharma, H. (2020). Effect of spray and freeze drying on physico-chemical, functional, moisture sorption and morphological characteristics of camel milk powder. *LWT* 134: 110117. https://doi.org/10.1016/j.lwt.2020.110117.

Du, Y., Zhang, M., Mujumdar, A.S. et al. (2022). Innovative applications of freeze-drying to produce compound formula instant foods: a review. *Drying Technology* 40 (13): 2583–2597. https://doi.org/10.1080/07373937.2021.1966795.

Duan, X., Zhang, M., and Mujumdar, A.S. (2007). Studies on the microwave freeze drying technique and sterilization characteristics of cabbage. *Drying Technology* 25 (10): 1725–1731. https://doi.org/10.1080/07373930701591044.

Duan, X., Zhang, M., Mujumdar, A.S., and Wang, R. (2010). Trends in microwave-assisted freeze drying of foods. *Drying Technology* 28 (4): 444–453. https://doi.org/10.1080/07373931003609666.

Duan, X., Yang, X., Ren, G. et al. (2016). Technical aspects in freeze-drying of foods. *Drying Technology* 34 (11): 1271–1285.

Dziki, D. (2020). Recent trends in pretreatment of food before freeze-drying. *Processes* 8 (12): 1661. https://doi.org/10.3390/pr8121661.

Feistel, R. and Wagner, W. (2007). Sublimation pressure and sublimation enthalpy of H2O ice Ih between 0 and 273.16K. *Geochimica et Cosmochimica Acta* 71 (1): 36–45. https://doi.org/10.1016/j.gca.2006.08.034.

Fennema, O. (1978). Cryogenic freezing of foods. *Advances in Cryogenic Engineering* 712–720.

Fikiin, K., & Anguel Fikiin. (1998). Individual Quick Freezing of Foods by Hydrofluidisation and Pumpable Ice Slurries. *International IIR Conference 'Advances in the Refrigeration Systems, Food Technologies and Cold Chain'At: Sofia (Bulgaria)Volume: IIR Proceedings Series 'Refrigeration Science and Technology'*, ISSN: 01511637, ISBN: 2913149006.

Fountain, D., Ralston, M., Higgins, N. et al. (1997). Liquid nitrogen freezers: a potential source of microbial contamination of hematopoietic stem cell components. *Transfusion* 37 (6): 585–591. https://doi.org/10.1046/j.1537-2995.1997.37697335152.x.

Franks, F. (1998). Freeze-drying of bioproducts: putting principles into practice. *European Journal of Pharmaceutics and Biopharmaceutics* 45 (3): 221–229. https://doi.org/10.1016/S0939-6411(98)00004-6.

Gaidhani, K.A., Harwalkar, M., Bhambere, D., and Nirgude, P.S. (2015). Lyophilization/freeze drying–a review. *World Journal of Pharmaceutical Research* 4 (8): 516–543.

Ganguly, A. and Alexeenko, A.A. (2012). Modeling and measurements of water–vapor flow and icing at low pressures with application to pharmaceutical freeze-drying. *International Journal of Heat and Mass Transfer* 55 (21–22): 5503–5513. https://doi.org/10.1016/j.ijheatmasstransfer.2012.05.021.

Garcia-Amezquita, L. E., Welti-Chanes, J., Vergara-Balderas, F. T., & Bermúdez-Aguirre, D. (2016). Freeze-drying: the basic process. In Caballero, B., Finglas, P.M., and Toldrá, F. Encyclopedia of Food and Health (pp. 104–109). Elsevier. https://doi.org/10.1016/B978-0-12-384947-2.00328-7.

George, R.M. (1993). Freezing proceseses used in the food industry. *Trends in Food Science & Technology* 4 (5): 134–138.

Gronwald, B.J., Kijak, K., Jezierska, K. et al. (2022). Influence of freeze-dried diet on oral hygiene indicators in strict isolation condition of an analog space mission.

International Journal of Environmental Research and Public Health 19 (3): 1367. https://doi.org/10.3390/ijerph19031367.

Hnin, K.K., Zhang, M., and Wang, B. (2021). Impact of different FD-related drying methods on selected quality attributes and volatile compounds of rose flavored yogurt melts. *Drying Technology* 39 (9): 1205–1218. https://doi.org/10.1080/07373937.2021.1871622.

I, S. and An, R. (2022). A review on freeze-drying: a stability enhancement technique. *Research Journal of Pharmacy and Technology* 4841–4846. https://doi.org/10.52711/0974-360X.2022.00813.

Izutsu, K. (2018). Applications of freezing and freeze-drying in pharmaceutical formulations. In: *Survival Strategies in Extreme Cold and Desiccation*, vol. 1081 (ed. M. Iwaya-Inoue, M. Sakurai, and M. Uemura), 371–383. Springer Singapore https://doi.org/10.1007/978-981-13-1244-1_20.

James, S.J. and James, C. (2014). Chilling and freezing. In: *Food Safety Management* (ed. Y. Motarjemi and H. Lelieveld), 481–510. Elsevier.

Jeeraruangrattana, Y., Smith, G., Polygalov, E., and Ermolina, I. (2020). Determination of ice interface temperature, sublimation rate and the dried product resistance, and its application in the assessment of microcollapse using through-vial impedance spectroscopy. *European Journal of Pharmaceutics and Biopharmaceutics* 152: 144–163. https://doi.org/10.1016/j.ejpb.2020.04.015.

Jouppila, K. and Roos, Y.H. (1997). Water sorption isotherms of freeze-dried milk products: applicability of linear and non-linear regression analysis in modelling. *International Journal of Food Science & Technology* 32 (6): 459–471. https://doi.org/10.1111/j.1365-2621.1997.tb02120.x.

Joy, P. and Lada, R. (2006). Crack development in individually quick frozen cut and peel carrots. *Journal of Food Science* 71 (9): E392–E397.

Karami, H., Shariatifar, N., Nazmara, S. et al. (2021). The concentration and probabilistic health risk of potentially toxic elements (PTEs) in edible mushrooms (wild and cultivated) samples collected from different cities of Iran. *Biological Trace Element Research* 199 (1): 389–400. https://doi.org/10.1007/s12011-020-02130-x.

Kawasaki, H., Shimanouchi, T., and Kimura, Y. (2019). Recent development of optimization of lyophilization process. *Journal of Chemistry* 2019: 1–14. https://doi.org/10.1155/2019/9502856.

Khadatkar, R.M., Kumar, S., and Pattanayak, S.C. (2004). Cryofreezing and cryofreezer. *Cryogenics* 44 (9): 661–678. https://doi.org/10.1016/j.cryogenics.2004.03.008.

Kobayashi, M., Kondo, K., Nakashima, C. et al. (2020). Design of self-dispersible microsponge containing cyclosporine through wet milling and drop freeze-drying processes to improve dissolution behavior. *Journal of Drug Delivery Science and Technology* 57: 101577. https://doi.org/10.1016/j.jddst.2020.101577.

Kyu, M.M. (2018). Design analysis of plate freezer. *International Journal of Research Publications* 18 (1): 11–11.

Lavelle, F., McKernan, C., Murphy, B., and Dean, M. (2023). Food safety and convenience meals: Consumers' actual and perceived behaviours – a mixed methods study. *British Food Journal* https://doi.org/10.1108/BFJ-05-2022-0429.

Lee, S., Choi, Y.-S., Jo, K. et al. (2021). Processing characteristics of freeze-dried pork powder for meat emulsion gel. *Food Science of Animal Resources* 41 (6): 997–1011. https://doi.org/10.5851/kosfa.2021.e51.

Liapis, A.I. and Bruttini, R. (2020). Freeze drying. In: *Handbook of Industrial Drying* (ed. A.S. Mujumdar), 309–343. CRC Press.

Liu, J., Viverette, T., Virgin, M. et al. (2005). A study of the impact of freezing on the lyophilization of a concentrated formulation with a high fill depth. *Pharmaceutical Development and Technology* 10 (2): 261–272. https://doi.org/10.1081/PDT-54452.

Lopez-Quiroga, E., Wang, R., Gouseti, O. et al. (2016). Crystallisation in concentrated systems: a modelling approach. *Food and Bioproducts Processing* 100: 525–534. https://doi.org/10.1016/j.fbp.2016.07.007.

Ma, Y., Yi, J., Jin, X. et al. (2022). Freeze-drying of fruits and vegetables in food industry: effects on phytochemicals and bioactive properties attributes – a comprehensive review. *Food Reviews International* 1–19: https://doi.org/10.1080/87559129.2022.2122992.

Marshall, R.T., Goff, H.D., and Hartel, R.W. (2003). The freezing process. In: *Ice Cream* (ed. R.T. Marshall, H.D. Goff, and R.W. Hartel), 171–206. US: Springer https://doi.org/10.1007/978-1-4615-0163-3_7.

Mediani, A., Hamezah, H.S., Jam, F.A. et al. (2022). A comprehensive review of drying meat products and the associated effects and changes. *Frontiers in Nutrition* 9: 1057366. https://doi.org/10.3389/fnut.2022.1057366.

Mehra, R., Kumar, H., Kumar, N. et al. (2021). Whey proteins processing and emergent derivatives: An insight perspective from constituents, bioactivities, functionalities to therapeutic applications. *Journal of Functional Foods* 87: 104760. https://doi.org/10.1016/j.jff.2021.104760.

Mehra, R., Kumar, S., Singh, R. et al. (2022). Biochemical, dielectric and surface characteristics of freeze-dried bovine colostrum whey powder. *Food Chemistry: X* 15: 100364. https://doi.org/10.1016/j.fochx.2022.100364.

Merivaara, A., Zini, J., Koivunotko, E. et al. (2021). Preservation of biomaterials and cells by freeze-drying: change of paradigm. *Journal of Controlled Release* 336: 480–498. https://doi.org/10.1016/j.jconrel.2021.06.042.

Messina, V., Sancho, A., Grigioni, G. et al. (2015). Evaluation of different bovine muscles to be applied in freeze-drying for instant meal. Study of physicochemical and senescence parameters. *Animal* 9 (4): 723–727. https://doi.org/10.1017/S1751731114002912.

Morgan, C.A., Herman, N., White, P.A., and Vesey, G. (2006). Preservation of micro-organisms by drying; a review. *Journal of Microbiological Methods* 66 (2): 183–193. https://doi.org/10.1016/j.mimet.2006.02.017.

Nowak, D. and Jakubczyk, E. (2020). The freeze-drying of foods—the characteristic of the process course and the effect of its parameters on the physical properties of food materials. *Foods* 9 (10): 1488. https://doi.org/10.3390/foods9101488.

Nualkaekul, S., Deepika, G., and Charalampopoulos, D. (2012). Survival of freeze dried lactobacillus plantarum in instant fruit powders and reconstituted fruit juices. *Food Research International* 48 (2): 627–633. https://doi.org/10.1016/j.foodres.2012.06.003.

Omah, E., Ajoiyi, R., and Nwankwo, C. (2015). Proximate composition and sensory properties of freeze-dried Nigerian soups. *Agro-Science* 14 (1): 37. https://doi.org/10.4314/as.v14i1.5.

Oyinloye, T.M. and Yoon, W.B. (2020). Effect of freeze-drying on quality and grinding process of food produce: a review. *Processes* 8 (3): 354. https://doi.org/10.3390/pr8030354.

Padma Ishwarya, S. and Anandharamakrishnan, C. (2015). Spray-freeze-drying approach for soluble coffee processing and its effect on quality characteristics. *Journal of Food Engineering* 149: 171–180. https://doi.org/10.1016/j.jfoodeng.2014.10.011.

Patel, S.M. and Pikal, M.J. (2011). Emerging freeze-drying process development and scale-up issues. *American Association of Pharmaceutical Scientists* 12 (1): 372–378. https://doi.org/10.1208/s12249-011-9599-9.

Patel, S.M., Doen, T., and Pikal, M.J. (2010). Determination of end point of primary drying in freeze-drying process control. *American Association of Pharmaceutical Scientists* 11 (1): 73–84. https://doi.org/10.1208/s12249-009-9362-7.

Persson, P.O. and Londahl, G. (1993). Freezing technology. *Frozen Food Technology* 20–58.

Prosapio, V. and Lopez-Quiroga, E. (2020). Freeze-drying technology in foods. *Foods* 9 (7): 920. https://doi.org/10.3390/foods9070920.

Reidy, G.A. and Heldman, D.R. (1972). Measurement of texture parameters of freeze-dried beef. *Journal of Texture Studies* 3 (2): 218–226. https://doi.org/10.1111/j.1745-4603.1972.tb00625.x.

Resch, G.P., Brandstetter, M., Pickl-Herk, A.M. et al. (2011). Immersion freezing of biological specimens: rationale, principles, and instrumentation. *Cold Spring Harbor Protocols* 2011 (7): pdb. top118.

Różyło, R. (2020). Recent trends in methods used to obtain natural food colorants by freeze-drying. *Trends in Food Science & Technology* 102: 39–50. https://doi.org/10.1016/j.tifs.2020.06.005.

Sailliol, A., Martinaud, C., Cap, A.P. et al. (2013). The evolving role of lyophilized plasma in remote damage control resuscitation in the French armed forces health service: evolving role of lyophilized plasma. *Transfusion* 53: 65S–71S. https://doi.org/10.1111/trf.12038.

Scholliers, P. (2015). Convenience foods. What, why, and when. *Appetite* 94: 2–6. https://doi.org/10.1016/j.appet.2015.02.017.

Silva, A.C.C. and Schmidt, F.C. (2022). Intensification of freeze-drying rate of coffee extract by vacuum freezing. *Innovative Food Science & Emerging Technologies* 78: 103022. https://doi.org/10.1016/j.ifset.2022.103022.

Silva, S.H., Neves, I.C.O., Meira, A.C.F.d.O. et al. (2021). Freeze-dried petit Suisse cheese produced with ora-pro-nóbis (Pereskia aculeata miller) biopolymer and carrageenan mix. *LWT* 149: 111764. https://doi.org/10.1016/j.lwt.2021.111764.

Silva-Espinoza, M.A., Ayed, C., Foster, T. et al. (2019). The impact of freeze-drying conditions on the physico-chemical properties and bioactive compounds of a freeze-dried orange puree. *Foods* 9 (1): 32. https://doi.org/10.3390/foods9010032.

Siow, C.R.S., Tang, D.S., Heng, P.W.S., and Chan, L.W. (2019). Probing the impact of HPMC viscosity grade and proportion on the physical properties of co-freeze-dried mannitol-HPMC tableting excipients using multivariate analysis methods. *International Journal of Pharmaceutics* 556: 246–262. https://doi.org/10.1016/j.ijpharm.2018.11.071.

Stratta, L., Capozzi, L.C., Franzino, S., and Pisano, R. (2020). Economic analysis of a freeze-drying cycle. *Processes* 8 (11): 1399. https://doi.org/10.3390/pr8111399.

Tan, M., Mei, J., and Xie, J. (2021). The formation and control of ice crystal and its impact on the quality of frozen aquatic products: a review. *Crystals* 11 (1): 68. https://doi.org/10.3390/cryst11010068.

Tarafdar, A., Shahi, N.C., Singh, A., and Sirohi, R. (2018). Artificial neural network modeling of water activity: a low energy approach to freeze drying. *Food and Bioprocess Technology* 11: 164–171.

Trelea, I., Fonseca, F., and Passot, S. (2016). Dynamic modelling of the secondary drying stage of freeze-drying reveals distinct desorption kinetics for bound water. *Drying Technology* 34: 335–345. https://doi.org/10.1080/07373937.2015.1054509.

Tsinontides, S.C., Rajniak, P., Pham, D. et al. (2004). Freeze drying—principles and practice for successful scale-up to manufacturing. *International Journal of Pharmaceutics* 280 (1–2): 1–16. https://doi.org/10.1016/j.ijpharm.2004.04.018.

Vargas-del-Río, L.M., García-Figueroa, A., Fernández-Quintero, A., and Rodríguez-Stouvenel, A. (2022). Spray-drying hen eggs: effects of the egg yolk to egg white ratio and sucrose addition on the physicochemical, functional, and nutritional properties of dried products and on their amino acid profiles. *Applied Sciences* 12 (9): 4516. https://doi.org/10.3390/app12094516.

Vesterlund, S., Salminen, K., and Salminen, S. (2012). Water activity in dry foods containing live probiotic bacteria should be carefully considered: a case study with lactobacillus rhamnosus GG in flaxseed. *International Journal of Food Microbiology* 157 (2): 319–321. https://doi.org/10.1016/j.ijfoodmicro.2012.05.016.

Visek, M., Joppolo, C.M., Molinaroli, L., and Olivani, A. (2014). Advanced sequential dual evaporator domestic refrigerator/freezer: system energy optimization. *International Journal of Refrigeration* 43: 71–79.

Witik, R.A., Phoutthasak, S., Collet, A., and Wyser, Y. (2019). Shelf life prediction of oxygen-sensitive products: the influence of moisture on prediction accuracy for freeze-dried coffee. *Packaging Technology and Science* 32 (7): 357–365. https://doi.org/10.1002/pts.2444.

Yamamoto, Y., Fong-in, S., and Kawai, K. (2021). Optimum physical properties of fruit puree for freeze-drying: effect of pulp content on freeze-concentrated glass transition temperature and yield stress of mango puree. *Journal of Food Engineering* 307: 110649. https://doi.org/10.1016/j.jfoodeng.2021.110649.

Yang, J., Mocking-Bode, H.C.M., van den Hoek, I.A.F. et al. (2022). The impact of heating and freeze or spray drying on the interface and foam stabilising properties of pea protein extracts: explained by aggregation and protein composition. *Food Hydrocolloids* 133: 107913. https://doi.org/10.1016/j.foodhyd.2022.107913.

Zarkadas, D.M. and Mitrakas, M. (1999). On the applicability of a plug flow immersion freezer: theoretical considerations. *LWT- Food Science and Technology* 32 (8): 548–552.

Zhang, Y., Guo, S., and Sun, Q. (2021). Research progress on lyophilization for pretreatment of emerging organic contaminants in environmental samples. *Chinese Journal of Chromatography* 39 (8): 827–834. https://doi.org/10.3724/SP.J.1123.2021.02034.

Zhang, Y., Zheng, Z., Liu, C. et al. (2022). A comparative study between freeze-dried and spray-dried goat milk on lipid profiling and digestibility. *Food Chemistry* 387: 132844. https://doi.org/10.1016/j.foodchem.2022.132844.

Index

Freeze Drying of Food Products: Fundamentals, Processes and Applications, First Edition.
Edited by Roji Balaji Waghmare, Manoj Kumar, and Parmjit Singh Panesar.

d

e